Progress in Probability
Volume 21

Probability in Banach Spaces 7

Proceedings of the Seventh International Conference

Ernst Eberlein
James Kuelbs
Michael B. Marcus
Editors

1990

Birkhäuser
Boston · Basel · Berlin

Ernst Eberlein
Institute for Mathematical Stochastics
7800 Freiburg
Federal Republic of Germany

James Kuelbs
Department of Mathematics
University of Wisconsin
Madison, WI 53706
USA

Michael B. Marcus
Department of Mathematics
Texas A & M University
College Station, TX 77843
USA

ISSN: 0892-063X

Library of Congress Cataloging-in-Publication Data
Probability in Banach spaces 7 : proceedings of the seventh international conference / Ernst Eberlein, James Kuelbs, Michael B. Marcus, editors.
p. cm. — (Progress in probability ; v. 21)
Includes bibliographical references.
ISBN 0-8176-3475-4 (alk. paper)
1. Probabilities—Congresses. 2. Banach spaces—Congresses. I. Eberlein, Ernst. II. Kuelbs, James. III. Marcus, Michael B. IV. Title: Probability in Banach spaces seven. V. Series: Progress in probability ; 21.
QA273.43.P772 1990
519.2—dc20 89-18631

Printed on acid-free paper

ISBN 0-8176-3475-4
ISBN 3-7643-3475-4

Camera-ready copy supplied by the authors.
Printed and bound by Edwards Brothers, Inc., Ann Arbor, Michigan.
Printed in the U.S.A.

9 8 7 6 5 4 3 2 1

Preface

The first international conference on Probability in Banach Spaces was held at Oberwolfach, West Germany, in 1975. It brought together European researchers who, under the inspiration of the Schwartz Seminar in Paris, were using probabilistic methods in the study of the geometry of Banach spaces, a rather small number of probabilists who were already studying classical limit laws on Banach spaces, and a larger number of probabilists, specialists in various aspects of the study of Gaussian processes, whose results and techniques were of interest to the members of the first two groups. This first conference was very fruitful. It fostered a continuing relationship among 50 to 75 probabilists and analysts working on probability on infinite-dimensional spaces, the geometry of Banach spaces, and the use of random methods in harmonic analysis.

Six more international conferences were held since the 1975 meeting. Two of the meetings were held at Tufts University, one at Sønderborg, Denmark, and the others at Oberwolfach. This volume contains a selection of papers by the participants of the Seventh International Conference held at Oberwolfach, West Germany, June 26–July 2, 1988. This exciting and provocative conference was attended by more than 50 mathematicians from many countries. These papers demonstrate the range of interests of the conference participants. In addition to the ongoing study of classical and modern limit theorems in Banach spaces, a branching out has occurred among the members of this group. Techniques that have been developed throughout the years are being applied to the study of empirical processes in statistics, to certain problems relating to diffusions, and to the study of other kinds of stochastic processes, such as Markov processes. This broadening of interests has loosened the tight focus of the group but it is still tied together by the similarity of the methods employed and the ability of the members to understand and contribute to each other's research. Arrangements have already been made for the eighth international conference in 1991. We look forward with great enthusiasm to further developments in this field.

Ernst Eberlein
James Kuelbs
Michael B. Marcus

Contents

On the functional law of the iterated logarithm for recurrent events

JON AARONSON AND MANFRED DENKER
UNIVERSITY OF TEL AVIV AND UNIVERSITY OF GÖTTINGEN

Let $X_1, X_2, \ldots$ be positive, independent and identically distributed random variables in the domain of attraction of an α–stable law where $0 < \alpha < 1$, the process being defined on the probability space $(\Omega, \mathcal{A}, P)$. It has been shown in [W, theorem 3.1] that the other law of the iterated logarithm holds, improving an old result of M. Lipschutz ([L2, theorem 1]). In [AD] we showed that this law can be obtained from a general upper and lower class result for conservative ergodic measure preserving transformations. The method of proof uses a representation of the process $(X_n)_{n\geq 1}$ as return time sequence (local time) for Ω embedded canonically (by a tower construction) into a conservative, ergodic, measure preserving discrete time flow $(X, \mathcal{F}, m, T)$ with σ–finite, infinite measure m, such that $m\,|_{\Omega} = P$. If N_n denotes the number of returns (under the flow T) to Ω up to time n for points in Ω, one observes that

$$N_n \geq m \Longleftrightarrow \sum_{k=1}^{m-1} X_k \leq n.$$

Consequently, the other law of the iterated logarithm follows from the law of the iterated logarithm for the random sequence $(N_n)_{n\geq 1}$.

A general result of this type has been obtained in [AD], special cases are dealt with in [CH], [L1], [L2], [JP] and [W]. Since part of this theorem will be basic to our discussion below we start with a description of it. Let T be a conservative, ergodic and measure preserving transformation on the σ–finite, non–atomic infinite measure space $(\mathcal{X}, \mathcal{F}, m)$. Recall that conservativity means that for any positive integrable function $f \neq 0$ the partial sums $\sum_{1\leq k\leq n} f \circ T^k$ diverge. The transformation T induces an operator $f \longrightarrow f \circ T$ on the space $L^\infty(m)$. Denote by $\hat{T}$ the corresponding dual operator restricted to the space $L^1(m)$. We shall say that a measurable set A is a *Darling–Kac set (DK–set)* if $0 < m(A) < \infty$ and if there exists a sequence $(a_n)_{n\geq 1}$ such that, uniformly on A,

$$\lim_{n\to\infty} \frac{1}{a_n} \sum_{k=1}^{n} \hat{T}^k 1_A = m(A).$$

It turns out that

$$(*) \qquad \lim_{n\to\infty} \frac{1}{a_n} \sum_{k=1}^{n} \hat{T}^k \circ f = \int_X f \, dm$$

for all $f \in L^1(m)$, and so the sequence a_n does not depend on A and is associated to T itself (see [A1]). It is called the *return sequence for* T. Given a measurable set B with $0 < m(B) < \infty$, define the *return time process* by $X_n = X_n(B) = \varphi \circ T_B^n$ $(n \geq 1)$ where

$$\varphi(x) = \min\{n \geq 1 : T^n(x) \in B\} \qquad x \in B,$$

and where T_B denotes the induced transformation $T_B(x) = T^{\varphi(x)}(x)$ $(x \in B)$. This process is defined on the probability space $(B, \mathcal{F} \cap B, m(B)^{-1} m \mid_B)$. Denote by $\mathcal{F}_0$ the σ–algebra generated by the return time process. Then $(X_n)_{n\geq 1}$ will be called

(a) *uniformly mixing from below* if for all $\mathcal{F}_0$–measurable sets C of positive probability

$$\liminf_{k\to\infty} \inf_{n\geq 1} \inf\{\frac{m(D \cap T_B^{-(n+k)} C)}{m(D)} : D \in \sigma(X_j : 0 \leq j < n)\} > 0$$

and

(b) *strongly mixing from below* if for every $\mathcal{F}_0$–measurable set C, there exists a non–negative sequence $(\alpha(n))_{n\geq 1}$ such that $\sum \alpha(n)/n < \infty$ and for $n \geq 1$, $k \geq 1$ and $D \in \sigma(X_j : 0 \leq j < n)$ we have

$$m(D \cap T^{-(n+k)}C) \geq m(D)m(C) - \alpha(k).$$

We not state the result from [AD] we are interested in.

THEOREM 1: Let $(\mathcal{X}, \mathcal{F}, m, T)$ be a conservative, ergodic, measure preserving system on a σ–finite, non–atomic infinite measure space. Assume that the return sequence for T is regularly varying with index $\alpha \in (0,1)$ and slowly varying function h. Moreover, suppose that there exists a DK–set A, such that the return time process is uniformly mixing from below or strongly mixing from below. Then, for every positive integrable function $f \in L^1_+(m)$,

$$\limsup_{n\to\infty} \frac{\sum_{k=1}^{n} f \circ T^k}{n^\alpha h(n/L_2(n))L_2(n)^{1-\alpha}} = K_\alpha \int_{\mathcal{X}} f\, dm \qquad \text{a.e.}$$

where $L_2(n) = \log\log n$ and $K_\alpha = \Gamma(1+\alpha)\alpha^{-\alpha}(1-\alpha)^{-(1-\alpha)}$.

By possibly renormalizing the measure, we may and do assume that the DK–set has measure one. Note that the theorem only has to be proved for $f = 1_A$ by Hopf's ratio ergodic theorem, and that the mixing conditions are weaker than the commonly used notions of uniform mixing and strong mixing.

The functional form of this theorem is the object of our discussion here. We are able to make a small unsatisfactory contribution towards it. So, the other purpose of this note is to explan this problem in greater detail and to give some ideas of the methology involved and to pose some questions.

Denote by K the set of all increasing, continuous functions $x(x)$ on $[0,1]$ satisfying the Hölder condition

$$|x(t) - x(s)| \leq K_\alpha |s - t|^\alpha$$

for all $0 \leq s,t \leq 1$. In particular, $x(1) \leq K_\alpha$. Denote by D the space of all functions defined on $[0,1]$ which are right continuous and have left limits, endowed with the supremum norm. We are able to prove

THEOREM 2: Let the assumptions of theorem 1 be satisfied. Define for $f \in L^1_+(m)$ the random functions

$$t \longrightarrow \frac{1}{n^\alpha h(n/L_2(n))L_2(n)^{1-\alpha}} \sum_{1 \le k \le nt} f \circ T^k \qquad (0 \le t \le 1).$$

Then, almost surely, the set of random functions is relatively compact in D and the set C_f of accumulation points is non–random and has the form

$$C_f = C \int_X f \, dm$$

where C does not depend on f and $C \subseteq K$.

REMARKS AND QUESTIONS:

1) Theorem 2 holds for pointwise dual ergodic transformations (i.e. those satisfying (*)) with regularly varying return sequences, index in $(0,1)$. The extra conditions in theorem 1 are not needed for the upper bound for the limsup there.

2) Under the assumption of theorem 2, what is the precise form of the limit set C? Is the limit set only dependent on the return sequence? If so, is it only dependent on the index of regular variation? (This is true for theorem 1.) The same question can be posed for pointwise dual transformation. In this case even the analogue to theorem 1 is unknown.

In case of i.i.d. positive random variables in the domain of attraction of a stable distribution of index $\alpha \in (0,1)$ and $f = 1_{[0,u]}$ theorem 2 has been proven in [W, theorem 3.2]. It turns out that C consists of all non–decreasing, absolutely continuous functions x such that

$$\int_0^1 (\dot{x}(t))^{1-\alpha} \, dt \le 1.$$

3) Two measure preserving transformations are similar, if they are both factors of the same measure preserving transformations (see [A1]). It is not hard to see, that if T is similar to

S in our class, and if theorem 2 holds for S, then theorem 2 holds for T and also $C(T) = C(S)$.

4) For which pointwise dual transformations does there exist a DK–set? Except for Markov shifts (which correspond to general random walks) and certain number theoretic transformations the existence of DK–sets is not known.

PROOF: By Hopf's ratio ergodic theorem for any two non-zero functions $f, g \in L^1_+(m)$

$$\lim_{n\to\infty} \frac{\sum_{k=1}^n f \circ T^k}{\sum_{k=1}^n g \circ T^k} = \frac{\int_X f\, dm}{\int_X g\, dm} \qquad \text{a.e.}$$

and hence the set C_f can be written in the form $C \int_X f\, dm$, where $C = C_{1_A}$. Next observe that the set of accumulation points C is non-random. Write $S_n(t)$ for the random function associated to 1_A at time t. Then $\|S_n - S_n \circ T\|_\infty \leq 1$ and consequently the limit set of a point $x \in X$ is invariant. Since the limit set consists of measures (increasing functions), it can be viewed as a subset of a standard Borel space. Consequently, this set must be constant a.e. by ergodicity. (In fact the space of measures is a polish space. Let B be a subset. Then $\{\lim_{k\to\infty} S_{n_k} \in B$ for some subsequence $(n_k)\}$ is invariant, and hence of measure zero or one by ergodicity.)

The only non–obvious assertion is the Hölder continuity of the limit functions. This is because the process $(1_A \cdot T^k)_{k\geq 1}$ on the probability space $(A, \mathcal{F} \cap A, m\,|_A)$ is not stationary, and hence the proof for theorem 1 does not carry over.

The proof of this will be complete proving

PROPOSITION: For every $0 \leq t \leq s + t \leq 1$

$$\limsup_{n\to\infty} \frac{S_n(t+s) - S_n(t)}{n^\alpha h(n/L_2(n))L_2(n)^{1-\alpha}} \leq K_\alpha s^\alpha \qquad \text{a.e.}$$

We shall prove this proposition via a sequence of lemmas. Fix s and t. We may assume that s and t are rational and that n is so large, that $ns,\ nt \in \mathbb{N}$.

Define $a(p,n) : A \longrightarrow \mathbb{Z}_+$ $(n, p \in \mathbb{N})$ by

$$a(0,n) = 1_{\{n \geq 1\}},$$

$$a(p+1,n)(x) = \sum_{k=1}^{n} 1_A(T^k(x))a(p,n-k)(T^k(x))$$

and $\gamma_p(q) \in \mathbb{N}$ by

$$\gamma_1(q) = \delta_{1,q}, \quad \gamma_{p+1}(q) = q(\gamma_p(q) + \gamma_p(q-1)).$$

A direct computation shows (see [AD])

LEMMA 1: For $sn \geq p \geq 1$ we have

$$(S_n(t+s,x) - S_n(t,x))^p = \sum_{q=1}^{p} \gamma_p(q)a(q,ns)(T^{tn}(x)).$$

Let $u_k = \int_{k-1}^{k} \alpha s^{\alpha-1} h(s)ds$.

LEMMA 2: There exist $\beta(k) \downarrow 1$ as $k \uparrow \infty$ such that for all $n \geq 1$ and a.e. $x \in \mathcal{X}$

$$\sum_{k=1}^{n} \hat{T}^k 1_A(x) \leq \sum_{k=1}^{n} \beta(k)u_k.$$

PROOF: Because of the uniform convergence of $a_n^{-1} \sum_{1 \leq k \leq n} \hat{T}^k 1_A$ on A, there exists a sequence $b(k) \downarrow 1$ such that for all $n \geq 1$ and $x \in A$:

$$\sum_{k=1}^{n} \hat{T}^k 1_A(x) \leq b(n)a_n.$$

Let $A_0 = A$ and $A_m = A \setminus \bigcup_{1 \leq k \leq m} T^{-k}A$ for $m \geq 1$. It is shown in [A2, p. 1045] that

(a) $$\sum_{n=0}^{\infty} \hat{T}^n 1_{A_n} = 1 \qquad \text{a.e. on } \mathcal{X}$$

(b) $$\sum_{n=0}^{N} \hat{T}^n 1_{A_n} = \sum_{n=0}^{N} \hat{T}^n \left(1_{A_n} \sum_{k=0}^{N-n} \hat{T}^k 1_A\right) \quad \text{a.e. on } \mathcal{X}$$

Consequently, for any $B \in \mathcal{F}$

$$\begin{aligned}
&\int_B \sum_{n=0}^{N} \hat{T}^n \left(1_{A_n} \sum_{k=0}^{N-n} \hat{T}^k 1_A\right) dm \\
&= \sum_{n=0}^{N} \int_{A_n} 1_B \circ T^n \left(1_{A_n} \sum_{k=0}^{N-n} \hat{T}^k 1_A\right) dm \\
&\leq \sum_{n=0}^{N} a_{N-n} b(N-n) \int_{A_n} 1_B \circ T^n \, dm \\
&= \sum_{n=0}^{N} a_{N-n} b(N-n) \int_B \hat{T}^n 1_{A_n} dm.
\end{aligned}$$

This implies together with (b) that

$$\sum_{n=0}^{N} \hat{T}^n 1_A \leq \sum_{n=0}^{n} a_{N-n} b(N-n) \hat{T}^n 1_{A_n}.$$

Since the return sequence is assumed to be regularly varying, by Karamata's theorem, $a_n(T) \sim \sum_{k=1}^{n} u_k$ as $n \to \infty$, and because of (a), the lemma follows.

Define $A(0,n) = 1_{\{n \geq 1\}}$ and

$$A(p+1, n) = \sum_{k \leq n} \beta(k) u_k A(p, n-k).$$

LEMMA 3: ([Ad, lemma 2.8]) For all $\beta > 1$ there exists t_β such that for all $n \geq t_\beta$ and $t_\beta \leq p \leq L_2(n)^2$,

$$\int_A a(p,n) dm \leq A(p,n) \leq \beta^p (\Gamma(1+\alpha) n^\alpha h(n/p))^p \frac{1}{\Gamma(1+\alpha p)}.$$

LEMMA 4: ([AD, lemmas 2.11, 2.12]) For all $\beta > 1$ there exists t_β such that for all $n \geq t_\beta$ and $t_\beta \leq p \leq L_2(n)^2$

$$\sum_{q=1}^{p} \gamma_p(q) \frac{1}{\Gamma(1+\alpha q)} (\beta\Gamma(1+\alpha)n^\alpha h(n/q))^q$$
$$\leq \beta^p \frac{\Gamma(1+\alpha)^p}{\alpha^{p\alpha} e^{(1-\alpha)p}} p^{(1-\alpha)p} n^{p\alpha} h(n/p)^p.$$

LEMMA 5: For all $\beta > 1$ there exists t_β such that for all $n \geq t_\beta$ and $t_\beta \leq p \leq L_2(n)^2$

$$\int_A a(p, ns) \circ T^{nt} dm \leq A(p, ns).$$

PROOF: By definition of $a(p, n)$ and the duality of $\hat{T}$ we have

$$\int_A a(p, ns) \circ T^{nt}\, dm$$
$$= \sum_{k=1}^{ns} \int_A (1_A a(p-1, ns-k)) \circ T^{nt+k}\, dm$$
$$= \sum_{k=1}^{ns} \int_{T^{-nt}A} a(p-1, ns-k) \circ T^{nt} (\hat{T}^k 1_A)\, dm$$
$$= \sum_{k=1}^{ns} \int_{T^{-nt}A} (\sum_{j=1}^{k} \hat{T}^j 1_A - \sum_{j=1}^{k-1} \hat{T}^j 1_A) a(p-1, ns-k) \circ T^{nt} dm$$
$$= \sum_{k=1}^{ns} \int_{T^{-nt}A} (a(p-1, ns-k) - a(p-1, ns-k-1)) \circ T^{nt}$$
$$\sum_{j=1}^{k} \hat{T}^j 1_A\, dm.$$

Since $a(p, n)$ is non-decreasing in n, we can apply lemmas 2

and 3 to obtain

$$\int_A a(p,ns)\circ T^{nt}\,dm$$
$$\leq \sum_{k=1}^{ns}\int_{T^{-nt}A}(a(p-1,ns-k)-a(p-1,ns-k-1))\circ T^{nt}$$
$$\sum_{j=1}^{k}\beta(j)u_j\,dm$$
$$=\sum_{k=1}^{ns}\int_{T^{-nt}A}\left(\sum_{j=1}^{k}\beta(j)u_j-\sum_{j=1}^{k-1}\beta(j)u_j\right)$$
$$a(p-1,ns-k)\circ T^{nt}\,dm$$
$$=\sum_{k=1}^{ns}\beta(k)u_k\int_{T^{-nt}}Aa(p-1,ns-k)\circ T^{nt}\,dm$$
$$=\sum_{k=1}^{ns}\beta(k)u_k\int_A a(p-1,ns-k)dm$$
$$\leq\sum_{k=1}^{ns}\beta(k)u_k A(p-1,ns-k)$$
$$=A(p,ns).$$

This proves the lemma.

Combining lemmas 1, 4 and 5 we obtain the following estimate for the moments of the partial sums.

LEMMA 6: For every $\beta > 1$ there exists $n_\beta \in \mathbb{N}$ such that for all integers $n \geq n_\beta$ and $n_\beta \leq p \leq L_2(n)^2$,

$$||S_n(t+s)-S_n(t))||_{L^p(A,\mu)} \leq \beta\,\frac{\Gamma(1+\alpha)}{\alpha^\alpha e^{(1-\alpha)}}\,p^{(1-\alpha)}(ns)^\alpha h(n/p).$$

PROOF OF THE PROPOSITION: Using lemma 6 with $p=[L_2(n)/(1-\alpha)]$ and Markov's inequality we have for $\beta > 1$

and n large enough that

$$m(A\cap\{S_n(t+s)-S_n(t)\ge\beta^2K_\alpha(ns)^\alpha h(\frac{n}{L_2(n)})L_2(n)^{1-\alpha}\})$$
$$\le\frac{\int_A(S_n(t+s)-S_n(t))^p\,dm}{(\beta^2K_\alpha(ns)^\alpha h(n/L_2(n))L_2(n)^{1-\alpha}\})^p}$$
$$\ll\left(\frac{h(n(1-\alpha)/L_2(n))}{\beta\exp(1-\alpha)h(n/L_2(n))}\right)^{L_2(n)/(1-\alpha)}$$
$$=\left(\frac{h(n(1-\alpha)/L_2(n))}{h(n/L_2(n))}\right)^{L_2(n)/(1-\alpha)}\exp(-L_2(n)(1+\frac{\log\beta}{1-\alpha})).$$

Since h is slowly varying, it follows that

$$\sum_{n=1}^{\infty}m(A\cap\{S_{[\gamma^n]}(t+s)-S_{[\gamma^n]}(t)$$
$$\ge\beta^2K_\alpha([\gamma^n]s)^\alpha h([\gamma^n]/L^2([\gamma^n]))L_2([\gamma^n])^{1-\alpha}\})<\infty$$

for every $\gamma>1$. The Borel–Cantelli lemma implies

$$\limsup_{n\to\infty}\frac{S_{[\gamma^n]}(t+s)-S_{[\gamma^n]}(t)}{([\gamma^n]s)^\alpha h([\gamma^n]/L_2([\gamma^n]))L_2([\gamma^n])^{1-\alpha}}\le\beta^2K_\alpha$$

for every $\beta,\gamma>1$, and this proves the proposition.

References

[A1] J. Aaronson: Rational ergodicity and a metric invariant for Markov shifts. Israel J. Math. **27**, (1977), 93–123

[A2] J. Aaronson: Random f–expansions. Ann. Probab. **14**, (1986), 1037–1057

[AD] J. Aaronson, M. Denker: Upper bounds for ergodic sums of infinite measure preserving transformations. To appear: Trans. Amer. Math. Soc.

[CH] K.L. Chung, G.A. Hunt: On the zeros of $\sum_1^n \pm 1$. Ann. of Math. (2) **50**, (1949), 385–400

[JP] N.C. Jain, W.E. Pruitt: An invariance principle for the local time of a recurrent random walk. Z. Wahrscheinlichkeitsth. verw. Geb. **66**, (1984), 141–156

[L1] M. Lipschutz: On strong laws for certain types of events connected with sums of independent random variables. Ann. of Math. **57**, (1953), 318–330

[L2] M. Lipschutz: On strong bounds for sums of independent random variables which tend to a stable distribution. Trans. Amer. Math. Soc. **81**, (1956), 135–154

[W] M.J. Wichura: Functional laws of the iterated logarithm for the partial sums of i.i.d. random variables in the domain of attraction of a completely asymmetric stable law. Ann. Propap. **2**, (1974), 1108–1138

Jon Aaronson, School of Mathematical Sciences Tel Aviv University, Ramat Aviv, 69978 Tel Aviv, Israel.

Manfred Denker, Institut für Mathematische Stochastik, Universität Göttingen, Lotzestr. 13, 3400 Göttingen, West Germany.

Sur la loi des grands nombres de Nagaev en dimension infinie

JEAN-CHRISTIAN ALT

1. Introduction

Cet article présente une condition nécessaire générale pour la loi forte des grands nombres (l.f.g.n.) dans un espace de Banach quelconque. Cette condition est la généralisation à la dimension infinie d'une condition précédemment énoncée par Nagaev pour des variables aléatoires réelles ([5]; voir aussi [6], théorème 2).

Rappelons brièvement le résultat initial de Nagaev, sous une forme légèrement différente, car nous remplaçons les restrictions de probabilités à des intervalles utilisées dans [5] par des troncations de variables aléatoires. Soit $(X_n)_n$ une suite des variables aléatoires réelles (v.a.r.) symétriques et indépendantes. Pour tous ε et $h > 0$ posons :

$$g_n(h,\varepsilon) = E\ \exp(hX_n 1_{\{|X_n| \leq \varepsilon n\}}),$$

et définissons $h_k(\varepsilon)$ comme étant la solution de l'équation :

$$\sum_{2^k < i \leq 2^{k+1}} g_i'(h,\varepsilon)/g_i(h,\varepsilon) = \varepsilon 2^{k+1}$$

(s'il n'y a pas de solution : $h_k(\varepsilon) = +\infty$).

Théorème 1. (S.V. Nagaev, [5]). *La suite $(X_n)_n$ vérifie la l.f.g.n. si et seulement si les deux conditions suivantes sont vérifiées :*

$$(1)\ \forall\ \ \varepsilon > 0\ \ \sum_{n \geq 1} P\{X_n \geq \varepsilon n\} < +\infty$$

$$(2)\ \forall\ \ \varepsilon \geq 0\ \ \sum_{k \geq 1} \exp\{-\varepsilon 2^{k+1} h_k(\varepsilon)\} + \infty. \qquad \text{(CN)}$$

Il nous sera utile d'opposer la l.f.g.n. de Prokhorov au théorème 1, dont elle constitue une conséquence facile, ainsi que l'a montré Nagaev dans [5].

Théorème 2. (You. V. Prokhorov, [7]). *Soit $(X_n)_n$ une suite de v.a.r. indépendantes et centrées, satisfaisant la condition :*

$$\exists \ C > 0, \quad \forall \quad n \in N \quad |X_n| \leq Cn/L_2 n \quad p.s.$$

$(L_2 x = ln[ln(xVe^e)])$.

Cette suite vérifie la l.f.g.n. si et seulement si :

$$\forall \quad \varepsilon > 0 \quad \sum_{k \geq 1} \exp(-\varepsilon \lambda_k^{-1}) < +\infty \qquad (CP)$$

où $\lambda_k = 2^{-2k} \sum_{2^k < 1 \leq 2^{k+1}} EX_i^2$.

Nous avons montré dans [2] que l'extension naturelle de la condition (CP) du théorème 2 pour des variables aléatoires $(X_n)_n$ à valeurs dans un espace de Banach séparable $(B, \| \ \|)$ se formule de la façon suivante :

$$\forall \quad \varepsilon > 0 \quad \sum_{k \geq 1} \exp(-\varepsilon \tilde{\lambda}_k^{-1}) < +\infty \qquad (CP')$$

où $\tilde{\lambda}_k = 2^{-2k} \sup \sum_{2^k < i \leq 2^{k+1}} E\langle X_i, x' \rangle^2; \quad \|x'\|_{B'} \leq 1$.

Similairement, nous montrerons dans cet article qu'en dimension infinie les analogues des quantités $h_k(\varepsilon)$ du théorème 1 sont les expressions:

$$\tilde{h}_k(\varepsilon) = \inf\{\tilde{h}_k(x', \varepsilon); \quad \|x'\| \leq 1\},$$

où $\tilde{h}_k(x', \varepsilon)$ est solution de :

$$\sum_{2^k < i \leq 2^{k+1}} \tilde{g}_i'(h, x', \varepsilon)/\tilde{g}_i(h, x', \varepsilon) = \varepsilon 2^{k+1},$$

et $\tilde{g}_n(h, x', \varepsilon) = E \ \exp(hx'(X_n) 1_{\{\|X_n\| \leq \varepsilon n\}})$, et qu'avec ces expressions on obtient une condition nécessaire pour la l.f.g.n. en dimension infinie, condition qui s'écrit :

$$\forall \quad \varepsilon > 0 \quad \sum_{k \geq 1} \exp\{-\varepsilon 2^{k+1} \tilde{h}_k(\varepsilon)\} < +\infty. \qquad (CN')$$

Nous étendons ainsi la nécessité de la condition (CN) du théorème 1 à n'importe quel espace de Banach; le théorème correspondant et sa démonstration font l'objet de la partie 3 de ce travail. Dans la seconde partie on trouvera une série de lemmes techniques et de remarques préparatoires. Par souci de complétude nous avons tenu à redonner dans la partie 4 la démonstration de la suffisance des conditions 1 et 2 du théorème 1 pour v.a. réelles. Utilisant un exemple de B. Heinkel, nous montrerons à la fin de cet article que la condition (CN') n'est en général pas suffisante pour la l.f.g.n. en dimension infinie. Mais nous prouverons dans la partie 5 que dans le cadre du théorème de Prokhorov, c'est à dire pour des v.a. indépendantes (X_n) vérifiant la condition de bornitude :

$$\exists \quad C > 0, \quad \forall \quad n \geq 1 \quad \|X_n\| \leq C n / L_2 n \qquad p.s.,$$

la condition (CN'), accompagnée d'une hypothèse de convergence en probabilité, entraîne la convergence presque sûre de la suite $X_1 + \cdots + X_n/n$.

2. Préliminaires techniques

Nous commençerons par rappeler sans démonstration un lemme qui aidera à comprendre les définitions qui suivent.

Lemme 3. *Soit X une v.a.r. bornée; on a :*

$$\lim_{a \to +\infty} \int X e^{aX} dP / \int e^{aX} dP = \sup ess\ X.$$

Soit X une v.a.r. symétrique, bornée et non dégénérée. Pour tout réel a nous poserons :

$$g(a) = g_X(a) = E(e^{aX}) \quad \text{et} \quad \varphi(a) = \varphi_X(a) = g'(a)/g(a).$$

La fonction φ est strictement croissante. Pour $a > 0$ nous définirons $h(X, a)$ de la façon suivante :

si $a <$ sup $essX, h(X, a)$ est l'unique solution de l'équation $\varphi(h) = a$.

si $a \geq$ sur $essX, h(X, a) = +\infty$.

Nous adopterons de plus la convention suivante, où $O_{\mathbb{R}}$ désigne toute v.a.r. presque sûrement nulle :

$$\forall \quad a > 0 \qquad h(O_{\mathbb{R}}, a) = +\infty.$$

Le lemme qui suit, implicitement contenu dans [5], est l'outil fondamental permettant de prouver la suffisance de la condition (CN) (cf. la démonstration du théorème 7).

Lemme 4. *Pour toute v.a.r. X symétrique, bornée et non dégénérée on a :*

$$\forall \quad a > 0 \quad P\{X \geq 2a\} \leq \exp\{-a\,h(X,a)\}.$$

Démonstration. On a d'une part pour tout $M \in \mathbf{R}$ et $h \geq 0$:

$$g(h) \geq E(1_{\{X \geq M\}} e^{hX}) \geq e^{hM} P\{X \geq M\}.$$

D'autre part :

$$ln[g(h)] = \int_0^h [ln\, g]'(x)dx = \int_0^h \varphi(x)dx \leq h\varphi(h),$$

soit : $g(h) \leq \exp\{h\varphi(h)\}$.
D'où : $e^{hM} P\{X \geq M\} \leq \exp\{h\varphi(h)\}$.
Si $h(X,a)$ est fini, on prend $M = 2a$ et $h = h(X,a)$; et l'inégalité à démontrer est triviale si $h(X,a) = +\infty$, car alors $a \geq \sup\, ess\, X$.

Remarque. L'inégalité du lemme précédent est à comparer à celle de Cramer, qui s'obtient à partir de l'inégalité de Tchebychev

$$P\{X \geq a\} \leq e^{-ah} E(e^{hX}) = e^{-ah} g(h)$$

en choisissant un réel h minimisant le membre de droite; ce réel est précisément égal à $h(X,a)$; il s'en suit que :

$$P\{X \geq a\} \leq \exp\{-a\,h(X,a) + ln\, g_X[h(X,a)]\}.$$

Pour une var. centrée X $g_X(h)$ est supérieur à 1. L'inégalité du lemme 4 est donc, au coefficient 2 près, plus fine que celle de Cramer.

Passons à présent à la généralisation aux espaces de Banach des expressions $h(X,a)$. Pour toute v.a. X symétrique et bornée à valeurs dans un espace de Banach séparable $(B, \|\ \|)$ nous définirons, en conservant les notations qui précèdent :

$$\forall \quad a > 0, \quad \forall \quad u \in B' \qquad h(X,a) = \inf\{h(u(X),a); \|u\| \leq 1\}.$$

Il n'est pas difficile de voir que dans le cas où l'espace de Banach considéré est la droite réelle, les deux définitions de $h(X,a)$ coïncident.

Nous conclurons cette partie en énonçant sous forme de lemme la minoration sur laquelle repose la démonstration de la nécessité de la condition (CN$'$); ce lemme est également implicitement contenu dans [5].

Lemme 5. *Soient $X_1, \dots, X_n$ des v.a.r. indépendantes symétriques et bornées. Posons : $Y = X_1 + \cdots + X_n$; pour tout $a > 0$ tel que $h = h(Y, a)$ est fini et pour tout $d > 0$ on a :*

$$P\{Y \geq a - d\} \geq e^{-(a+d)h}[1 - d^{-2} \sum_{i=1}^{n} \int X_i^2 e^{hX_i} dP / \int e^{hX_i} dP].$$

Démonstration. Désignons par μ_i la loi de X_i et posons :

$$\mu = \mathcal{L}(Y) = *_{1 \leq i \leq n} \mu_i, \pi = \prod_{1 \leq i \leq n} E\ e^{hX_i}, \nu_i = (e^{h\cdot} / E e^{hX_i}) \mu_i$$

et $\nu = *_{1 \leq i \leq n} \nu_i$.

Soient $Z_1, \dots, Z_n$ des variables aléatoires indépendantes de lois respectives $\nu_1, \dots, \nu_n$. Posons $Z = Z_1 + \cdots + Z_n$. On vérifie facilement les égalités:

$$E(Z) = \sum_{1 \leq i \leq n} \varphi_{X_i}(h) = \varphi_Y(h) = a.$$

D'où : $E(Z - a)^2 \leq \sum_{1 \leq i \leq n} E(Z_i^2)$.

Il en résulte :

$$P\{|Z - a| > d\} \leq d^{-2} \sum_{1 \leq i \leq n} E(Z_i^2),$$

ou de façon équivalente :

$$\nu([a - d, a + d]) \geq 1 - d^2 \sum_{1 \leq i \leq n} E(Z_i^2). \tag{1}$$

On a d'autre part :

$$\begin{aligned} P\{Y \geq a - d\} &= \mu([a - d, +\infty[) \\ &= \int_{a-d}^{+\infty} \pi e^{-hx} d\nu(x) \\ &\geq \pi \int_{a-d}^{a+d} e^{-hx} d\nu(x) \\ &\geq e^{-(a+d)h} \nu([a - d, a + d]), \end{aligned} \tag{2}$$

π étant supérieur à 1 en conséquence de l'inégalité de Jensen et de la symétrie des variables X_i.

La conclusion résulte alors de (1), (2) et de l'égalité :

$$E(Z_i^2) = \int X_i^2 e^{hX_i} dP / \int e^{hX_i} dP.$$

3. Nécessité de la condition (CN′)

Nous présentons ici l'extension à la dimension infinie de la nécéssité de la condition (CN) du théorème 1. On se donne:

— une suite $(X_n)_n$ de v.a. symétrique et indépendantes à valeurs dans un espace de Banach séparable B

— une suite croissante de réels strictement positifs $(a_n)_n$

— une suite strictement croissante d'entiers $(n_k)_k$.

Pour tout réel $\delta > 0$ nous noterons :

$$X_n^\delta = X_n 1_{\{\|X_n\| \le \delta a_n\}}$$

et

$$T_k^\delta = \sum_{i \in I_k} X_i^\delta, \quad \text{avec} \quad : I_k =]n_{k-1}, n_k].$$

Enfin, avec les notations des préliminaires, nous poserons pour tout $\varepsilon > 0$:

$$h_k(\varepsilon) = h(T_k^\varepsilon, \varepsilon a_{n_k}).$$

Le théorème énonçant la condition générale pour la l.f.g.n. en dimension infinie est alors le suivant :

Théorème 6. *Soient $(Xn), (a_n)$ et (n_k) les suites définies ci-dessus. Si la suite $(X_1 + \cdots + X_n / a_n)_n$ converge presque sûrement vers 0, alors la condition suivante est satisfaite :*

$$\forall \quad \varepsilon > 0 \qquad \sum_{k \ge 1} \exp\{-\varepsilon a_{n_k} h_k(\varepsilon)\} < +\infty \qquad (CN')$$

Démonstration. Désignons par B_1' la boule unité fermée du dual de B. Pour $u \in B_1', \varepsilon > 0, \delta > 0, k \in N$ posons : $h_k(u, \delta, \varepsilon) = h(u(T_k^\delta), \varepsilon a_{n_k})$.

Première étape. Soit $\delta > 0$; posons : $S_n = X_1 + \cdots + X_n$ et $S_n^\delta = X_1^\delta + \cdots + X_n^\delta$. Si S_n / a_n converge p.s. vers 0, on a par le lemme de Borel-Cantelli :

$$\sum_{n \ge 1} P\{\|X_n\| > \delta a_n\} < +\infty.$$

Par conséquent S_n^δ/a_n converge également p.s. vers 0. Un lemme de de Acosta ([1], lemme 3.1) entraîne alors :

$$\lim_{n\to\infty} E(\|S_n^\delta\|^2/a_n^2) = 0,$$

donc aussi :

$$\lim_{k\to\infty} E(\|T_k^\delta\|^2 a_{n_k}^2) = 0. \tag{3}$$

Fixons $\varepsilon > 0$ et $u \in B_1'$. Posons $\delta = \varepsilon/2$ et $Y_i = u(X_i^\delta)$, et choisissons $\eta = \eta(\varepsilon)$ tel que :

$$(1+\eta)\delta \leq (\delta+\varepsilon)/2, \quad \eta < \varepsilon \quad \text{et} \quad 2(\varepsilon-\eta)^2 > (\delta+\varepsilon)\varepsilon. \tag{4}$$

Soit enfin $c = c(\varepsilon)$ vérifiant : $(1-e^{-2c})^{-1} \leq 1 + \eta/2$. Pour tous $i \in I_k$ et $a > 0$ tel que $ca^{-1} \leq \delta a_i$ on a :

$$\begin{aligned} E(1_{\{ca^{-1}\leq Y_i \leq \delta a_i\}} Y_i e^{aY_i}) &\leq (1-e^{-2c})^{-1} E(1_{\{ca^{-1}\leq Y_i\leq \delta a_i\}} Y_i [e^{aY_i} - e^{-aY_i}]) \\ &\leq (1+\eta/2) E(1_{\{|Y_i|\leq \delta a_i\}} Y_i e^{aY_i}), \end{aligned} \tag{5}$$

la symétrie des variables X_i ayant été utilisée pour la dernière inégalité.

Soit $k \in N$; posons : $h = h_k(u,\delta,\varepsilon)$, et supposons h fini. Pour $i \in I_k$ on a :

$$A = E(Y_i^2 e^{hY_i}) = E(1_{\{|Y_i|\leq \delta a_i\}} Y_i^2 e^{hY_i}).$$

Distingons deux cas :

(a) $ch^{-1} \leq \delta a_i$: tenant compte de (5) il vient :

$$\begin{aligned} A &\leq e^c E(1_{\{-\delta a_i \leq Y_i \leq ch^{-1}\}} Y_i^2) + \delta a_{n_k} E(1_{\{ch^{-1}\leq Y_i \leq \delta a_i\}} Y_i e^{hY_i}) \\ &\leq e^c E(Y_i^2) + (1+\eta/2)\delta a_{n_k} E(Y_i e^{hY_i}) \end{aligned}$$

(b) $ch^{-1} > \delta a_i$: par symétrie des variables X_i on a :

$$\begin{aligned} A &\leq e^c E(Y_i^2) \\ &\leq e^c E(Y_i^2) + (1+\eta/2)\delta a_{n_k} E(Y_i e^{hY_i}). \end{aligned}$$

Remarquons qu'en conséquence de l'inégalité de Jensen: $E(e^{hY_i}) \geq 1$; d'où:

$$\begin{aligned} \sum_{i\in I_k} \frac{E(Y_i^2 e^{hY_i})}{E(e^{hY_i})} &\leq \sum_{i\in I_k} \{e^c E(Y_i^2) + (1+\eta/2)\delta a_{n_k} \frac{E(Y_i e^{hY_i})}{E(e^{hY_i})}\} \\ &\leq e^c E(u^2(T_k^\delta)) \quad + (1+\eta/2)\delta a_{n_k} \varphi_{u(T_k^\delta)}(h) \\ &\leq e^c E(\|T_k^\delta\|^2) \quad + (1+\eta/2)\delta a_{n_k}^2 \varepsilon. \end{aligned}$$

D'après l'inégalité (3) il existe un entier $k_0 = k_0(\varepsilon)$ tel que :

$$\forall \quad k \geq k_0 \quad \sum_{i \in I_K} E(Y_i^2 e^{hY_i})/E(e^{hY_i}) \leq (1+\eta)\varepsilon\delta a_{n_k}^2$$
$$\leq (\delta+\varepsilon)\varepsilon a_{n_k}^2/2.$$

On applique alors le lemme 5 aux variables $Y_i, i \in I_k$, avec $a = \varepsilon a_{n_k}$ et $d = a_{n_k}(\varepsilon - \eta)$ pour obtenir :

$$P\{u(T_k^\delta) \geq \eta a_{n_k}\} \geq e^{-h(2\varepsilon-\eta)}a_{n_k}[1-(\varepsilon-\eta)^{-2}a_{n_k}^{-2}\sum_{i\in I_k} E(Y_i^2 e^{hY_i})/E(e^{hY_i})]$$
$$\geq e^{-2\varepsilon h a_{n_k}}[1-\varepsilon(\delta+\varepsilon)/2(\varepsilon-\eta)^2],$$

avec $1-\varepsilon(\delta+\varepsilon)/2(\varepsilon-\eta)^2 > 0$ par (4).

On a donc finalement prouvé :

$$\forall \varepsilon > 0, \exists \eta = \eta(\varepsilon) > 0, \exists \alpha(\varepsilon) > 0, \exists k_0(\varepsilon), \forall u \in B_1', \forall k \geq k_0(\varepsilon)$$

$$\alpha(\varepsilon)P\{u(T_k^{\varepsilon/2}) \geq \eta a_{n_k}\} \geq \exp\{-2\varepsilon a_{n_k} h_k(u, \varepsilon/2, \varepsilon)\}$$
$$\Rightarrow \alpha(\varepsilon)P\{\|T_k^{\varepsilon/2}\| \geq \eta a_{n_k}\} \geq \exp\{-2\varepsilon a_{n_k} h_k(u, \varepsilon/2, \varepsilon)\} \quad (6)$$

Deuxième étape. Fixons à nouveau $\varepsilon > 0$ et $u \in B_1'$; soit $\delta = \varepsilon/4$. Supposons $h = h(u(T_k^\varepsilon), \varepsilon a_{n_k})$ fini et distingons comme précédemment deux cas :

(a) $\sum_{i\in I_k} E(1_{\{a_i\delta < \|X_i\| \leq a_i\varepsilon\}} u(X_i^\delta) e^{hu(X_i^\delta)}) \geq \varepsilon a_{n_k}/2$.

Alors :

$$\varepsilon a_{n_k}/2 \leq \sum_{i\in I_k} \varepsilon a_{n_k} e^{h\varepsilon a_{n_k}} P\{a_i\delta < \|X_i\| \leq \varepsilon\alpha_i\},$$

et par conséquent :

$$\exp\{-\varepsilon h a_{n_k}\} \leq 2\sum_{i\in I_k} P\{\delta a_i \leq \|X_i\|\}. \quad (7)$$

(b) $\sum_{i\in I_k} E(1_{\{\delta a_i < \|X_i\| \leq \varepsilon a_i\}} u(X_i^\delta) e^{hu(X_i^\delta)}) < \varepsilon a_{n_k}/2$.

Alors *à fortiori*

$$\sum_{i\in I_k} E(1_{\{\delta a_i < \|X_i\| \leq \varepsilon a_i\}} u(X_i^\delta) e^{hu(X_i^\delta)})/E(e^{hu(X_i^\varepsilon)}) < \varepsilon a_{n_k}. \quad (8)$$

L'application $\delta \to E\exp\{hu(X_i^\delta)\}$ étant croissante pour tout i on a :

$$\sum_{i\in I_k} E(u(X_i^\delta)e^{hu(X_i^\delta)})/E(e^{hu(X_i^\delta)}) \geq \sum_{i\in I_k} E(u(X_i^\delta)e^{hu(X_i^\delta)})/E(e^{hu(X_i^\varepsilon)}). \tag{9}$$

D'autre part :

$$\begin{aligned} E(u(X_i^\delta)e^{hu(X_i^\delta)}) &= E(1_{\{\|X_i\|\leq\varepsilon a_i\}}u(X_i)e^{hu(X_i)}) \\ &\quad -E(1\{\delta a_i < \|X_i\| \leq \varepsilon a_i\}u(X_i)e^{hu(X_i)}) \\ &= E(u(X_i^\varepsilon)e^{hu(X_i^\varepsilon)}) - E(1_{\{\delta a_i<\|X_i\|\leq\varepsilon a_i\}}u(X_i)e^{hu(X_i)}). \end{aligned} \tag{10}$$

De (8), (9), (10) et de la définition de h il résulte :

$$\sum_{i\in I_k} E(u(X_i^\delta)e^{hu(X_i^\delta)})/E(e^{hu(X_i^\delta)}) \geq \varepsilon a_{n_k} - \varepsilon a_{n_k}/2 = \varepsilon a_{n_k}/2.$$

On en déduit l'inégalité :

$$h_k(u(T_k^\varepsilon), \varepsilon a_{n_k}) \geq h_k(u, \varepsilon/4, \varepsilon/2) = h(u(T_k^{\varepsilon/4}), \varepsilon a_{n_k}/2). \tag{11}$$

En conséquence de (6), (7), et (11) :

$$\forall \quad \varepsilon > 0, \quad \exists \ \alpha > 0, \quad \exists \ k_0 \in N, \quad \forall \ k \geq k_0, \quad \forall \ u \in B_1'$$

$$\begin{aligned} &\exp\{-\varepsilon a_{n_k} h_k(u(T_k^\varepsilon), \varepsilon a_{n_k})\} \\ &\qquad \leq \max\{2\sum_{i\in I_k} P\{\|X_i\| \geq \varepsilon a_i/4\}, e^{-\varepsilon a_{n_k} h_k(u,\varepsilon/4,\varepsilon/2)}\} \\ &\qquad \leq \max\{2\sum_{i\in I_k} P\{\|X_i\| \geq \varepsilon a_i/4\}, \alpha(\varepsilon/2)P\{\|T_k^{\varepsilon/4}\| \geq \eta a_{n_k}\}\}. \end{aligned}$$

Il en résulte finalement que le terme $\exp\{-\varepsilon a_{n_k} h_k(\varepsilon)\}$ est majoré par le membre de droite de l'inégalité précédente, ce qui implique la convergence de la série

$$\sum_{k\geq 1} \exp\{-\varepsilon a_{n_k} h_k(\varepsilon)\}$$

et conclut la démonstration du théorème 6.

4. Suffisance de la condition (CN)

Il nous a paru intéressant de faire figurer dans cet article la démonstration, adaptée de [5], de la partie du théorème 1 relative à la suffisance de la condition (CN) pour des variables aléatoires à valeurs réelles.

Comme précédemment nous considérerons :

—une suite $(X_n)_n$ de v.a.r. symétriques et indépendantes

—une suite croissante de réels strictement positifs $(a_n)_n$

—une suite strictement croissante d'entiers $(n_k)_k$.

Nous supposerons de plus satisfaite l'hypothèse suivante :

$$\exists\ M > 1,\ \exists\ N > 0, \forall\ k \geq 1 \qquad Ma_{n_k} \leq a_{n_{k+1}} \leq Na_{n_k+1}. \tag{12}$$

Nous adopterons des notations similaires à celles de la partie 3 :

$$X_n^\delta = X_n 1_{\{|X_n| \leq \delta a_n\}},$$

$$T_k^\delta = \sum_{i \in I_k} X_i^\delta, \qquad (I_k =]n_{k-1}, n_k])$$

et

$$h_k(\varepsilon) = h(T_k^\varepsilon, \varepsilon a_{n_k}).$$

Le théorème énonçant la suffisance de la condition (CN) est le suivant:

Théorème 7. (Nagaev [5]). *Pour que $X_1 + \cdots + X_n/a_n$ converge presque sûrement vers 0 il suffit que les deux conditions ci-dessous soient vérifiées :*

(1) $\forall \quad \varepsilon > 0 \qquad \sum_{n\geq 1} P\{|X_n| \geq \varepsilon a_n\} < +\infty$

(2) $\forall \quad \varepsilon > 0 \qquad \sum_{k\geq 1} \exp\{-\varepsilon a_{n_k} h_k(\varepsilon)\} < +\infty.$

Démonstration. Du fait de la symétrie des variables et des inégalités (12), $X_1 + \cdots + X_n/a_n$ converge p.s. vers 0 si et seulement si la suite $(a_{n_k}^{-1} \sum_{i\in I_k} X_i)_k$ converge p.s. vers 0; cette dernière convergence équivaut à :

$$\forall \quad \varepsilon > 0 \sum_{k\geq 1} P\{a_{n_k}^{-1} |\sum_{i \in I_k} X_i| \geq \varepsilon\} < +\infty \tag{13}$$

Tenant compte de la symétrie des variables et de l'hypothèse (1), (13) équivaut à :

$$\forall \quad \varepsilon > 0 \sum_{k \geq 1} P\{T_k^{\varepsilon/2} \geq \varepsilon a_{n_k}\} < +\infty. \tag{14}$$

Le lemme 2 montre que :

$$P\{T_k^{\varepsilon/2} \geq \varepsilon a_{n_k}\} \leq exp\{-\varepsilon a_{n_k} h(T_k^{\varepsilon/2}, \varepsilon a_{n_k}/2)/2\}$$
$$= exp\{-\frac{1}{2}\varepsilon a_{n_k} h_k(\varepsilon/2)\}.$$

(14) est alors une conséquence immédiate de l'hypothèse (2).

5. Suffisance de la condition (CN′)

La partie précédente soulève le problème de l'extension du théorème 7 à la dimension infinie, autrement dit celui d'une éventuelle suffisance de la condition (CN′) du théorème 6. Nous apportons ici une réponse partielle en prouvant que dans le cadre de l.f.g.n. de Prokhorov cette condition est effectivement suffisante.

Rappelons au préalable la généralisation aux espaces de Banach du théorème 2 démontrée dans [3].

Soit $(X_n)_n$ une suite de v.a. indépendantes et centrées à valeurs dans un espace de Banach séparable B; et soient $(A_n)_n$ et $(b_n)_n$ deux suites croissantes de réels strictement positifs. Soit d'autre part un suite croissante d'entiers pour laquelle il existe un réel $M > 1$ tel que :

$$\forall \quad k \geq 1 \qquad M a_{n_k} \leq a_{n_{k+1}} \leq M^3 a_{n_k+1}. \qquad (H)$$

Posons :

$$I_k =]n_{k-1}, n_k],$$
$$\sigma_k^2 = \sup\{\sum_{I \in I_k} E\langle X_i, x'\rangle^2; \|x'\| \leq 1\},$$

et

$$S_n = X_1 + \cdots + X_n.$$

Le résultat généralisant le théorème 2 à la dimension infinie est le suivant :

Théorème 8. ([3], théorème 7). *En plus des hypothèses précédentes supposons vérifiées les conditions :*

(1) $\forall \quad n \geq 1 \qquad \|X_n\| \leq b_n \qquad$ p.s.

(2) $\exists \quad \gamma > 0 \qquad \sum_{k \geq 1} \exp(-\gamma a_{n_k}/b_{n_k}) < +\infty$

(3) S_n/a_n *converge en probabilité vers* 0.

Sous ces hypothèses la suite $(S_n/a_n)_n$ *converge p.s. vers* 0 *si et seulement si* :

$$\forall \quad \varepsilon > 0 \sum_{k \geq 1} \exp(-\varepsilon[a_{n_k}/\sigma_k]^2) < +\infty. \qquad (CP')$$

C'est dans le cadre que nous venons de préciser que sera énoncée la suffisance de la condition (CN′) sous forme d'équivalence entre les conditions (CN′) et (CP′). Cette équivalence peut être considérée comme une extension à la dimension infinie du corollaire 2 (et de sa démonstration) de [5]. Elle montre que dans le théorème 8 la condition (CP′) peut être remplacée par la condition (CN′); cette dernière est donc, dans le cadre restrictif de ce théorème, une condition suffisante pour la l.f.g.n. en dimension infinie. Le théorème qui suit donne l'énoncé précis de cette équivalence.

Théorème 9. *Soit $(X_n)_n$ une suite de v.a. symétriques et indépendantes à valeurs dans B; soient $(a_n)_n$ et $(b_n)_n$ deux suites croissantes de réels strictement positifs et $(n_k)_k$ une suite d'entiers vérifiant (H). Supposons satisfaites les deux conditions :*

$$(1) \quad \forall \quad n \geq 1 \qquad \|X_n\| \leq b_n \qquad \text{p.s.}$$

$$(2) \quad \exists \quad \gamma > 0 \qquad \sum_{k \geq 1} \exp(-\gamma a_{n_k}/b_{n_k}) < +\infty.$$

Les conditions suivantes sont alors équivalentes :

$$(CP') \quad \forall \quad \varepsilon > 0 \qquad \sum_{k \geq 1} \exp(-\varepsilon[a_{n_k}/\sigma_k]^2) < +\infty$$

$$(CN') \quad \forall \quad \varepsilon > 0 \qquad \sum_{k \geq 1} \exp(-\varepsilon h_k(\varepsilon) a_{n_k}) < +\infty.$$

Démonstration. Introduisons en premier lieu les notations complémentaires suivantes :

$$Y_i = Y_i(\varepsilon, u) = u(X_i^\varepsilon), \qquad (X_i^\varepsilon = X_i \, 1_{\{\|X_i\| \leq \varepsilon a_i\}})$$

$$T_k^\varepsilon = \sum_{i \in I_k} X_i^\varepsilon,$$

et

$$g_i^\varepsilon(a, u) = g_{Y_i}(a) = E(e^{aY_i}).$$

Puis :

$$\Phi_k^\varepsilon(a, u) = \varphi_{u(T_k^\varepsilon)}(a) = \sum_{i \in I_k} g'_{Y_i}(a)/g_{Y_i}(a),$$

$$h_k(\varepsilon, u) = h(u(T_k^\varepsilon), \varepsilon a_{n_k}), \qquad (\text{solution de: } \Phi_k^\varepsilon(h, u) = \varepsilon a_{n_k})$$

et

$$h_k(\varepsilon) = h(T_k^\varepsilon, \varepsilon a_{n_k}). \qquad (= \inf\{h(u(T_k^\varepsilon), \varepsilon a_{n_k}); \|u\| \leq 1\})$$

Donnons nous un réel $\varepsilon > 0$ et fixons $\delta > 0$ vérifiant :

$$\delta < 1, \quad \varepsilon' = \varepsilon\delta < 1/2e, \quad \gamma \leq 1/2e\delta. \tag{15}$$

Posons : $h_k = b_{n_k}^{-1}$. Comme a_n/b_n tend vers l'infini quand n croît, et tenant compte de (H), il existe un entier k_0 tel que :

$$\forall\ k \geq k_0, \quad \forall\ \ i \in I_k, \qquad \varepsilon a_i \geq \varepsilon' a_i \geq b_{n_k} \geq b_i. \tag{16}$$

Dans la suite nous supposerons $k \geq k_0$.

Soit $a \geq 0$ et $u \in B_1'$. En conséquence de (16) :

$$\forall\ i \in I_k \qquad Y_i' = Y_i(\varepsilon', u) = u(X_i^{\varepsilon'}) = u(X_i) \qquad p.s.$$

Donc :

$$\begin{aligned} g'_{Y_i'}(a) &= E(u(X_i)e^{au(X_i)}) \\ &= E(\sum_{j\geq 0} a^j [u(X_i)]^{j+1}/j!). \end{aligned}$$

Par symétrie de la variable X_i les puissances impaires de $u(X_i)$ sont d'espérance nulle. D'où la minoration :

$$\forall \quad a \geq 0 \qquad g'_{Y_i'}(a) \geq a\, E\langle X_i, u\rangle^2. \tag{17}$$

Si de plus $0 \leq a \leq h_k (= b_{n_k}^{-1})$, alors presque sûrement :

$$|aY_i'| = a|u(X_i)| \leq b_{n_k}^{-1} b_i \leq 1.$$

Par conséquent :

$$\forall\ i \in I_k, \quad \forall\ a \leq h_k \qquad 1 \leq g_{Y_i'}(a) = E\ e^{aY_i'} \leq e. \tag{18}$$

Preuve de (CN′) ⇒ (CP′). Par définition de σ_k^2 il existe $u \in B_1'$ tel que :

$$\sigma_k^2 \leq 2 \sum_{i\in I_k} E\langle X_i, u\rangle^2. \tag{19}$$

Fixons un tel élément u et distingons deux cas :

$$(a)\ \ h_k(\varepsilon', u) \geq h_k. \qquad\qquad (\Leftrightarrow \Phi_k^{\varepsilon'}(h_k, u) \leq \varepsilon' a_{n_k}) \tag{20}$$

Utilisant (17) et (18) on obtient immédiatement :

$$\forall\ i \in I_k \quad g'_{Y'_i}(h_k)/g'_{Y'_i}(h_k) \geq e^{-1} h_k E\langle X_i, u\rangle^2.$$

D'où :

$$\Phi_k^{\varepsilon'}(h_k, u) \geq e^{-1} h_k \sum_{i \in I_k} E\langle X_i, u\rangle^2. \tag{21}$$

Des inégalités (15), (19) (20) et (21) résulte alors :

$$\varepsilon' a_{n_k} \geq (2e)^{-1} h_k \sigma_k^2 = (2e)^{-1} \sigma_k^2 / b_{n_k}$$

$$\Rightarrow -\varepsilon a_{n_k}^2/\sigma_k^2 \leq -(2e\delta)^{-1} a_{n_k}/b_{n_k}$$

$$\leq -\gamma a_{n_k}/b_{n_k}. \tag{22}$$

(*b*) $h_k(\varepsilon', u) < h_k$.

Un calcul similaire au précédent montre que :

$$\Phi\varepsilon'_k(h_k(\varepsilon', u), u) \geq e^{-1} h_k(\varepsilon', u) \sum_{i \in I_k} \langle X_i, u\rangle^2.$$

Comme par définition $\Phi\varepsilon'_k(h_k(\varepsilon', u), u) = \varepsilon' a_{n_k}$, on obtient par (19) :

$$\varepsilon' a_{n_k} \geq (2e)^{-1} h_k(\varepsilon', u) \sigma_k^2.$$

D'où :

$$-\varepsilon' a_{n_k}^2/\sigma_k^2 \leq -(2e)^{-1} a_{n_k} h_k(\varepsilon', u).$$

Puisque $\varepsilon > \varepsilon'$ et $\varepsilon' < (2e)^{-1}$, il vient finalement :

$$-\varepsilon a_{n_k}^2/\sigma_k^2 \leq -\varepsilon' a_{n_k} h_k(\varepsilon', u)$$

$$\leq -\varepsilon' a_{n_k} h_k(\varepsilon'). \tag{23}$$

Les inégalités (22) et (23) prouvent que si la condition (CN′) est vérifiée la condition (CP′) l'est également.

Preuve de (CP′) ⇒ (CN′) . Soit $u \in B'_1$ tel que : $h_k(\varepsilon, u) \leq 2\, h_k(\varepsilon)$. Nous distinguerons à nouveau deux cas :

(a) $\varepsilon h_k(\varepsilon, u) > 2\gamma h_k$.

Dans ce cas on a immédiatement :

$$2\varepsilon\, h_k(\varepsilon) > 2\gamma\, h_k = 2\gamma / b_{n_k};$$

d'où :

$$-\varepsilon a_{n_k} h_k(\varepsilon) < -\gamma a_{n_k}/b_{n_k}. \tag{24}$$

(b) $\varepsilon h_k(\varepsilon, u) \le 2\gamma h_k$.

Soit $a \ge 0$; on a en vertu de l'inégalité de Jensen :

$$\forall\ i \in I_k \qquad g_{Y_i}(a) \ge 1.$$

D'autre part, tenant compte des inégalités (16) :

$$\begin{aligned} g'_{y_i}(a) &= E[u(X_i)e^{au(X_i)}] \\ &= \sum_{j\ge 0} a^j\, E[u(X_i)]^j/j! \\ &\le E[u(X_i)]^2(a + a^3 b_i^2/3! + a^5 b_i^4/5!\ + \cdots) \\ &\le a\, e^{ab_i}\, E\langle X_i, u\rangle^2. \end{aligned}$$

Prenant $a = h_k(\varepsilon, u)$ et remarquant que :

$$b_i h_k(\varepsilon, u) \le 2\gamma b_i/\varepsilon b_{n_k} \le 2\gamma/\varepsilon,$$

on obtient :

$$g'_{Y_i}[h_k(\varepsilon, u)] \le h_k(\varepsilon, u) e^{2\gamma/\varepsilon} E\langle X_i, u\rangle^2.$$

Il en résulte :

$$\begin{aligned} \varepsilon a_{n_k} = \Phi_k^\varepsilon(h_k(\varepsilon, u), u) &< e^{2\gamma/\varepsilon} h_k(\varepsilon, u)\sigma_k^2 \\ &\le e^{2\gamma/\varepsilon} h_k(\varepsilon)\sigma_k^2. \end{aligned}$$

D'où finalement :

$$-\varepsilon a_{n_k} h_k(\varepsilon) \le -\varepsilon^2 (2e^{2\gamma/\varepsilon})^{-1} a_{n_k}^2/\sigma_k^2. \tag{25}$$

Les inégalités (24) et (25), ainsi que l'hypothèse (2), montrent alors que la condition (CP′) entraîne la condition (CN′), ce qui achève la démonstration du théorème 9.

6. Un exemple

Nous présentons pour conclure un exemple prouvant que la condition (CN′) n'est pas en général suffisante pour la l.f.g.n. ; plus précisément nous montrerons que le théorème 2 ne s'étend pas tel quel à la dimension infinie.

L'exemple, dû à B. Heinkel, est tiré de [4] et s'adapte facilement au présent cadre. Rappelons la façon dont cet exemple est construit. On considère d'abord pour tout entier n une suite $\varepsilon_1, \ldots, \varepsilon_{2^n}$ de v.a. de Rademacher indépendantes et une suite $\eta_1, \ldots, \eta_{2^n}$ de v.a. indépendantes entre elles et indépendantes des ε_i, vérifiant :

$$\forall\ i = 1, \ldots, 2^n \qquad \begin{array}{l} P\{\eta_i = 1\} = 2^{-n} \\ P\{\eta_i = 0\} = 1 - 2^{-n}. \end{array}$$

On divise l'ensemble d'entiers $\{1, 2, \ldots, 2^n\}$ en s intervalles consécutifs $B_1, \ldots, B_s$ de longueur $[(L_2 n)^4]$, sauf peut-être le dernier. Notant par $(e_n)_{n \geq 1}$ la base canonique de l^2, on pose pour tout $i = 1, \ldots, s$ et $k \in B_i$:

$$Y_k = (2^n \varepsilon_k \eta_k / L_2 n) e_i.$$

On désigne alors par $(X_n)_n$ une suite de v.a. indépendantes à valeurs dans l^2 telles que pour tout entier n et $k \in I_n =]2^n, 2^{n+1}]$ on ait : $\mathcal{L}(X_k) = \mathcal{L}(Y_{k-2^n})$.

Il est montré dans [4] que la suite $(X_n)_n$ vérifie la loi faible des grands nombres, mais non la loi forte. De plus, par construction des variables X_n, il est aisé de constater que pour tout $\varepsilon > 0$ on a :

$$\sum_{n \geq 1} P\{\|X_n\| \geq \varepsilon n\} < +\infty.$$

Pour mettre en défaut la généralisation du théorème 7 à la dimension infinie (avec $a_n = n$ et $n_k = 2^k$) il ne reste plus qu'à démontrer que la condition (CN′) est satisfaite. Pour cela fixons un réel $\varepsilon > 0$ et remarquons que pour n assez grand, par définition des X_n :

$$T = T^\varepsilon = \sum_{k \in I_n} X_k 1_{\{\|X_k\| \leq \varepsilon k\}} = \sum_{k \in I_n} X_k.$$

Soit $u = (a_1, \ldots, a_s, 0, 0, \ldots)$ un élément de la boule unité de l^2. Posons pour tout $h > 0$:

$$g_n(h) = E\exp[hu(T)];$$

on voit sans peine que :

$$\begin{aligned} \varphi_n(h) = g_n'(h)/g_n(h) &= \sum_{i=1}^{s} \frac{2^n a_i}{L_2 n} \sum_{k \in B_i} \frac{E(\varepsilon_k \eta_k \exp\{h a_i \varepsilon_k \eta_k 2^n / L_2 n\})}{E\exp\{h a_i \varepsilon_k \eta_k 2^n / L_2 n\}} \\ &= \sum_{i=1}^{s} a_i \sum_{k \in B_i} \frac{2^n}{L_2 n} \frac{sh(h a_i 2^n / L_2 n)}{1 - 2^{-n} + 2^{-n} ch(h a_i 2^n / L_2 n)} \\ &= \sum_{i=1}^{s} a_i b_i. \end{aligned}$$

Comme $\lim_{h\to+\infty} \varphi_n(h) \sim_{n\to+\infty} 2^n(L_2n)^3 \sum_{i=1}^{s}|a_i| \geq 2^n(L_2n)^3$, on a pour n assez grand : $\lim_{h\to+\infty} \varphi_n(h) > \varepsilon 2^{n+1}$. L'équation $\varphi_n(h) = \varepsilon 2^{n+1}$ admet donc une solution finie $h_n(u,\varepsilon)$.

Fixons à présent $h = 2^{-n}(\log n)^2$, et supposons que tous les a_i sont positifs ou nuls, ce qui ne restreint pas la généralité. Posons :

$$A = \{i \in [1,s] \mid a_i \leq L_2n(\log n)^{-2}\}\text{et } \ B = [1,s] - A.$$

Ces notations permettent d'écrire :

$$\varphi_n(h) = \sum_{i\in A} a_i b_i + \sum_{i\in B} a_i b_i.$$

Tenant compte de l'inégalité :

$$\forall\ x \in [0,1] \qquad sh(x) \leq ex,$$

on obtient :

$$\begin{aligned}\sum_{i\in A} a_i b_i &\leq \sum_{i\in A} e(\log n/L_2n)^2 a_i^2[(L_2n)^4] \\ &\leq (\log n)^3.\end{aligned}$$

Remarquons qu'il y a au plus $(\log n)^4/(L_2n)^2$ indices i dans B. Il en résulte que :

$$\begin{aligned}\sum_{i\in B} a_i b_i &\leq \sum_{i\in B} a_i[(L_2n)^4](L^2n)^{-1} sh\{(\log n)^2/L_2n\} \\ &\leq (\log n)^4 L_2n sh\{(\log n)^2/L_2n\} \\ &< \exp{(\log n)^2}.\end{aligned}$$

D'où :

$$\begin{aligned}\varphi_n(2^{-n}(\log n)^2) &\leq \exp\{2(\log n)^2\} \\ &\leq \varepsilon 2^{n+1}.\end{aligned}$$

On en déduit la minoration :

$$h_n(u,\varepsilon) \geq 2^{-n}(\log n)^2.$$

Le choix particulier de la forme linéaire u optimise toutes les situations possibles; par conséquent :

$$h_n(\varepsilon) = \inf\{h_n(u,\varepsilon); \quad ||u|| \leq 1\} \geq 2^{-n}(\log n)^2.$$

La minoration précédente implique que la condition (CN′) est satisfaite :

$$\forall\ \varepsilon > 0 \qquad \sum_{n\geq 1} \exp\{-\varepsilon 2^{n+1} h_n(u)\} < +\infty,$$

ce qui achève de prouver que le théorème 7 ne s'étend pas à la dimension infinie.

REFERENCES

[1] De Acosta, A., *Inequalities for B-valued random vectors with applications to the strong law of large numbers*, Ann. Probability, **9** (1981), 157–161.

[2] Alt, J.-C., *La loi des grands nombres de Prokhorov dans les espaces de type p*, Ann. Inst. Henri Poincaré, **23** (1987), 561–574.

[3] Alt, J.-C., *Une forme générale de la loi forte des grands nombres pour des variables aléatoires vectorielles*, à paraître dans les actes de la 4ème Conférence internationale sur la théorie des probabilités dans les espaces vectoriels, Lancut (Pologne), juin 1987.

[4] Heinkel, B., *A law of large numbers for random vectors having large norms*, prepublication.

[5] Nagaev, S. V., *On necessary and sufficient conditions for the strong law of large numbers*, Theor. Prob. Appl., **17** (1972), 573–581.

[6] Nagaev, S. V. et Volodin, N.A., *A remark on the strong law of large numbers*, Theor. Prob. Apppl., **22** (1977), 810-813.

[7] Prokhorov, Y.V., *Some remarks on the strong law of large numbers*, Theor. Prob. Appl., **4** (1959), 204–208.

Jean-Christian Alt
Université Louis Pasteur
Département de Mathématique
7, rue René Descartes
67084 Strasbourg cédex
France

HYPERACCURACY OF BOOTSTRAP BASED PREDICTION

Chongen Bai[1]
Department of Economics
Harvard University
Cambridge, MA 02138

Peter J. Bickel[2]
Department of Statistics
University of California, Berkeley
Berkeley, CA 94720

Richard A. Olshen[1,3]
Department of Mathematics
University of California, San Diego
La Jolla, CA 92093

1. INTRODUCTION

This paper grew out of attempts to understand a bootstrap-based nonparametric percentile-t-like method for forming prediction intervals. Our interest in these intervals derives from a study of the development of walking in children. The particular application is to prediction of a "test case" zeroth order Fourier coefficient in a random coefficient model. See Sutherland et al. (1988), Olshen et al. (1989), and the discussion by Bai and Olshen (1988) of Hall's (1988) paper for background. Our main message here is summarized in a heuristic that not only "explains" an anomaly regarding this type of bootstrap-based prediction, but also clarifies the hyperaccuracy of certain percentile-t-based confidence intervals, and more. (We learned the term "hyperaccuracy" from Wing Wong.) We also discuss the behavior of the "naive" method of using the percentiles of the empirical distribution of the learning sample as prediction bounds (suggested to us by M. Pollak).

The development of our approach begins with data $Z, X_1, \ldots, X_n$ that are assumed to be independent with common distribution F and density $F' = f$. (We denote both the cumulative distribution and its corresponding probability measure with the same notation.) The X's are termed a ***learning sample*** and Z a ***test case***. In the context of prediction, on the basis of the learning sample, one wishes to compute a reasonable interval in which the test case lies with preassigned (unconditional) probability, say $\alpha < 1$. Of course, if F is Gaussian then this is an easy exercise, whether or not the common variance σ^2 is known. However, typically F is unknown. One of our main

[1]Research supported in part by National Science Foundation Grants DMS 85-05609 and DMS 87-22306.

[2]Research supported in part by Office of Naval Research Contract N00014-80-C-0163.

[3]Research supported in part by the John Simon Guggenheim Memorial Foundation.

results is that a bootstrap process yields intervals for which the preassigned probability is quite accurately attained. In order to proceed with the details, we define

$$\bar{X} = n^{-1}\sum_{i=1}^{n} X_i \quad \text{and} \quad S^2 = n^{-1}\sum_{i=1}^{n}(X_i - \bar{X})^2.$$

Throughout, we require that F has a positive density f; additional assumptions are required to complete the arguments.

Define $t_{\alpha,p} = t_{\alpha,p}(n)$ by

$$P\left\{\frac{\bar{X} - Z}{S} \le t_{\alpha,p}\right\} = \alpha \,. \tag{1}$$

Denote $P\{\cdot\,|X_1,\ldots,X_n\}$ by $P^*\{\,\cdot\,\}$ and bootstrapped $\bar{X}$ and Z, respectively, by X^* and Z^*. Our assumptions entail that the empirical distribution of the learning sample (almost surely) has jumps of height only n^{-1}, so it is possible to define $t^*_{\alpha,p} = t^*_{\alpha,p}(n)$ that satisfies

$$P^*\left\{\frac{\bar{X}^* - Z^*}{S^*} \le t^*_{\alpha,p}\right\} = \alpha + O_p(n^{-1})\,. \tag{2}$$

While we do not know $t_{\alpha,p}$, we can substitute for it in (1) an estimate $t^*_{\alpha,p}$ derived by an empirical bootstrap process. The estimation can be done not only in principle, but, what matters, to within arbitrarily high accuracy with arbitrarily high probability by doing sufficiently many simulations. So in what follows we concentrate only on the theoretical bootstrap quantities denoted by asterisks. Two basic questions arise. How close is

$$P\left\{\frac{\bar{X} - Z}{S} \le t^*_{\alpha,p}\right\} \tag{3}$$

to α? How close is $t^*_{\alpha,p}$ to $t_{\alpha,p}$? Answers to these questions are most interesting when they are compared and contrasted to the corresponding questions for confidence intervals.

These questions can be reframed in terms of (say) the "$1 - \alpha$" lower prediction bound

$$\underline{\hat{Z}} \equiv \bar{X} - St^*_{\alpha,p}\,.$$

How close is $P[Z \ge \hat{Z}]$ to $1 - \alpha$? How close is $\underline{\hat{Z}}$ to the theoretical (but uncomputable) bound

$$\underline{Z} \equiv \bar{X} - St_{\alpha,p}\,?$$

In the context of confidence intervals, we introduce the unknown mean μ of F and define $t_{\alpha,c} = t_{\alpha,c}(n)$ by

$$P\left\{\frac{n^{\frac{1}{2}}(\bar{X} - \mu)}{S} \le t_{\alpha,c}\right\} = \alpha \,. \tag{4}$$

Note that in this more widely discussed setting, the role of the unknown parameter μ replaces that of the test case Z in (1). The $n^{\frac{1}{2}}$ term in the numerator of (4) renders its scale comparable to that of the numerator of (1). We define $t^*_{\alpha,c}$ by

$$P^*\left\{\frac{n^{\frac{1}{2}}(\bar{X}^*-\bar{X})}{S^*}\le t^*_{\alpha,c}\right\}=\alpha+O_p(n^{-1}), \tag{5}$$

and again ask two questions. How close is

$$P\left\{\frac{n^{\frac{1}{2}}(\bar{X}-\mu)}{S}\le t^*_{\alpha,c}\right\} \tag{6}$$

to α? How close is $t^*_{\alpha,c}$ to $t_{\alpha,c}$?

Hall (1988) indicates that under suitable conditions

$$t^*_{\alpha,c}-t_{\alpha,c}=O_p(n^{-1}) \tag{7}$$

and

$$(6)-(4)=O(n^{-1}); \tag{8}$$

see also Singh (1981). Bai and Olshen (1988) demonstrate that, perhaps surprisingly,

$$O_p(n^{-1})\neq t^*_{\alpha,p}-t_{\alpha,p}=O_p(n^{-\frac{1}{2}}). \tag{9}$$

However, despite (9), rigorous arguments establish that

$$(3)-(1)=O(n^{-\frac{3}{4}+\gamma}) \quad \text{for all} \quad \gamma>0, \tag{10}$$

and heuristic arguments suggest that even more is true. ((10) requires less in the way of assumptions than does (8).)

2. HEURISTICS

To see why the behavior given in (8), (9), (10), is not unexpected, consider the "naive" $1-\alpha$ lower prediction bound $F_n^{-1}(\alpha)$, the Fisher consistent (or "bootstrap") estimate of the theoretical percentile $F^{-1}(\alpha)$.

$$F_n^{-1}(\alpha)=X_{([n\alpha])}$$

where $X_{(1)}\le\cdots\le X_{(n)}$ are the order statistics of the training sample and $[n\alpha]$ is the largest integer $\le n\alpha$. If F is continuous, then

$$P\{Z\ge X_{([n\alpha])}\}=1-\frac{[n\alpha]}{n+1}=1-\alpha+O(n^{-1}),$$

while if F has a continuous density, then

$$\mathcal{L}(n^{\frac{1}{2}}(F_n^{-1}(\alpha)-F^{-1}(\alpha))\to N\left(0,\frac{\alpha(1-\alpha)}{f^2(F^{-1}(\alpha))}\right),$$

so that $F_n^{-1}(\alpha)=F^{-1}(\alpha)+O_p(n^{-\frac{1}{2}})$.

Our heuristics are based on looking at the limiting behavior of the joint distribution of $(Z, \sqrt{n}\,(t^*_{\alpha,p} - t_{\alpha,p}))$ in the prediction case and $\left(\frac{\sqrt{n}\,(\bar{X}-\mu)}{S}, \sqrt{n}\,(t^*_{\alpha,c} - t_{\alpha,c})\right)$ for confidence bounds. Specifically, suppose that (U_n, V_n) converges in law to (U, V) (say) and that $\sup_n E(V_n)^2 < \infty$. Suppose also that both U_n and $U_n + \frac{V_n}{\sqrt{n}}$ admit asymptotic expansions of the form

$$P\{U_n \leq t\} = A(t) + n^{-\frac{1}{2}}B(t) + O(n^{-1}) \tag{11}$$

$$P\left\{U_n + \frac{V_n}{\sqrt{n}} \leq t\right\} = A(t) + n^{-\frac{1}{2}}C(t) + O(n^{-1}), \tag{12}$$

where $A(t)$ is necessarily $P\{U \leq t\}$, and $O(n^{-1})$ is taken uniform in t. Suppose A, B, C are all continuously differentiable with bounded derivatives a, b, c.

Lemma 1. *Under these conditions,*

$$B(t) = C(t) = 0 \quad \textit{for all} \quad t$$

if and only if (13)

$$E(V|U) = 0 \quad a.s.$$

Proof. Let Ψ be an infinitely differentiable function with compact support. Then by an integration by parts argument,

$$E\Psi(U_n) = \int \Psi(u)a(u)\,du + n^{-\frac{1}{2}} \int \Psi(u)b(u)\,du + O(n^{-1}) \tag{14}$$

$$E\Psi\left(U_n + \frac{V_n}{\sqrt{n}}\right) = \int \Psi(u)a(u)\,du + n^{-\frac{1}{2}} \int \Psi(u)c(u)\,du + O(n^{-1}). \tag{15}$$

But

$$\begin{aligned} E\Psi\left(U_n + \frac{V_n}{\sqrt{n}}\right) &= E\Psi(U_n) + n^{-\frac{1}{2}}E(V_n\Psi'(U_n)) + O(n^{-1}) \\ &= E\Psi(U_n) + n^{-\frac{1}{2}}E(V\Psi'(U)) + o(n^{-\frac{1}{2}}) \end{aligned} \tag{16}$$

since $\sup_n EV_n^2 < \infty$.

Therefore, $B = C$ if and only if for all such Ψ

$$E(V\Psi'(U)) = 0.$$

Since the $\Psi'(U)$ are dense the lemma follows.

Now take

$$U_n = \frac{\sqrt{n}\,(\bar{X} - \mu)}{S}; \; V_n = \sqrt{n}\,(t_{\alpha,c} - t^*_{\alpha,c}).$$

If $E(X)^4 < \infty$ and F is absolutely continuous, then, by Hall (1988) for instance, (11) and (12) hold with U normal, $V \equiv 0$ (in view of (7)). If instead of using $\frac{\sqrt{n}(\bar{X}-\mu)}{S}$ as our pivot we had used $\sqrt{n}(\bar{X}-\mu)$, then the same argument implies that $\sqrt{n}$ times the difference between the bootstrap and true percentiles tends jointly with $\sqrt{n}(\bar{X}-\mu)$ to a joint normal distribution with nonzero correlation. By Lemma 1, in this case where the difference between the estimated and true percentiles is of the order $n^{-\frac{1}{2}}$, the difference between the probabilities of coverage must also be of order $n^{-\frac{1}{2}}$. On the other hand, in the prediction situation we can show that $U_n \equiv Z$ and $V_n \equiv \sqrt{n}(\hat{\underline{Z}} - \underline{Z})$ do have a limiting joint distribution, but with necessarily independent components U, V. Moreover, V has a nondegenerate normal distribution with mean 0. So even though $\hat{\underline{Z}} - \underline{Z}$ is of order $n^{-\frac{1}{2}}$, if the regularity conditions (11) and (12) hold, we expect the probability of coverage of $\hat{\underline{Z}}$ to be $1 - \alpha + O(n^{-1})$ because $E(V|U) = E(V) = 0$. This approach can successfully but unnecessarily be applied to $F_n^{-1}(\alpha)$. Unfortunately we have not been able to verify (12) since the structure of V_n is hard to handle. However, in what follows we give a more indirect argument.

3. RESULTS

The remainder of the paper contains indirect arguments for (10). Our exposition is simplified by taking σ as known. The arguments are essentially the same for the scenario of the previous section in which the statistic in (1) and (3) is "studentized." Thus here we take $t_{\alpha,p}$ and $t^*_{\alpha,p}$ defined by

$$P\left\{\frac{\bar{X}-Z}{\sigma} \le t_{\alpha,p}\right\} = \alpha \tag{17}$$

and

$$P^*\left\{\frac{\bar{X}^*-Z^*}{S} \le t^*_{\alpha,p}\right\} = \alpha + O_p(n^{-1}) \tag{18}$$

rather than the definitions of (1) and (2). We shall establish the

Theorem. *Suppose that*

(i) $E(|Z|^{10}) < \infty$,

(ii) $f = F' \neq 0$, *and that*

(iii) $f' = F''$ *exists and is bounded.*

Then

$$\mathcal{L}(n^{\frac{1}{2}}(t^*_{\alpha,p} - t_{\alpha,p})) \to N(0,\tau^2), \tag{19}$$

where

$$\tau^2 = \operatorname{Var}\left(\frac{1(Z \le -t_\alpha)}{f(-t_\alpha)} + Z - \frac{t_\alpha}{2}Z^2\right).$$

However,

$$P\left\{\frac{\bar{X}-Z}{\sigma} \le t^*_{\alpha,p}\right\} = \alpha + O(n^{-\frac{3}{4}+\gamma}) \tag{20}$$

for all $\gamma > 0$.

This theorem appears with only 4 moments required in the recent Ph.D. dissertation of Bai (1988). His arguments are quite different from ours. They are based on large deviation results for empirical probabilities, and permit ready extension to multivariate scenarios. We believe that the arguments given here also could be extended to the vector case and our moment conditions improved.

It is also our hope, so far unachieved, that they can be used to change the error bound in (20) from $O(n^{-\frac{3}{4}+\gamma})$, to the highly plausible $O(n^{-1+\gamma})$, $\gamma > 0$. In any case we believe that our present approach is somewhat more transparent.

Without loss of generality in what follows, we take $E(Z) = 0$, and $E(Z^2) = 1$. Always (i), (ii), and (iii) are in force. Subscripted constants $c_1, c_2, \ldots$ are universal; their exact values are immaterial to our conclusions. We denote the empirical distribution of the learning sample $X_1, \ldots, X_n$ by F_n, and the signed normalized empirical measure $n^{\frac{1}{2}}\{F_n - F\}$ by ν_n. For convenience, we write t_α for $t_{\alpha,p}$ and t_α^* for $t_{\alpha,p}^*$.

By definition,

$$E^*(F_n(\bar{X}^* - St_\alpha^*)) = 1 - \alpha + O(n^{-1}) \tag{21}$$

$$F_n(-\hat{t}_\alpha) = 1 - \alpha + O(n^{-1}), \tag{22}$$

where we write $O(n^{-1})$ for random quantities whose absolute values are uniformly bounded by a/n for some a. Subtracting (22) from (21) and centering yields

$$|E^*((F(\bar{X}^* - St_\alpha^*) - F(-\hat{t}_\alpha))| = n^{-\frac{1}{2}}|E^*(\nu_n(-\hat{t}_\alpha, \bar{X}^* - St_\alpha^*))| + O(n^{-1}). \tag{23}$$

Let

$$A_n = \{|\bar{X}^*| \leq 2n^{-\frac{1}{2}+\epsilon}\}$$

and

$$T_n = \{|\hat{t}_\alpha - t_\alpha| \leq n^{-\frac{1}{2}+\epsilon}, \quad |\bar{X}| \leq n^{-\frac{1}{2}+\epsilon}, \quad |S-1| \leq \frac{1}{2},$$
$$m_4 \leq 2\mu_4, \quad \sup |\nu_n(-\infty, a)| \leq n^\epsilon\},$$

where

$$m_j = n^{-1}\sum_{i=1}^{n}(X_i - \bar{X})^j, \quad \mu_j = E(Z^j) \quad \text{and} \quad 0 < \epsilon < \frac{1}{2}.$$

Lemma 2. *On T_n, $|t_\alpha^* - \hat{t}_\alpha| \leq c_5 \max\{n^{-4\epsilon}, n^{-\frac{1}{2}+\epsilon}\}$.*

Proof. From (23) it follows that

$$1_{T_n}|E^*((F(\bar{X}^* - St_\alpha^*) - F(-\hat{t}_\alpha))1_{A_n})|$$
$$= n^{-\frac{1}{2}}1_{T_n}|E^*(\nu_n(-\hat{t}_\alpha, \bar{X}^* - St_\alpha^*)1_{A_n})| + \Delta_{n1}, \tag{24}$$

where, since on T_n, $\sup|\nu_n(-\infty, a)| \leq n^\epsilon$,

$$|\Delta_{n1}| \leq c_1(n^{-\frac{1}{2}+\epsilon} + P^*\{|\bar{X}^*| \geq 2n^{-\frac{1}{2}+\epsilon}\}).$$

On T_n

$$P^*\{|\bar{X}^*| \geq 2n^{-\frac{1}{2}+\epsilon}\} \leq P^*\{n^{\frac{1}{2}}|\bar{X}^* - \bar{X}| \geq n^{\epsilon}\} \leq c_2 m_4\, n^{-4\epsilon} \tag{25}$$

so that $|\Delta_{n1}| \leq c_3(n^{-\frac{1}{2}+\epsilon} + n^{-4\epsilon})$. Similarly,

$$1_{T_n} E^*(F_n(\bar{X}^* - tS) - F_n(-\hat{t}_\alpha)) = 1_{T_n} E^*(F(\bar{X}^* - tS) - F(-\hat{t}_\alpha)) + R_n(t)\,,$$

where $\sup_t |R_n(t)| \leq n^{-\frac{1}{2}+\epsilon}$. And there is a $\gamma > 0$ such that if $|t + \hat{t}_\alpha| \leq \gamma$, then

$$|1_{T_n} E^*((F(\bar{X}^* - tS) - F(-\hat{t}_\alpha))1_{A_n})| \geq \frac{1}{2} f(-t_\alpha)(t + \hat{t}_\alpha + n^{-\frac{1}{2}+\epsilon})\,.$$

Since

$$1_{T_n}|E^*((F(\bar{X}^* - tS) - F(-\hat{t}_\alpha))1_{A_n^c})| \leq c_4 n^{-4\epsilon}\,,$$

we conclude from (21) and (22) that on T_n,

$$|-t_\alpha^* - (\hat{t}_\alpha)| \leq c_5 \max\{n^{-\frac{1}{2}+\epsilon},\, n^{-4\epsilon}\}\,.$$

We now bring embedding arguments to bear upon our study of the difference between $\bar{X}^* - St_\alpha^*$ and $-\hat{t}_\alpha$. In particular, we use this extension of Theorem 3 of Komlós, Major, and Tusnády: after augmenting the underlying probability space if necessary, it is possible to define a sequence of Brownian bridges $\{B_n(t): 0 \leq t \leq 1\}$ for which

$$P\{\sup_{0 \leq t \leq 1} |n(F_n(F^{-1}(t)) - t) - n^{\frac{1}{2}} B_n(t)| > a \log n + x\} \leq K e^{-\lambda x}\,,$$

where a, K, and λ are positive constants.

Lemma 3. $1_{T_n}(\bar{X} - St_\alpha^* + \hat{t}_\alpha)$

$$\begin{aligned} &= n^{-\frac{1}{2}}(f(-t_\alpha))^{-1}(1 + o(1))1_{T_n}\{E^*((B_n(F(\bar{X}^* - St_\alpha^*) - B_n(F(-\hat{t}_\alpha))1_{A_n}) \\ &\quad + c_9 n^{-2\epsilon} + c_{10} n^{-1+\delta} M_{n1}\} \quad \textit{whenever} \quad \delta > 0, \end{aligned}$$

where

$$\sup_n E(|M_{n1}|^r) < \infty \quad \textit{for all} \quad r \geq 0\,.$$

Proof. $1_{T_n}(E^*((F(\bar{X}^* - St_\alpha^*) - F(-\hat{t}_\alpha))1_{A_n}))$

$$= (1 + o(1)) f(-t_\alpha)(E^*(\bar{X}^* 1_{A_n}) - St_\alpha^* - (-\hat{t}_\alpha))1_{T_n} \tag{26}$$

according to Lemma 2. Now $E^*(\bar{X}^* 1_{A_n}) = \bar{X} - r_n$, where in view of (25),

$$|r_n| \leq [E^*((\bar{X}^*)^2)]^{\frac{1}{2}}[P^*\{\bar{X}^* \in A_n^c\}]^{\frac{1}{2}} = \left[\frac{S^2}{n} + (X)^2\right]^{\frac{1}{2}} [P^*\{\bar{X}^* \in A_n^c\}]^{\frac{1}{2}} \leq c_6 n^{-\frac{1}{2}-\epsilon}\,. \tag{27}$$

From (23), (24), and (26) we obtain

$$\begin{aligned} &1_{T_n}|\bar{X} - St_\alpha^* - (-\hat{t}_\alpha)| \\ &\quad = (f(t_\alpha))^{-1}(1 + o(1))n^{-\frac{1}{2}}|E^*(\nu_n(-\hat{t}_\alpha, \bar{X}^* - St_\alpha^*)1_{A_n})|1_{T_n} + \Delta_{n2}\,, \end{aligned} \tag{28}$$

where

$$|\Delta_{n2}| \le c_7(|r_n| + E^*(\sup_a |\nu_n(-\infty, a)| 1_{A_n})) 1_{T_n}$$

$$\le c_8(n^{-\frac{1}{2}-\epsilon} + n^{\epsilon} P^*\{\bar{X}^* \in A_n^c\}) 1_{T_n} \le c_9 n^{-2\epsilon}. \tag{29}$$

From the cited extension of the theorem of Komlós, Major, and Tusnády, (28), and (29), we have

$$1_{T_n}(\bar{X} - St_\alpha^* - (-\hat{t}_\alpha)) = 1_{T_n} n^{-\frac{1}{2}}\{E^*((B_n(\bar{X} - St_\alpha^*) - B_n(F(-\hat{t}_\alpha)))1_{A_n})\} + \Delta_{n3} + \Delta_{n2},$$

where

$$\sup_{n>1} \left(\frac{n}{\log n}\right)^r E(|\Delta_{n3}|^r) < \infty \quad \text{for all} \quad r \ge 0.$$

Lemma 3 follows.

The next two lemmas are stated without proof. Lemma 4 is a consequence of arguments concerning Lévy's modulus of continuity for Brownian motion. It is not sharp, but is adequate for our purposes. One can prove it from (25), page 541 of the book by Shorack and Wellner (1986). Lemma 5 follows from the work of Singh (1981).

Lemma 4. *If B is a Brownian bridge, $0 < a \le \frac{1}{2}$, and*

$$M_n = \sup\left\{\frac{|B(t) - B(u)|}{|t-u|^{1/2}} : an^{-1+\delta} < |t-u| \le \frac{1}{2}\right\},$$

then

$$\sup_n n^{-\gamma} E(|M_n|^r) < \infty \quad \textit{for all} \quad r, \gamma > 0.$$

Lemma 5. *Let M_{n3}/n be*

$$\sup_x \left| P^*\left(\frac{\sqrt{n}(\bar{X}^* - \bar{X})}{S} \le x\right) - \Phi(x) - \frac{m_3}{6\sqrt{n}} p_2(x)\varphi(x) \right| 1(|S-1| \le 1/2) 1(m_4 \le 2\mu_4),$$

where $p_2(x) = x^2 - 1$ and φ is the standard normal density function. Then

$$\sup_n E\{|M_{n3}|^r\} < \infty \quad \textit{for all} \quad r \ge 0.$$

The next lemma is the heart of our proof of the theorem.

Lemma 6.

$$1_{T_n}|E^*((B_n(F(\bar{X}^* - St_\alpha^*) - B_n(F(-\hat{t}_\alpha))1_{A_n})|$$

$$\le 1_{T_n}\left\{n^{\gamma}(M_{n4} + M_{n2})[|\bar{X} - St_\alpha^* - (-\hat{t}_\alpha)|^{\frac{1}{2}} + c_{11} n^{-\frac{1}{4}}] + \frac{M_{n4}M_{n3}}{n}\right\}, \tag{30}$$

where

$$\sup_n E(|M_{nj}|^r) < \infty \quad \textit{for all} \quad j, r \ge 0 \quad \textit{and all} \quad \gamma > 0.$$

Proof. On $T_n A_n$,

$$f(-t_\alpha)(1-\theta_n)|\bar{X}^* - St_\alpha + \hat{t}_\alpha| \le |F(\bar{X}^* - St_\alpha) - F(-\hat{t}_\alpha)| \le f(-t_\alpha)(1+\theta_n)|\bar{X}^* - St_\alpha^* + \hat{t}_\alpha|, \tag{31}$$

where the θ_n are constants tending to 0. Let

$$C_n = \{dn^{-1+\delta} \le |\bar{X}^* - St_\alpha^* + \hat{t}_\alpha| \le 1/2\}.$$

According to Lemma 4 and (31), on T_n

$$|E^*((B_n(F(\bar{X} - St_\alpha^*)) - B_n(F(-\hat{t}_\alpha)))1_{A_n}1_{C_n})|$$

$$\le \sup\left\{\frac{|B_n(t) - B_n(u)|}{|t-u|^{\frac{1}{2}}} : dn^{-1+\delta} \le |t-u| \le \frac{1}{2}\right\} E^*\{|\bar{X}^* - St_\alpha^* + \hat{t}_\alpha|^{\frac{1}{2}}\} \tag{32}$$

$$\le M_n[E^*(\bar{X}^* - St_\alpha^* + \hat{t}_\alpha)^2]^{\frac{1}{4}}, \tag{33}$$

$$= M_n\left\{(\bar{X} - St_\alpha^* + \hat{t}_\alpha)^2 + \frac{S^2}{n}\right\}^{\frac{1}{4}}$$

$$\le 2M_n\{|\bar{X} - St_\alpha^* + \hat{t}_\alpha|^{\frac{1}{2}} + n^{-\frac{1}{4}}\}.$$

On the other hand,

$$|E^*((B_n(F(\bar{X}^* - St_\alpha^*)) - B_n(F(-\hat{t}_\alpha)))1_{A_n}1_{C_n^c})|1_{T_n}$$

$$\le 1_{T_n} \sup_{0\le t\le 1} |B_n(t)|\{2^{-\frac{1}{2}}E^*(|\bar{X}^* - St_\alpha^* + \hat{t}_\alpha|^{\frac{1}{2}}1(|\bar{X}^* - St_\alpha^* + \hat{t}_\alpha| \ge \frac{1}{2}))$$

$$+ P^*\{|\bar{X}^* - St_\alpha^* + \hat{t}_\alpha| \le dn^{-1+\delta}\}\}. \tag{34}$$

By Lemma 5, if $S\Delta^* = n^{\frac{1}{2}}(\hat{t}_\alpha - St_\alpha^* + \bar{X})$, then

$$1_{T_n}P^*\left\{\frac{-dn^{-\frac{1}{2}+\delta}}{S} - \Delta^* \le \frac{(\bar{X}^* - \bar{X})n^{1/2}}{S} \le \frac{dn^{-\frac{1}{2}+\delta}}{S} - \Delta^*\right\}$$

$$\le 1_{T_n}\left\{\frac{dn^{-\frac{1}{2}+\delta}}{S}\left(1 + \frac{|m_3|}{n^{\frac{1}{2}}}\right) + \frac{M_{n3}}{n}\right\}$$

$$< 1_{T_n}\left(c_{11}n^{-\frac{1}{2}+\delta} + \frac{M_{n3}}{n}\right). \tag{35}$$

Further, as in (32),

$$E^*(|\bar{X}^* - St_\alpha^* + \hat{t}_\alpha|1(|\bar{X}^* - St_\alpha^* + \hat{t}_\alpha| \ge 1/2)) \le 2\{|\bar{X} - St_\alpha^* + \hat{t}_\alpha|^{\frac{1}{2}} + n^{-\frac{1}{4}}\}. \tag{36}$$

By setting $M_{n2} = M_n n^{-\gamma}$ and $M_{n4} = \sup_{0\le t\le 1} |B_n(t)|$, and combining (32) through (36), we infer that the left hand side of (30) is bounded above by

$$1_{T_n}\left\{n^{\gamma}(M_{n2}+M_{n4})|[\bar{X} - St^*_\alpha + \hat{t}_\alpha|^{\frac{1}{2}} + c_{11}n^{-\frac{1}{4}}] + \frac{M_{n4}M_{n3}}{n}\right\},$$

as was asserted.

Lemma 7. *For* $j = 1, 2, E(|St^*_\alpha - t_\alpha|^j 1_{T^c_n}) = O(n^{-\frac{3}{4}})$.

Proof. From (2) (with S^* replaced by S) and elementary considerations it follows that $X_{(1)} - X_{(n)} \le St^*_\alpha < X_{(n)} - X_{(1)}$, where $X_{(i)}$ is the ith order statistic of the learning sample. Therefore, $|St^*_\alpha - t_\alpha| \le 2|X|_{(n)} + |t_\alpha|$, where $|X|_{(n)}$ is the largest absolute value of the learning sample. So Hölder's inequality implies that

$$E(|St^*_\alpha - t_\alpha|^j 1_{T^c_n}) \le c_{12}E^{1/p}(|X|^{jp}_{(n)} + 1)P^{1/q}\{T^c_n\}. \tag{37}$$

An elementary argument gives

$$E(|X|^{\ell}_{(n)}) \le nE(|Z|^{\ell}). \tag{38}$$

Moreover, if $E(|Z|^\ell) < \infty$ for $\ell \ge 4$, then

$$P\{|\bar{X}| \ge n^{-\frac{1}{2}+\epsilon}\} \le c_{13}n^{-\ell\epsilon}, \quad P\{|S-1| \ge 1/2\} \le c_{14}n^{-\ell/4},$$

and for $\ell \ge 8$

$$P\{m_4 \ge 2\mu_4\} \le c_{15}n^{-\ell/8}.$$

Also, since under our assumptions

$$P\{\sup_a |\nu_n(-\infty, a)| \ge n^\epsilon\} \le c_{16}n^{-r\epsilon}$$

and

$$P\{|\hat{t}_\alpha - t_\alpha| \ge n^{-\frac{1}{2}+\epsilon}\} \le c_{17}n^{-r\epsilon}$$

for all $r \ge 0$, we conclude that

$$P^{1/q}\{T^c_n\} \le c_{18}(n^{-\ell/8q} + n^{-\ell\epsilon/q}). \tag{39}$$

For $\ell = 10$ take $p = \ell$ if $j = 1$, $p = \ell/2$ if $j = 2$. Substitute (38) and (39) into (37) to conclude the proof of Lemma 7.

We turn now to the proof of the theorem. From Lemmas 2, 3, and 6 it follows that

$$\begin{aligned}&1_{T_n}|\bar{X} - St^*_\alpha + \hat{t}_\alpha| \\ &\quad\le 1_{T_n}c_{19}\left\{n^{-\frac{1}{2}+\gamma}(M_{n4}+M_{n2})[|\bar{X} - St^*_\alpha + \hat{t}_\alpha|^{\frac{1}{2}} + c_{11}n^{-\frac{1}{4}}] + \frac{M_{n4}M_{n3}}{n}\right\}.\end{aligned}$$

Note that $0 \le x \le K_1 + (K_2/x)$ implies that $x \le K_1 + K_2 + 1$. Put $x = n^{\frac{3}{8}-\frac{\gamma}{2}}|\bar{X} - St^*_\alpha + \hat{t}_\alpha|^{\frac{1}{2}}$ to see that

$$\begin{aligned} &1_{T_n}|\bar{X} - St^*_\alpha + \hat{t}_\alpha| \\ &\le c_{20} n^{-\frac{3}{4}+\gamma}\left(n^{-\frac{1}{8}+\frac{\gamma}{2}}(M_{n2} + M_{n4}) + M_{n4}M_{n3}n^{-\frac{1}{4}-\gamma} + c_{21}\right)^2 1_{T_n}\,. \end{aligned} \tag{40}$$

Since $P(T_n^c) = o(n^{-1})$,

$$\bar{X} - St^*_\alpha + \hat{t}_\alpha = O_p(n^{-\frac{3}{4}+\gamma}) \quad \text{for all} \quad \gamma > 0\,;$$

(19) follows from (40) and the Bahadur (1966) representation of $\hat{t}_\alpha$.

To prove (20), write

$$\begin{aligned} P\{Z > \bar{X} - St^*_\alpha\} = E((1 - F(\bar{X} - St^*_\alpha)) &= \alpha - (\bar{X} - St^*_\alpha + t_\alpha)f(-t_\alpha) \\ &+ O(E((\bar{X} - St^*_\alpha + t_\alpha)^2)\,. \end{aligned} \tag{41}$$

$$\begin{aligned} |E(\bar{X} - St^*_\alpha + t_\alpha)| &\le E(|\bar{X} - St^*_\alpha + \hat{t}_\alpha|) + |E(\hat{t}_\alpha - t_\alpha)| \\ &= E(|\bar{X} - St^*_\alpha + \hat{t}_\alpha|) + O(n^{-1})\,. \end{aligned} \tag{42}$$

We need only show that $E(|\bar{X} - St^*_\alpha + \hat{t}_\alpha|) = O(n^{-\frac{3}{4}+\gamma})$ and $E((\bar{X} - St^*_\alpha + t_\alpha)^2) = O(n^{-3/4+\gamma})$ to complete the description of (41). An application of (40) suffices on T_n. For the remaining part, write

$$|\bar{X} - St^*_\alpha + \hat{t}_\alpha| \le |\bar{X}| + |\hat{t}_\alpha - t_\alpha| + |t_\alpha - St^*_\alpha| \tag{43}$$

and multiply both sides by $1_{T_n^c}$. The first two terms can be dealt with in a straightforward manner, and the third by Lemma 7, so $E(|\bar{X} - St^*_\alpha + \hat{t}_\alpha|)$ is bounded as required; $E((\bar{X} - St^*_\alpha + t_\alpha)^2)$ is bounded by a similar argument.

REFERENCES

Bahadur, R. (1966). A note on quantiles in large samples. Ann. Math. Statist. **37**, 577–580.

Bai, C. (1988). *Asymptotic Properties of Some Sample Reuse Methods for Prediction and Classification.* Ph.D. Dissertation, Department of Mathematics, University of California, San Diego.

Bai, C. and Olshen, R. A. (1988). Discussion of "Theoretical comparison of bootstrap confidence intervals," by Peter Hall. Ann. Statist. **16**, 953–956.

Hall, P. (1988). Theoretical comparison of bootstrap confidence intervals (with discussion). Ann. Statist., **16**, 927–985.

Komlós, J., Major, P., and Tusnády, G. (1975). An approximation of partial sums of independent RV's and the sample DFV I. Z. Wahrscheinlichkeitstheorie verw. Gebiete **32**, 111–131.

Olshen, R. A., Biden, E. N., Wyatt, M. P., and Sutherland, D. H. (1989). Gait analysis and the bootstrap. Ann. Statist. **17**, to appear.

Shorack, G. R. and Wellner, J. A. (1986). *Empirical Processes With Applications to Statistics*, Wiley, New York.

Singh, K. (1981). On the asymptotic accuracy of Efron's bootstrap. Ann. Statist. **9**, 1187–1195.

Sutherland, D. H., Olshen, R. A., Biden, E. N., and Wyatt, M. P. (1988). *The Development of Mature Walking*. MacKeith Press, London.

REPRESENTATION OF BANACH SPACE VALUED MARTINGALES AS STOCHASTIC INTEGRALS

Egbert Dettweiler

If $M = (M_t)_{t\geq o}$ is a real-valued, continuous local martingale whose quadratic variation is absolutely continuous relative to Lebesgue measure, then by a theorem of Doob ([7],p.449), M is the stochastic integral of a certain function relative to a Brownian motion-on a possibly extended probability space. This theorem is also well-known in the d-dimensional case (see [12], th.4.5.2). The classical methods of proof can be used without major difficulties to obtain the same theorem for Hilbert space valued, continuous local martingales, but they do not work beyond that case. For continuous local martingales with values in a real separable Banach space, we give a complete different proof of Doob's theorem. The main application is the representation of Banach space valued Itô processes (defined as in [12]) as solutions of infinite-dimensional stochastic differential equations. This result then can be used to obtain a uniqueness theorem for the so-called martingale problem (in the sense of [12]) in the Banach space case.

1. BANACH SPACE VALUED LOCAL MARTINGALES AND GAUSSIAN PROCESSES

Let $(\Omega,\mathcal{F},P)$ be a probability space with a right continuous filtration $(\mathcal{F}_t)_{t\geq o}$. We assume that this probability space is rich enough to serve as the domain of definition for all random functions which we will consider. For this reason, we will not always mention the underlying probability space of a process. The same remark holds for the filtration. Talking about adaptedness of a process without mentioning the filtration will always mean adaptedness relative to the filtration $(\mathcal{F}_t)_{t\geq o}$.

Let E denote a real, separable Banach space, and suppose that $Y= (Y(t))_{t\geq o}$ is an E-valued, continuous local martingale relative to the filtration $(\mathcal{F}_t)$. The continuity implies that Y is also locally bounded and especially locally p-integrable for every $p\geq 1$. If τ is a stopping time, Y^τ denotes as usual the stopped process $Y(\tau \wedge .)$. Let Π denote the family of all finite partitions of $\mathbb{R}_+$. For $\Delta\in\Pi$ with $\Delta=(o=t_o< t_1 <...< t_n)$ we put $|\Delta|=\max\ (t_{k+1}-t_k)$. For every

$\Delta=(o=t_o <...< t_n) \in \Pi$ and every $t \geq o$ we define

$$V_\Delta(t) = \sum_{k=o}^{n-1} (Y^t(t_{k+1})-Y^t(t_k))^{\otimes 2},$$

where $x^{\otimes 2}$ denotes the element $x\otimes x$ of the tensor product $E\otimes E$. Let $E\otimes_\varepsilon E$ denote the tensor product $E\otimes E$ provided with the ε-topology. The dual $(E\otimes_\varepsilon E)'$ of $E\otimes_\varepsilon E$ is the space of integral bilinear forms on $E\times E$ (see[13]). Then one can prove (cf. [4], lemma 2.9) that for every $\varphi\in(E\otimes_\varepsilon E)'$ the net $(\langle V_\Delta,\varphi\rangle)_{\Delta\in\Pi}$ of $C(\mathbb{R}_+)$-valued random vectors converges in probability to a limit which we denote by

$$<[Y],\varphi> = (<[Y](t),\varphi>)_{t\geq o}.$$

For $\varphi = x'\otimes x'$ ($x'\in E'$, the dual of E), $<[Y], x'\otimes x'>$ is just the quadratic variation of the one-dimensional, continuous local martingale $<Y,x'>$. Since for every $t\geq o$,

$$[Y](t): (E\otimes_\varepsilon E)' \longrightarrow \mathcal{L}^o(\Omega,\mathcal{F}_t, P)$$

is a cylindrical process (or linear random function, see [10]), we call $[Y] = ([Y](t))_{t\geq o}$ the *cylindrical tensor quadratic variation of* Y. In case that every $[Y](t)$ is an $E\tilde{\otimes}_\varepsilon E$-valued random vector (see [10] for general conditions that a cylindrical process in a genuine random vector), the $E\tilde{\otimes}_\varepsilon E$-valued process $[Y]$ is simply called the *(tensor) quadratic variation of* Y.

Typical examples of continuous local martingales with a nice quadratic variation are connected with Gaussian processes and stochastic integration in Banach spaces. Let G(E) denote the family of all Gaussian measures on E. For $\rho\in G(E)$ the linear operator $Q=Q(\rho)\in E\tilde{\otimes}_\varepsilon E \subset L(E',E)$, defined by

$$<Q(x'),y'>=\int <x,x'><x,y'>\rho(dx) \quad \text{for all } x', y'\in E'.$$

is called the *covariance operator* of ρ. We denote by Cov(E) the cone of all covariance operators of Gaussian measures on E. Every $Q\in \mathrm{Cov}(E)$ has a factorization $Q=S\circ S^t$ (S^t=transposed operator of S), where $S\in L(H,E)$ is a bounded linear operator from a real, separable Hilbert space into E (see e.g.[1] for these well known facts on Gaussian measures). If $Q = S\circ S^t$, there is a classical method to construct from S a symmetric Gaussian independent increment process $Z=(Z(t))$ with values in E such that Q is the covariance operator of the distribution ρ of Z(1). Q is also called the covariance operator of the Gaussian process Z. The construction of Z is as follows. Let $(e_k)_{k\geq 1}$ be a complete orthonormal system of H (shortly:CONS) and let $\xi=(\xi_k)$ be an infinite-dimensional Brownian motion. Then Z is given by

$$Z(t)=\sum_{k=1}^{\infty} S(e_k)\,\xi_k(t) \qquad \text{for } t\geq o,$$

where the series is a.s. convergent and also in the p-th mean for every $p\geq 1$ (see [1], p.143). A Gaussian (symmetric, independent increment) process Z with covariance operator Q is an elementary example of a continuous (local) martingale with existing quadratic variation, which is just given by $[Z](t) = t\,Q$.

Let us denote by $L_g(H,E)$ the subspace of all $S\in L(H,E)$ such that $Q=S\circ S^t\in \mathrm{Cov}(E)$. On $L_g(H,E)$ we now introduce a norm, which will turn $L_g(H,E)$ into a Banach space. Let (η_k) be an i.i.d. sequence of standard normal distributed random variables (for short: standard normal sequence) and define

$$|||S||| = \left(\mathbb{E}\|\sum_{k=1}^{\infty} S(e_k)\,\eta_k\|^2\right)^{1/2} \text{ for } S \,\varepsilon\, L_g(H,E),$$

where $\mathbb{E}$ denotes the expectation relative to P. The definition of $|||S|||$ does not depend on the special choice of (e_k) and (η_k) since

$$|||S|||^2 = \int \|x\|^2 \rho(dx) ,$$

where ρ is the Gaussian measure with covariance operator Q. Clearly, $|||\bullet|||$ is a norm on $L_g(H,E)$ and $L_g(H,E)$ is a Banach space relative to $|||\bullet|||$. In case E=H , $|||\bullet|||$ is just the classical Hilbert-Schmidt norm of a Hilbert-Schmidt operator.

For the Banach space $G = L_g(H,E)$ we call $S : \mathbb{R}_+ \times \Omega \to G$ an *elementary function*, if for some $\Delta = (t_j)_{0 \le j \le n}, \in \Pi$,

(1.1) $$S = \sum_{j=0}^{n-1} S_j\, 1_{[t_j,\, t_{j+1}[} \qquad \text{with} \quad S_j \in \mathcal{L}^2(\Omega, \mathcal{F}_{t_j}, P; G) \qquad \text{for } 0 \le j \le n-1 .$$

Let $\xi = (\xi_k)$ be an infinite-dimensional Brownian motion and (e_k) be a fixed CONS of H, and let $\mathcal{E}(G)$ denote the space of all elementary functions. For $S \in \mathcal{E}(G)$ of the form (1.1) we define the *stochastic integral process*

$$Y(S) = \int_0^{\cdot} [S(r) \,,\, d\xi(r)]$$

of S relative to ξ by

(1.2) $$Y(S)(t) = \int_0^t [S(r), d\xi(r)] := \sum_{j=0}^{n-1} \sum_{k=1}^{\infty} S_j(e_k)(\xi_k^t(t_{j+1}) - \xi_k^t(t_j))$$

for every $t \ge 0$. It is easy to see that Y(S) is an E-valued, continuous square integrable martingale with quadratic variation [Y(S)] given by

(1.3) $$[Y(S)](t) = \int_0^t Q(r)dr \qquad \text{with} \qquad Q = \sum_{j=0}^{n-1} (S_j \circ S_j^t)\, 1_{[t_j, t_{j+1}[}$$

Our next aim is to extend the above notion of a stochastic integral for elementary functions to larger classes of G-valued functions. Let us suppose for the moment that the Banach space E is 2-smoothable, i.e. has an equivalent 2-uniformly smooth norm. By a theorem of Pisier ([9]), E is 2-smoothable if and only if there exists a constant C such that for every E-valued, square integrable martingale $(M_n)_{n \ge 1}$ the inequality

(1.4) $$\sup_{n \ge 1} \mathbb{E}\, \|M_n\|^2 \le C \sum_{k=0}^{\infty} \mathbb{E}\, \|M_{k+1} - M_k\|^2 \quad (M_0 := 0)$$

holds. From (1.4) one can derive a nice inequality for the stochastic integral of an elementary function $S \in \mathcal{E}(G)$ of the form (1.1). For every $t \geq o$ one has

$$
\begin{aligned}
(1.5)\quad & \mathbb{E}\| \int_0^t [S(r), d\xi(r)] \|^2 \\
&= \mathbb{E}\| \sum_{j=0}^{n-1} \sum_{k=1}^{\infty} S_j(e_k)\, (\xi_k^t(t_{j+1}) - \xi_k^t(t_j)) \|^2 \\
&\leq C \sum_{j=0}^{n-1} \mathbb{E}\| \sum_{k=1}^{\infty} S_j(e_k)\, (\xi_k^t(t_{j+1}) - \xi_k^t(t_j)) \|^2 \\
&= C \sum_{j, t_j<t} (t \wedge t_{j+1} - t_j)\, \mathbb{E}\| \sum_{k=1}^{\infty} S_j(e_k)\, \eta_{j,k} \|^2 \qquad \left(\text{where } \eta_{j,k} := \frac{\xi_k(t \wedge t_{j+1}) - \xi_k(t_j)}{(t \wedge t_{j+1} - t_j)^{1/2}} \right) \\
&= C \sum_{j, t_j<t} (t \wedge t_{j+1} - t_j)\, \mathbb{E}\left[\mathbb{E}\left(\| \sum_{k=1}^{\infty} S_j(e_k)\, \eta_{j,k} \|^2 \mid \mathcal{F}_{t_j} \right) \right] \\
&= C \sum_{j, t_j<t} (t \wedge t_{j+1} - t_j)\, \mathbb{E}|||S_j|||^2 \\
&= C\mathbb{E} \int_0^t |||S(r)|||^2 dr \,,
\end{aligned}
$$

where the last but one line follows from the fact that $(\eta_{j,k})_{k \geq 1}$ is standard normal and independent from $\mathcal{F}_{t_j}$ for every j.

Analogously to classical stochastic integration theory (following e.g. the arguments in [11], ch. 2 or [12], ch.4.3.) one can use inequality (1.5) to extend the stochastic integral to all functions $S: \mathbb{R}_+ \times \Omega \longrightarrow G = L_g(H,E)$, which are progressively measurable and for which

$$\int_0^t |||S(r)|||^2 dr < \infty \qquad \text{P-a.s.} \qquad \text{for all } t \geq o.$$

Let $I^2_{loc}(G)$ denote this space of G-valued, progressively measurable functions, and denote by $I^2(G)$ the subspace of all $S \in I^2_{loc}(G)$ with the stronger property $\mathbb{E}\int_0^t |||S(r)|||^2 dr < \infty$ for all $t \geq o$. For every $S \in I^2_{loc}(G)$ the stochastic integral process $Y(S) = \int_0^{\cdot} [S(r), d\xi(r)]$ is a continuous, local martingale (even a square integrable martingale in case $S \in I^2(G)$) with quadratic variation $\int_0^{\cdot} Q(r)\, dr$, where $Q = S \circ S^t$.

One can prove (see [2], th.3.2) that the validity of (1.5) for all $S \in \mathcal{E}(G)$ is necessary and sufficient for E to be 2-smoothable.

To get an essential extension of the stochastic integral beyond $\mathcal{E}(G)$ also in case that E is not necessarily 2-smoothable, we proceed as follows (see [3] for the motivation behind). Let F denote a second Banach space, and call now an operator $R \in L(F,E)$ 2-smoothable, if there is a constant C such that for every F-valued, square-integrable martingale $(M_n)_{n\geq 1}$ the inequality

$$(1.6) \qquad \sup_{n\geq 1} \mathbb{E} \| R \circ M_n \|^2 \leq C \sum_{k=o}^{\infty} \mathbb{E} \| M_{k+1} - M_k \|^2 \qquad (M_o := 0)$$

holds. From (1.6) one can derive now the following inequality, valid for all $S \in \mathcal{E}(L_g(H,F))$:

$$(1.7) \quad \mathbb{E} \| \int_0^t [R \circ S(r) \, , \, d\xi(r) \,] \|^2 \leq C \, \mathbb{E} \int_0^t ||| S(r) |||^2 \qquad \text{for all} \quad t \geq 0 \, .$$

Again, it follows from (1.7) that now for every $S \in I^2_{loc}(L_g(H,F))$ and every 2-smoothable $R \in L(F,E)$ there is a stochastic integral process $Y(T) = \int_0^t [T(r), d\xi(r)]$ for the function $T = R \circ S$ relative to ξ, and that Y(T) is an E-valued, continuous local martingale with quadratic variation $[Y(T)] = \int_0^{\bullet} Q(r)dr$ where $Q = T \circ T^t$. Again, Y(T) is even a square integrable martingale in case that $S \in I^2(Lg(H,F))$. That 2-smoothable operators occur naturally in connection with stochastic integration on general Banach spaces, was already indicated in [3], where we discussed the stochastic integrability of Banach space valued functions relative to a one-dimensional Brownian motion. In a forthcoming paper we will prove even stronger results than in [3], showing that under reasonable assumptions the existence of a stochastic integral of the type $\int_0^t [\, T(r) \, , \, d\xi(r) \,]$ necessarily implies that T is of the form $T(.) = R \circ S(.)$ with a fixed 2-smoothable operator R.

2. REPRESENTATION OF A LOCAL MARTINGALE AS A STOCHASTIC INTEGRAL

The above discussion shows that on any real, separable Banach space there exist non-trivial examples of continuous local martingales, whose quadratic variation is of the form $\int_0^{\bullet} Q(r) \, dr$ where Q is a progressively measurable function with values in Cov(E). Now we turn to the converse problem: Suppose that $Y = (Y(t))_{t\geq o}$ is a given E-valued, continuous local martingale such that Y has a quadratic variation $[Y]$ of the form $[Y] = \int_0^{\cdot} Q(r)\, dr$ with $Q\colon \mathbb{R}_+ \times \Omega \longrightarrow \mathrm{Cov}(E)$ progressively measurable. What can be said about the structure of Y? More precisely: Is Y of the stochastic integral type described above? In case $E = \mathbb{R}$ a solution of this problem is given already in [7]. For $E = \mathbb{R}^d$ we refer to [12] (ch.4, cf. also [8]).

For the formulation of the main results we need some preparations.

Suppose that $(\Omega', \mathcal{F}', P')$ is a second probability space with a filtration $(\mathcal{F}'_t)_{t\geq o}$. Then we will call the product space $(\Omega^*, \mathcal{F}^*, P^*) = (\Omega\times\Omega', \mathcal{F}\otimes\mathcal{F}', P\otimes P')$ an *extended probability space* of $(\Omega, \mathcal{F}, P)$ and the filtration $(\mathcal{F}_t^*)_{t\geq o}$ defined by $\mathcal{F}_t^* = \mathcal{F}_t \otimes \mathcal{F}'_t$ for all $t \geq o$, an *extended filtration* of $(\mathcal{F}_t)$. Any random function X defined on Ω can also be viewed as a random function on Ω^* by putting $X(\omega,\omega') = X(\omega)$ for all $(\omega,\omega') \in \Omega^*$, and if X is a process on Ω adapted to $(\mathcal{F}_t)$, then X as a process on Ω^* is obviously adapted to $\mathcal{F}_t^*$.

Suppose that $S\colon \mathbb{R}_+ \times \Omega \longrightarrow L_g(H,E)$ is a progressively measurable function such that the associated covariance function $Q = S \circ S^t$ has the weak integrability property that

$$\mathbb{E}\int_0^t < Q(r)(x')\, ,\, x' > dr < \infty \qquad \text{for all } t\geq 0 \quad \text{and all } x' \in E'$$

Let (e_k) be a CONS of H and $\xi = (\xi_k)$ be an infinite-dimensional Brownian motion. Then the following L^2-isometry holds:

$$(2.1) \qquad \mathbb{E}\left(\sum_{k=1}^{\infty} \int_o^t <S(r)(e_k), x'> d\xi_k(r)\right)^2$$

$$= \mathbb{E}\int_0^t \sum_{k=1}^{\infty} < S(r)(e_k)\, ,\, x' >^2 dr$$

$$= \mathbb{E}\int_0^t \| S^t(r)(x') \|^2 dr = \mathbb{E}\int_0^t < Q(r)(x')\, ,\, x' > dr \qquad \text{for all } x' \in E' .$$

If, more generally,

$$\int_o^t <Q(r)(x'), x'> dr < \infty \text{ a.s.}$$

for all $t\geq o$ and all $x' \in E'$, one can show with the aid of (2.1) by using suitable stopping times that also in that case

(2.2) $< \int_o^t [S(r), d\xi(r)], x'> := \sum_{k=1}^{\infty} \int_o^t <S(r)(e_k), x'> d\xi_k(r)$

exists for all $t \geq o$ and all $x' \in E'$, and we will say shortly that the stochastic integral process $\int_0^{\bullet} [S(r), d\xi(r)]$ *exists in the weak sense.*

We will call a family $(Q_i)_{i \in I} \subset Cov(E)$ *quasi-injective*, if there exists a linear independent sequence (y_n') in E', such that every Q_i is injective on $L[(y_n')]$, the linear hull of (y_n'), and such that $L^*[(y_n')]$, the weak* sequential closure of $L[(y_n')]$, is the whole dual space E'. The sequence (y_n') we will shortly call a *basic sequence* for $(Q_i)_{i \in I}$. Let us indicate the importance of the notion of quasi-injectivity. Suppose that $(Q_i)_{i \in I}$ is quasi-injective and that $Q_i = S_i \circ S_i^t$ for every $i \in I$ with $S_i \in L_g(H,E)$. Let (y_n') be a basic sequence for (Q_i) and denote by $< .,. >_i$ the bilinear form on $E' \times E'$, defined by $<x',y'>_i = < Q_i(x'),y'>$ for $x',y' \in E'$. Since every Q_i is injective on $L[(y_n')]$ the sequence (y_n') can be orthonormalized by the Gram-Schmidt process relative to any of the bilinear forms $<.,.>_i$. Let $(z'_{n,i})_{n \geq 1}$ denote the orthonormalized sequence obtained from (y'_n) relative to $<.,.>_i$. Then the sequence $(f_{n,i})_{n \geq 1}$ in H, defined by $f_{n,i} = S_i^t(z'_{n,i})$, is orthogonal in H for every $i \in I$. Let H_i denote the Hilbert subspace of H having $(f_{n,i})_{n \geq 1}$ as a CONS. Then it is not difficult to see that $(S_i^t(E'))^- = H_i = (Ker S_i)^{\perp}$. Furthermore, $Q_i = \sum_{n \geq 1} x_{n.i}^{\otimes 2}$, where $x_{n,i} = Q_i(z'_{n,i})$. Let P_i denote the orthogonal projection from H onto H_i and define for every $m \geq 1$ the operator $\overline{S}_i^m \in L(E,H_i)$ by

(2.3) $$\overline{S}_i^m = \sum_{k=1}^{m} < x, z'_{k,i} > f_{k,i}$$

Then it is easy to see that the sequence $(\overline{S}_i^m \circ S_i)_{m \geq 1}$ of operators in L(H) converges strongly to P_i, and we will call the sequence $(\overline{S}_i^m)_{m \geq 1}$ an *approximate quasi-inverse* of S_i.

The above construction is mainly important because of the following observation. Suppose that the index space I has an additional structure. For example, let us suppose that I is a measurable space and that the family $(S_i)_{i \in I}$ depends measurably on i. Then the Gram-Schmidt process shows that also the sequences $(z'_{n,i})_{n \geq 1}$ and $(f_{n,i})_{n \geq 1}$ depend measurably on i, and it follows that also (P_i) and $(\overline{S}_i^m)$ are measurable (see[6] for the origin of these ideas).

Now we can formulate our main theorem.

(2.4). **Theorem.** Let E be a separable Banach space and H be a separable Hilbert space (both real) and suppose that $S: \mathbb{R}_+ \times \Omega \to L_g(H,E)$ is progressively measurable. Let $Q: \mathbb{R}_+ \times \Omega \to Cov(E)$ denote the associated covariance function given by $Q = S \circ S^t$. Assume that $Y=(Y(t))$ is a given E-valued, continuous local martingale with cylindrical quadratic variation $[Y]$ of the form $[Y](t)=\int_0^t Q(r)dr$. Finally, let (e_k) be a fixed CONS of H. Then there exists an infinite-dimensional Brownian motion $\xi=(\xi_k)$ on a possibly extended probability space Ω^* such that (cf. (2.2))

$$(2.5)\qquad <\int_0^{\cdot} [S(r)\ ,\ d\xi(r)]\ ,\ x'> = <Y(t)\ ,\ x'> \qquad P^*\text{-a.s.}$$

for all $t \geq o$ and all $x'\in E'$. If $(Q(s,\omega))_{s\geq o,\ \omega\in\Omega}$ is quasi-injective and if $S(s,\omega)$ is injective for all $s \geq o$ and all $\omega\in\Omega$, then a Brownian motion $\xi=(\xi_k)$ exists even on Ω such that (2.5) holds, and if S has the property that the stochastic integral process $\int_0^{\cdot} [S(r)\ ,\ d\xi(r)]$ exists in the sense of Chapter1, then $\int_0^{\cdot} [S(r)\ ,\ d\xi(r)] = Y$ P-a.s..

Proof. (1) Let us first suppose that Q is quasi-injective, and let (y_n') be a fixed basic sequence for Q. As in the remarks preceding the theorem let $(z'_n(t,\omega))_{n\geq 1}$ denote the orthonormalized sequence obtained from (y_n') by the Gram-Schmidt process relative to the bilinear form on $E'\times E'$ given by $Q(t,\omega)$ and put $f_n(t,\omega)=S^t(t,\omega)(z'_n(t,\omega))$. Let $H(t,\omega)$ denote the Hilbert subspace having $(f_n(t,\omega))_{n\geq 1}$ as CONS, and let $P(t,\omega)$ be the orthogonal projection from H onto $H(t,\omega)$. It follows that $P: \mathbb{R}_+\times\Omega\longrightarrow L(H)$ is progressively measurable. As above, every $S(t,\omega)$ has an approximate quasi-inverse $(\overline{S}^m(t,\omega))_{m\geq 1}$ given by

$$(2.6)\qquad \overline{S}^m(t,\omega)(x) = \sum_{k=1}^{m} < x\ ,\ z'_k(t,\omega) > f_k(t,\omega)$$

and every $\overline{S}^m : \mathbb{R}_+\times\Omega\longrightarrow L(E,H)$ is also progressively measurable.

The following auxiliary construction is not necessary in case that $H(t,\omega) = H$ for every $t \geq o$ and $\omega\in\Omega$, i.e. in case that the operators $S(t,\omega)$ are all injective.

We take a second probability space $(\Omega',\mathcal{F}',P')$ with a right continuous filtration $(\mathcal{F}'_t)$ and assume that on Ω' there exists an infinite dimensional Brownian motion $\zeta= (\zeta_k)$ adapted to $(\mathcal{F}_t')$. Since E is separable, it is easy to see that there exists an injective covariance operator $Q_0\in Cov(E)$. Orthonormalize the given basic sequence (y_n') relative to Q_0 and let $(z'_{n,o})$ denote the orthonormalized sequence. Define $S_0(e_n)=Q_0(z'_{n,o})$. Then it is easy to see that this defines a linear operator $S_0\in L(H,E)$ such that $Q_0=S_0 o S_0^t$ and S_0 is injective. Let $(\overline{S}_0^m)_{m\geq 1}$ denote the approximate quasi-inverse of S_0, which is similar to (2.6) given by

$$(2.7)\qquad \overline{S}_0^m(x) = \sum_{k=1}^{m} < x\ ,\ z'_{k,0}> e_k \qquad \text{for } x\in E\ ,\ m\geq 1\ ,$$

and it follows from the injectivity of S_0 that in this case the sequence $(\overline{S}_0^m o S_0)_{m\geq 1}$ converges strongly to the identity on H.

Finally, we put

$$Z_o(t) = \sum_{k\geq 1} S_o(e_k)\,\zeta_k(t) \qquad \text{for every } t\geq o.$$

Then Z_o is an E-valued Gaussian process on Ω' (symmetric, with ind. increments) with $[Z_o](t) = t\,Q_o$.

As above let $\Omega^* = \Omega \times \Omega'$ denote the extended probability space and consider all random functions $Y, S, \ldots, Z_o, \ldots$ as random functions on Ω^*. As usual, we will suppress the variable ω and just write $S(r)$, $P(r), \ldots$ instead of $S(r,\omega)$, $P(r,\omega), \ldots$ Finally, let us denote by P_n ($n \in \mathbb{N}$) the projection from H onto the n-dimensional subspace generated by $e_1, \ldots, e_n$.

For every $n,m \in \mathbb{N}$ we define

$$(2.8) \qquad Z_{n,m}(t) = \int_0^t P_n \overline{S}^m(r)\, dY(r) + \int_0^t P_n (I-P(r))\, \overline{S}_o^{\,m}\, dZ_o(r)$$

(I = identity on H). Since the integrands map the integrators into the finite dimensional Hilbert space $P_n(H)$, both stochastic integrals on the right side of (2.8) exist as a consequence of the special structure of the quadratic variations of Y and Z_o.

For $m,m' \in \mathbb{N}$ the quadratic variation of $Z_{n,m}-Z_{n,m'}$ is given by

$$[Z_{n,m}-Z_{n,m'}](t) = \int_0^t P_n(\overline{S}^m(r)- \overline{S}^{m'}(r))S(r)(P_n(\overline{S}^m(r)-\overline{S}^{m'}(r))\, S(r))^t dr$$

$$+ \int_0^t P_n (I-P(r))\, (\overline{S}_o^{\,m}-\overline{S}_o^{\,m'})\, S_o (P_n(I-P(r))\, (\overline{S}_o^{\,m} - \overline{S}_o^{\,m'}) S_o)^t dr,$$

since Y and Z_o are independent. It follows from $\lim_{m\to\infty} \overline{S}^m(r)S(r)=P(r)$ and $\lim_{m\to\infty} \overline{S}_o^{\,m} S_o = I$ strongly that $\lim_{m',m\to\infty} [Z_{n,m}-Z_{n,m'}](t) = 0$. This implies (see e.g. [8], ch.I) that there exists a ($P_n(H)$-valued) continuous local martingale Z_n such that $\lim_{m\to\infty} Z_{n,m} = Z_n$ in probability uniformly on every bounded interval. Z_n has the quadratic variation

(2.9) $$[Z_n](t) = \lim_{m\to\infty} [Z_{n,m}](t) = \int_0^t P_n P(r)\, dr + \int_0^t P_n(I - P(r))\, dr = tP_n$$

and it follows from Levy's theorem that there exists an n-dimensional Brownian motion $\xi_n = (\xi_k^n)_{1\le k\le n}$ on Ω^* such that

$$Z_n(t) = \sum_{k=1}^{n} e_k\, \xi_k^n(t) \;.$$

Since $P_k Z_n = Z_k$ for $k\le n$, it follows that $\xi_k = \xi_k^n$ does not depend on n. Hence we have proved that there exists an infinite dimensional Brownian motion $\xi=(\xi_k)$ such that

(2.10) $$Z_n(t) = \sum_{k=1}^{n} e_k \xi_k(t) \qquad \text{for every } n \in \mathbb{N}$$

Now let us prove the identity (2.5). Let $\tau \le t$ be a stopping time such that Y^τ is square integrable. Then for every $x' \in E$ also $< \int_0^\tau [S(r),d\xi(r)]\, , x' >$ is square integrable, since $< \int_0^{\cdot} [S(r),d\xi(r)]\, , x' >$ has the same quadratic variation as $<Y,x'>$. We get

$$\mathbb{E}\left(<\int_0^\tau [S(r),d\xi(r)]\, , x'> - <Y(\tau)\, , x'>\right)^2$$

$$= \quad \mathbb{E}\left(\sum_{k\ge 1}\int_0^\tau <S(r)(e_k)\, , x'>d\xi_k(r) - <Y(\tau), x'>\right)^2$$

$$= \quad \lim_{n\to\infty} \mathbb{E}\left(\sum_{k=1}^{n}\int_0^\tau <S(r)(e_k)\, , x'> d\xi_k(r) - <Y(\tau)\, , x'>\right)^2$$

$$= \quad \lim_{n\to\infty} \mathbb{E}<\int_0^\tau S(r)dZ_n(r) - Y(\tau)\, , x'>^2$$

$$= \quad \lim_{n\to\infty}\lim_{m\to\infty} \mathbb{E}<\int_0^\tau S(r)dZ_{n,m}(r) - Y(\tau)\, , x'>^2$$

$$= \quad \lim_{n\to\infty}\lim_{m\to\infty} \mathbb{E}<\int_0^\tau S(r)P_n \bar{S}^{\,m}(r)dY(r) + \int_0^\tau S(r)P_n(I-P(r))\, \bar{S}_0^{\,m}\, dZ_0(r) - Y(\tau)\, , x'>^2$$

$$= \lim_{n\to\infty}\lim_{m\to\infty} \mathbb{E}<\int_0^\tau S(r)P_n \bar{S}^m(r)dY(r) - Y(\tau) , x'>^2$$

$$+ \lim_{n\to\infty}\lim_{m\to\infty} \mathbb{E}<\int_0^\tau S(r)P_n(I-P(r)) \bar{S}_0^m dZ_0(r) , x'>^2$$

(since Y and Z_0 are independent)

$$= \lim_{n\to\infty}\lim_{m\to\infty} \mathbb{E}\int_0^\tau \| (S(r)P_n \bar{S}^m(r)S(r) - S(r))^t (x') \|_H^2 dr$$

$$+ \lim_{n\to\infty}\lim_{m\to\infty} \mathbb{E}\int_0^\tau \| (S(r)P_n(I-P(r)) \bar{S}_0^m S_0)^t (x') \|_H^2 dr$$

$$= \lim_{n\to\infty}\lim_{m\to\infty} \mathbb{E}\int_0^\tau \| ((\bar{S}^m(r)S(r))P_n - I)(S(r)^t (x')) \|_H^2 dr$$

$$+ \lim_{n\to\infty}\lim_{m\to\infty} \mathbb{E}\int_0^\tau \| ((\bar{S}_0^m S_0)(I - P(r)) P_n)(S(r)^t (x')) \|_H^2 dr$$

(since $\bar{S}^m(r)S(r)$, $\bar{S}_0^m S_0$, P_n , $P(r)$ are orth.projections)

$$= \lim_{n\to\infty} \mathbb{E}\int_0^\tau \| (P(r) P_n - I)(S(r)^t (x')) \|_H^2 dr$$

$$+ \lim_{n\to\infty} \mathbb{E}\int_0^\tau \| ((I - P(r)) P_n)(S(r)^t (x')) \|_H^2 dr$$

(by Lebesgue's theorem, since $\lim_{m\to\infty} \bar{S}^m(r)S(r) = P(r)$ and

$\lim_{m\to\infty} \bar{S}_0^m S_0 = I$ strongly)

$$= 2\, \mathbb{E}\int_0^\tau \| (I - P(r))(S(r)^t(x')) \|_H^2 d\, r$$

(again by Lebesgue's theorem, since $\lim_{n\to\infty} P_n = I$ strongly)

By assumption, $S(r)^t(E') \subset H(r)$. This shows that the last integral must be zero, and we have proved

$$(2.11)\quad E\left(<\int_0^\tau [S(r), d\,\xi(r)], x'> - <Y(\tau), x'>\right)^2 = 0$$

for every $x' \in E'$ and every stopping time $\tau \le t$ such that $Y(\tau)$ is square integrable. By assumption on Y, there exists a sequence (τ_n) of stopping times such that $Y(\tau_n \wedge t)$ is square integrable for every $n \in \mathbb{N}$ and such that $\lim_{n\to\infty} P[\tau_n > t] = 1$, and it follows from (2.11) that (2.5) holds.

(2) Now suppose that Q is not quasi-injective and dim $H=\infty$. Let $(\Omega'',\mathcal{F}'',P'')$ be another probability space with a right continuous filtration $(\mathcal{F}_t'')_{t\geq o}$ and suppose that Y_o is an H-valued, symmetric Gaussian, independent increment process on Ω'' with an injective covariance operator $K\in Cov(H)$. Let $\Omega^1=\Omega\times\Omega''$ denote the extended probability space and consider Y,Y_o etc. as extended to Ω^1. Let Y_1 denote the $E\oplus H$-valued process $Y_1=(Y,Y_o)$. Then Y_1 is a continuous, local martingale with quadratic variation $[Y_1]$ given by

$$(2.12) \qquad [Y_1](t)=\int_o^t Q_1(r)\,dr \ ,$$

where $Q_1: \mathbb{R}_+\times\Omega^1 \longrightarrow Cov\,(E\oplus H)$ is defined by

$$(2.13) \qquad Q_1\,(t,\omega) = (Q(t,\omega),K) \quad \text{for } t\geq o,\ \omega\in\Omega^1.$$

It was proved in [6] (lemma 2) that Q_1 is now quasi-injective. If $K=R\circ R^t$ with $R\in L(H)$, then Q_1 has the factorization

$$(2.14) \qquad Q_1=S_1\circ S_1^t \qquad \{\text{ with } \ S_1=(S,R):\mathbb{R}_+\times\Omega^1\to L_g(H\oplus H,E\oplus H)\,\}$$

through the Hilbert space $H_1=H\oplus H$.

Let (e_k) denote a fixed CONS of H. Then the two sequences $((e_k,o))$ and $((o,e_k))$ together define a CONS (f_j) of H_1. Now we can make use of part (1) of the proof. So there exists an extended probability space $\Omega^*=\Omega^1\mathrm{x}\Omega'$ and an infinite-dimensional Brownian motion $\xi=(\xi_k)$ on Ω^* such that for all $t\geq o$ and all $(x',\eta)\in E'\oplus H$,

$$(2.15) \qquad <Y_1(t),(x',\eta)> = <\int_0^t [S_1(r),d\xi(r)],(x',\eta)> \qquad P^*\text{-a.s.}\ ,$$

where the right hand side is more explicitely given by

$$(2.16) \qquad <Y_1(t),(x',\eta)> = \sum_{j\geq 1}\ \int_0^t <S_1(r)(f_j),(x',\eta)>\, d\xi_j(r)\ .$$

Now let $P:E\oplus H\longrightarrow E$ denote the canonical projection. Then we obtain for every $x'\in E'$ and $t\geq o$,

$$<Y(t),x'> = <P\,Y_1(t),x'> = <Y_1(t),P^tx'> = <Y_1(t),(x',o)>$$

$$= \sum_{j\geq 1}\ \int_0^t <(S(r),R)(f_j)\ ,\ (x',0)>\, d\xi_j(r) = \sum_{k\geq 1}\ \int_0^t <S(r)(e_k)\ ,\ x'>\, d\xi_k(r)$$

P^*-a.s., i.e. (2.5) also holds in the general situation.

For the theorem above we already assumed that the covariance function Q is of the form $Q=S \circ S^t$ with S progressively measurable. Following the ideas in [6] one can prove the following: Suppose that $Q : \mathbb{R}_+ \times \Omega \longrightarrow Cov(E)$ is progressively measurable. Then there exists a progressively measurable $S : \mathbb{R}_+ \times \Omega \longrightarrow L_g(H,E)$ such that $Q=S \circ S^t$. Let us outline the proof of this statement. Again the proof is done in two steps, first assuming that Q is quasi-injective. As in the proof of the theorem we take a basic sequence (y_n') in E' for Q and orthonormalize (y_n') to $(z_n'(t,\omega))$ relative to the bilinear form $Q(t,\omega)$ on E'xE'. It was proved in [6] that

$$(t_k)_{k\geq 1} \rightarrow \sum_{k\geq 1} t_k \, Q(t,\omega)\,(z_k\,(t,\omega))$$

with $(t_k) \in \ell^2$ defines a continuous linear operator $S(t,\omega): \ell^2 \longrightarrow E$ such that $Q(t,\omega)=S(t,\omega) \circ S(t,\omega)^t$. The Gram-Schmidt procedure shows that $S:\mathbb{R}_+ \times \Omega \longrightarrow L(H,E)$ is progressively measurable. Moreover, in this case we have $\overline{(S(t,\omega)^t(E'))}=\ell^2$. If Q is not quasi-injective, we proceed similar to step (2) of the proof of theorem (2.4). Take an injective covariance operator $K \in Cov(\ell^2)$ and enlarge E to $E \oplus \ell^2$. Then $(Q,K) : \mathbb{R}_+ \times \Omega \longrightarrow Cov(\ell^2)$ is progressively measurable and quasi-injective. Hence $(Q,K)=\overline{S} \circ \overline{S}^t$ with $\overline{S} : \mathbb{R}_+ \times \Omega \longrightarrow L(\ell^2, E \oplus \ell^2)$ progressively measurable. If $P:E \oplus \ell^2 \longrightarrow E$ denotes the canonical projection, then $S = P\overline{S}$ is progressively measurable and has the property $Q = S \circ S^t$. With these remarks the following corollary follows from the theorem (2.4).

(2.17) **Corollary.** Let $Q:\mathbb{R}_+ \times \Omega \longrightarrow Cov(E)$ be progressively measurable and let Y be a given E-valued, continuous local martingale having a quadratic variation [Y] given by $[Y](t)= \int_0^t Q(r)dr$. Then there exist a progressively measurable $S:\mathbb{R}_+ \times \Omega \longrightarrow L(\ell^2,E)$ and an infinite dimensional Brownian motion $\xi= (\xi_k)$ on a possibly extended probability space such that

$$(2.18) \qquad <\int_0^t [S\,(r), d\xi\,(r)], x'> = <Y(t), x'> \quad \text{a.s.}$$

for all $t \geq o$, $x' \in E'$. If Q is quasi-injective, then ξ can even be defined on the original probability space Ω. Moreover (see[6]), if Q is in addition a.s. continuous, then also S can be taken to be a.s. continuous.

3. *APPLICATION:* UNIQUENESS OF THE MARTINGALE PROBLEM ON BANACH SPACES

Let $C_E=C(\mathbb{R}_+,E)$ denote the space of all continuous functions $f:\mathbb{R}_+ \longrightarrow E$ with the topology of uniform convergence on bounded intervals. We will call a function F defined on $\mathbb{R}_+\times C_E$ *predictable*, if $F(t,f)=F(t,f^t)$ for all $t\geq o$ and $f\in C_E$, where $f^t=f(t\wedge .)$. By P_E we will shortly denote the set of all Radon probability measures on C_E. For the description of the martingale problem we need the following data.

Let $R\in L(F,E)$ be a fixed 2-smoothable operator defined on a second (real, separable) Banach space F, and let $\mu:\mathbb{R}_+\times C_E\longrightarrow G(F)$ be continuous and predictable. Put $\rho=R\circ\mu$. Then $\rho:\mathbb{R}_+\times C_E\longrightarrow G(E)$ is also continuous and predictable, and we will call ρ shortly a *smooth field of Gaussian measures on* E. Suppose further that $G:\mathbb{R}_+\times C_E\longrightarrow E$ is another continuous and predictable function, which we will call shortly a *smooth vector field on* E. Let $K:\mathbb{R}_+\times C_E\longrightarrow Cov(F)$ be the covariance function associated to μ, i.e. $K(t,f)$ is the covariance operator of $\mu(t,f)$ for all $t\geq o$, $f\in C_E$. Then the covariance function $Q:\mathbb{R}_+\times C_E\longrightarrow Cov(E)$ associated ρ is given by $Q=R\circ K\circ R^t$.

We will now call a predictable function $\mathbb{P}:\mathbb{R}_+\times C_E\longrightarrow P_E$ a *solution of the martingale problem for* ρ and G, if the following conditions hold for the canonical process $X:\mathbb{R}_+\times C_E\longrightarrow E$, defined by $X(t)\,(f) = f\,(t)$:

(3.1) $$\mathbb{P}_{t,f}\,[X^t=f^t] = 1 \quad \text{for all } (t,f)\in\mathbb{R}_+\times C_E.$$

(3.2) For every $(s,f)\in\mathbb{R}_+\times C_E$ the process $Y = (Y(t))_{t\geq s}$, defined by

$$Y(t) = X(t) - X(s) - \int_s^t G(r,X)\,dr$$

is a local $\mathbb{P}_{s,f}$-martingale relative to the canonical filtration $(\mathcal{B}_t)_{t\geq s}$ on $C_E(\mathcal{B}_t=\sigma\{X(r):o\leq r\leq t\})$ with quadratic variation $[Y]$ given by

$$[Y]\,(t) = \int_s^t Q(r,X)\,dr.$$

The single measure $\mathbb{P}_{s,f}$ is also called a *solution of the martingale problem for* ρ *and* G *starting from* (s,f).

In case that $\rho(t,f)$ and $G(t,f)$ only depend on the value $f(t)$ of f at time t and not on the whole past before, one can prove (see[4]) that the above formulation of the martingale problem is the same as given in [12] for the case $E = \mathbb{R}^d$.

Suppose now that ρ and G fulfill the following additional *growth condition:*

(3.3) For every $t\geq o$ there exists a constant $K=K(t)$ such that for all $f\in C_E$ and all $s\in[0,t]$,

$$\left(\int_F \|y\|^4 \ \mu(s,f)\,(dy)\right)^{1/4} + \| G\ (s,f) \| \le K(1 + \|f^s\|)$$

(where 4 could be replaced by any p>2).

Then it is proved in [5] that (3.3) implies that there *exists* a solution of the martingale problem for ρ and G. As in [12] the idea is to construct from ρ and G a net $(\mathbb{P}^{\Delta}_{s,f})_{\Delta\in\Pi}$ of rather simple measures on C_E(see the proof of the theorem below) and to prove that this net is relatively compact and that any limit point gives a solution $\mathbb{P}_{s,f}$ of the martingale problem for ρ and G starting from (s,f).

In [5] we also proved a uniqueness theorem under rather strong conditions on ρ, G and the Banach space E, using the known representation theorem for martingales ([12], theorem 4.5.2) in the finite-dimensional case. With the aid of theorem 2.4 we can now prove a more general uniqueness theorem.

For this uniqueness result we impose a kind of *local Lischitz condition* on ρ and G. We assume that the covariance function K has a factoritation $K=S\circ S^t$ with a continuous and predictable function $S:\mathbb{R}_+\times C_E\rightarrow L_g(H,F)$ such that the following condition holds for S and G:

(3.4) For all $(t,f)\in\mathbb{R}_+\times C_E$ there exists constants η and C such that for all $g\in C_E$ with

$\|g^t-f^t\|<\eta$ we have for any $o\le s\le t$,

$|||S(s,g)-S(s,f)||| + \|G(s,g)-G(s,f)\|\le C\,\|g^s-f^s\|$.

Of course, $\|f^s\|$ means the sup-norm of the stopped function f^s, and $|||.|||$ denotes the norm on $L_g(H,F)$ introduced in chapter 1.

(3.5) **Theorem.** Under conditions (3.3) and (3.4) there is a unique solution $\mathbb{P}$ of the martingale problem for ρ and G.

Proof: As already mentioned, the growth condition (3.3.) implies that there exists a solution of the martingale problem for ρ and G, and it remains to prove uniqueness. Let $(s_o,f_o)\in\mathbb{R}_+\times C_E$ and let $\mathbb{P}^1$ and $\mathbb{P}^2$ denote two solutions of the martingale problem for ρ and G starting from (s_o,f_o). By theorem (2.4) there exist possibly extended probability spaces $(\Omega^i,\mathcal{F}^i,P^i)$ (i=1,2) of $(C_E,\mathcal{B},\mathbb{P}^i)$ (where $\mathcal{B}=\sigma\{X(t):t\ge o\}$), extended filtrations $(\mathcal{F}^i_t)$, and infinite dimensional Brownian motions $\xi^i=(\xi^i_k)_{k\ge1}$ on Ω^i, such that the martingale Y of (3.2), extended to Ω^i, is of the form

$$(3.6)\quad Y(t)=\int_{s_o}^{t}[T(r,X),d\xi^i(r)]=\sum_{k\geq 1}\int_{s_o}^{t}T(r,X)(e_k),d\xi_k^i(r)\qquad (t\geq s_o),$$

where $T=R\circ S$ and (e_k) denotes a fixed CONS of H. Since we have two extended probability spaces, we will denote more carefully by X_1, X_2 the extensions of the canonical process X to Ω^1 and Ω^2 resp. By the definition of the martingale Y, it follows from (3.6) that X_1 and X_2 are solutions of the stochastic integral equations

$$(3.7.i)\quad X_i(t)=f_0(s_o)+\int_{s_o}^{t}G(r,X_i),dr+\int_{s_o}^{t}[T(r,X_i),d\xi^i(r)]\qquad (t\geq s_o),$$

for i=1,2. Let us remark that the stochastic integrals of (3.6), (3.7.i) exist in the strong sense of chapter 1, since S is continuous and predictable relative to the norm $|||.|||$ on $L_g(H,F)$.

Let t_o denote in the following a fixed but arbitrary time point $t_o>s_o$, and denote by Π_o the family of all finite partitions of $[s_o,t_o]$. For $\Delta=(s_o<s_1<...<s_n=t_o)\in\Pi_o$ we define on Ω^i (i=1,2) a stochastic process by induction. We put

$$X_i^\Delta(t)=\begin{cases} f_0(t) & \text{for } t\in[0,s_0]\\ X_i^\Delta(s_k)+\int_{s_k}^{t}G(r,(X_i^\Delta)^{s_k})\,dr+\int_{s_k}^{t}[T(r,(X_i^\Delta)^{s_k}),d\xi^i(r)] & \\ \qquad\text{for } t\in\,]s_k,s_{k+1}]\quad (k=0,1,...,n-1). \end{cases}$$

With the aid of the function

$$\varphi^\Delta=\sum_{k=0}^{n-1}s_k\,1_{]s_k,s_{k+1}]}$$

we can write X_i^Δ in the closed form

$$(3.8)\quad X_i^\Delta(t)=f_0(s_o)+\int_{s_o}^{t}G(r,X_i^\Delta\circ\varphi^\Delta)\,dr+\int_{s_o}^{t}[T(r,X_i^\Delta\circ\varphi^\Delta),d\xi^i(r)]\ .$$

It is easy to see that X_1^Δ and X_2^Δ have the same distribution $\mathbb{P}^\Delta$ on $C([o,t_o],E)$ and one can prove (see[5]) that the condition (3.3) implies the uniform tightness of the net $(\mathbb{P}^\Delta)_{\Delta\in\Pi_o}$. Let $\mathbb{P}^i(t_o)$ (i=1,2) denote the measure on $C([o,t_o],E)$ given by the solution $\mathbb{P}^i$ of the martingale problem, and let us consider the process X_i on Ω^i of (3.7) as restricted to $[o,t_o]$. Then the uniform tightness of the family

$\{\mathbb{P}^{\Delta}:\Delta\in\Pi_o\}\cup\{\mathbb{P}^1(t_o),\ \mathbb{P}^2(t_o)\}$ implies that for every $\varepsilon>o$ there exists a compact subset K of $C([o,t_o],E)$ such that

$$(3.9)\qquad P^i[X_i^{\Delta}\notin K]<\varepsilon \quad\text{and}\quad P^i[X_i\notin K]<\varepsilon$$

for all $\Delta\in\Pi_o$ and i=1,2. The compact set K can and will be assumed (see[4]) to have the property that $f\in K$ implies $f^t\in K$ for $t\le t_o$, and we may also assume that the filtrations on Ω^i are right continuous. Then

$$\tau_i^{\Delta}=\begin{cases} t_o\ , & \text{if } X_i^{\Delta}\in K\\ \inf\{\ t\ge o : (X_i^{\Delta})^t\notin K\} & \text{else}\end{cases}$$

and

$$\tau_i=\begin{cases} t_0\ , & \text{if } X_i^{\Delta}\in K\\ \inf\{\ t\ge 0 : X_i^t\notin K\} & \text{else}\end{cases}$$

are stopping times (see[4],or[5]), and we put $\sigma_i^{\Delta}:=\tau_i^{\Delta}\wedge\tau_i$.

Now we compare the stopped processes $X_i^{\Delta}(\sigma_i^{\Delta}\wedge.)$ and $X_i(\sigma_i^{\Delta}\wedge.)$, using the fact that on the compact set K we get from the local Lipschitz condition (3.4) a uniform Lipschitz condition: There exists a $C\ge o$ such that

$$(3.10)\qquad |||S(t,g)-S(t,f)|||\ +\ ||G(t,g)-G(t,f)||\ \le C\ ||g^t-f^t||$$

for all $f,g\in K$ and $o\le t\le t_o$. Let $\mathbb{E}_i$ denote the expectation relative to P^i. Since by

our assumption on K every path of the stopped processes $X_i^{\Delta}(\sigma_i^{\Delta}\wedge.)$ and $X_i(\sigma_i^{\Delta}\wedge.)$ belongs to K, the integrals in the following inequalities are all finite. For every $t\in[s_o,t_o]$ we have

$$\mathbb{E}_i \sup_{s_0\le s\le t}\|X_i(s\wedge\sigma_i^{\Delta})-X_i^{\Delta}(s\wedge\sigma_i^{\Delta})\|^2$$

$$\le\ 2\,\mathbb{E}_i \sup_{s_0\le s\le t}\left\|\int_{s_0}^{s\wedge\sigma_i^{\Delta}}(G(r,X_i)-G(r,X_i^{\Delta}\circ\varphi^{\Delta}))\,dr\right\|^2$$

$$+\ 2\,\mathbb{E}_i \sup_{s_0\le s\le t}\left\|\int_{s_0}^{s\wedge\sigma_i^{\Delta}}[(T(r,X_i)-T(r,X_i^{\Delta}\circ\varphi^{\Delta}))\ ,\ d\xi^i(r)]\right\|^2$$

$$\le\ 2t\,\mathbb{E}_i\int_{s_0}^{t}\|G(r\,,X_i^{\sigma_i^{\Delta}})-G(r\,,(X_i^{\Delta}\circ\varphi^{\Delta})^{\sigma_i^{\Delta}})\ \|^2dr$$

$$+ \quad 4D\ \mathbb{E}_i \int_{s_0}^{t} |||S(r, X_i^{\sigma_i^\Delta}) - S(r, (X_i^\Delta \circ \varphi^\Delta)^{\sigma_i^\Delta})|||^2\, dr$$

(this follows from Doob's L^p-inequality together with (1.7))

$$\leq \ C^2(2t+4D)\,\mathbb{E}_i \int_{s_0}^{t} \| X_i^{\sigma_i^\Delta \wedge r} - (X_i^\Delta \circ \varphi^\Delta)^{\sigma_i^\Delta \wedge r} \|^2\, dr$$

$$= \ K_0 \int_{s_0}^{t} \mathbb{E}_i \sup_{s_0 \leq u \leq r} \| X_i(\sigma_i^\Delta \wedge u) - (X_i^\Delta \circ \varphi^\Delta)(\sigma_i^\Delta \wedge u) \|^2\, dr \qquad (\text{with } K_0 = C^2(2t+4D))$$

$$\leq \ 2K_0 \int_{s_0}^{t} \mathbb{E}_i \sup_{s_0 \leq u \leq r} \| X_i(\sigma_i^\Delta \wedge u) - X_i^\Delta(\sigma_i^\Delta \wedge u) \|^2\, dr \ + 2K_1(\Delta) \ ,$$

$$\text{with} \quad K_1(\Delta) = 2K_0 \int_{s_0}^{t_0} \mathbb{E}_i \sup_{s_0 \leq u \leq r} \| X_i(\sigma_i^\Delta \wedge u) - (X_i^\Delta \circ \varphi^\Delta)(\sigma_i^\Delta \wedge u) \|^2\, dr$$

Gronwall's inequality now implies that

$$(3.11) \qquad \mathbb{E}_i \sup_{s_0 \leq s \leq t_0} \| X_i(\sigma_i^\Delta \wedge s) - X_i^\Delta(\sigma_i^\Delta \wedge s) \|^2 \ \leq \ C_1 K_1(\Delta)$$

for a certain constant C_1. Since $X_i^\Delta(\sigma_i^\Delta \wedge .) \in K$ for all Δ and K as a compact subset of $C([o,t_o],E)$ is equicontinuous, we can finally find for a given $\delta > o$ a partition $\Delta(\delta)$ such that $K_1(\Delta) \leq C_1^{-1}\delta$ for all partitions $\Delta \geq \Delta(\delta)$. So we have from (3.11),

$$(3.12) \qquad \mathbb{E}_i \sup_{s_0 \leq s \leq t_0} \| X_i(\sigma_i^\Delta \wedge s) - X_i^\Delta(\sigma_i^\Delta \wedge s) \|^2 \ \leq \ \delta \qquad \text{for all } \Delta \geq \Delta(\delta)$$

where, of course, $\Delta(\delta)$ depends on K and hence on ε.

Let us prove that $\lim_\Delta X_i^\Delta = X_i$ in probability on $C([0,t_0],E)$.

For every $\eta > 0$ we get

$$P^i[\, \| X_i - X_i^\Delta \| \geq \eta \,] = P^i[\, \sup_{s_o \leq s \leq t_o} \| X_i(s) - X_i^\Delta(s) \| \geq \eta \,]$$

$$\leq P^i[\, \sigma_i^\Delta < t_o \,] + P^i[\, \sigma_i^\Delta = t_o \ \text{and} \ \sup_{s_o \leq s \leq t_o} \| X_i(s) - X_i^\Delta(s) \| \geq \eta \,]$$

$$\leq P^i[\, \sigma_i^\Delta < t_o \,] + P^i[\, \sup_{s_o \leq s \leq t_o} \| X_i(\sigma_i^\Delta \wedge s) - X_i^\Delta(\sigma_i^\Delta \wedge s) \| \geq \eta \,]$$

$$\leq P^i[\, \sigma_i^\Delta < t_o \,] + \eta^{-2}\, \mathbb{E}_i \sup_{s_o \leq s \leq t_o} \| X_i(\sigma_i^\Delta \wedge s) - X_i^\Delta(\sigma_i^\Delta \wedge s) \|^2$$

for every $\Delta \in \Pi_o$. First we choose now a compact K as above such that

$$P^i [\sigma_i^\Delta < t_o] \leq P^i [X_i \notin K] + P^i [X_i^\Delta \notin K] \leq \eta/2 ,$$

and then we choose a Δ_o, depending on η, such that for all $\Delta \geq \Delta_o$,

$$\mathbb{E}_i \sup_{s_0 \leq s \leq t_0} \| X_i(\sigma_i^\Delta \wedge s) - X_i^\Delta(\sigma_i^\Delta \wedge s) \|^2 \leq \frac{\eta^3}{2}$$

which is possible by (3.12). Then

$$P^i [\| X_i - X_i^\Delta \| \geq \eta] \leq \eta \text{ for all } \Delta \geq \Delta_o ,$$

i.e., $\lim_\Delta X_i^\Delta = X_i$ in probability, since $\eta > 0$ was arbitrary. Since convergence in probability implies the weak convergence of the distributions, we have finally

$$\lim_\Delta \mathbb{P}^\Delta = \mathbb{P}^i (t_o) \qquad \text{for } i = 1,2,$$

which proves $\mathbb{P}^1(t_o) = \mathbb{P}^2(t_o)$ and hence $\mathbb{P}^1 = \mathbb{P}^2$ since t_o was arbitrary. This finishes the proof that the martingale problem for ρ and G has a unique solution. ∎

Using similar arguments as in the above proof we get the following approximation method for solutions of the martingale problem.

(3.13) **Corollary.** Let ρ and G have the properties (3.3) and (3.4). Suppose that $(\Omega, \mathcal{F}, P)$ is a probability space with a right continuous filtration $(\mathcal{F}_t)$, on which there exists an infinite dimensional Brownian motion $\xi = (\xi_k)$ adapted to $(\mathcal{F}_t)$. Define as in the proof of the theorem for a given $(s_o, f_o) \in \mathbb{R}_+ \times C_E$ and $\Delta = (s_o < s_1 < \ldots < s_n)$

$$X^\Delta(t) = \begin{cases} f_o(t) & \text{for } t \leq s_o \\ f_o(s_o) + \int_{s_o}^t G(r, X^\Delta \circ \varphi^\Delta)\, dr + \int_{s_o}^t [T(r, X^\Delta \circ \varphi^\Delta), d\xi(r)] & (t > s_o) . \end{cases}$$

Then the net (X^Δ) of C_E-valued random vectors converges in probability to a limit X, which is a solution of the stochastic integral equation

$$X(t) = f_o(s_o) + \int_{s_o}^t G(r, X)\, dr + \int_{s_o}^t [T(r, X), d\xi(r)]$$

for $t \geq s_o$, and the distribution $\mathbb{P}_{s_o, f_o}$ of X is the unique solution of the martingale problem for ρ and G stanting from (s_o, f_o).

(3.14) **Remark.** In [5] the existence of solutions of the martingale problem was proved under the additional assumption that the Banach space E is reflexive. With slightly more general but essentially the same arguments one can show that the existence follows without the assumption of reflexivity.

REFERENCES

[1] A.ARAUJO, E.GINÉ: The central limit theorem for real and Banach valued random variables. New York: Wiley 1980.

[2] E.DETTWEILER: Banach space valued processes with independent increments and stochastic integration. In: Probability in Banach spaces IV, Lecture Notes in Math. 990, 54-83. Berlin-Heidelberg-New York: Spinger 1983.

[3] E.DETTWEILER: Stochastic integration of Banach space valued functions. In:L.Arnold, P.Kotelenez: Stochastic Space-Time Models and Limit Theories, D.Reidel Publ. Comp., 53-79 (1985).

[4] E.DETTWEILER: On the martingale problem for Banach space valued stochastic differential equations. To appear in: Journal of Theoretical Probability.

[5] E.DETTWEILER: The martingale problem on Banach spaces. To appear in: Expositiones Mathematicae.

[6] E.DETTWEILER, G.LITTLE: Continuous factorizations of covariance operators and Gaussian processes. Studia Math. 85, 91-106 (1987).

[7] J.L.DOOB: Stochastic Processes. New York: Wiley 1952.

[8] I.I.GIHMAN, A.V.SKOROHOD: The Theory of Stochastic Processes, Vol. III. Berlin-Heidelberg-New York: Springer 1979.

[9] G.PISIER: Martingales with values in uniformly convex spaces. Israel J. Math. 20, 326-350 (1975).

[10] L.SCHWARTZ: Radon measures on arbitrary tolopogical spaces and cylindrical measures. London: Oxford University Press 1973.

[11] A.V.SKOROHOD: Studies in the theory of random processes. Addison-Wesley 1965.

[12] D.W.STROOCK, S.R.S.VARADHAN: Multidimensional Diffusion Processes. Berlin-Heidelberg-New York: Springer 1979.

[13] F.TREVES: Topological Vector Spaces, Ditributions and Kernels. New York-London: Academic Press 1967.

Egbert Dettweiler
Mathematisches Institut
der Universitat Tübingen
Auf der Morgenstelle 10
74 Tübingen
West Germany

NONLINEAR FUNCTIONALS OF EMPIRICAL MEASURES AND THE BOOTSTRAP

R. M. Dudley

Room 2-245, M.I.T., Cambridge, MA 02139, USA

1. Introduction. This paper aims at extending the theory of differentiable statistical functionals in connection with empirical processes and the bootstrap.

A statistical functional is a function $T = T(P)$ whose arguments are probability measures P on a measurable space $(X,\tilde{A})$. In the original formulation of Mises (1936, 1947) and in most of the subsequent literature, X is the line $\mathbb{R}$ and P is represented by its cumulative distribution function F, so one had functionals $T(F)$ and studied the asymptotic behavior of $T(F_n)$ for empirical distribution functions F_n. More generally, Fortet and Mourier (1954) and Fortet (1958, pp. 196-198) considered statistical functionals on Banach spaces. Huber (1981, pp. 34ff.) considers differentiability with respect to metrics which metrize convergence of laws.

Some of the simplest statistical functionals are those of the form $T(P) = \int g\, dP$ for a measurable function g. If g is unbounded then one restricts attention to a class $\tilde{P}$ of laws P for which g is integrable, and for central limit theorems (as in most of this paper), also square-integrable. Functionals of this form are linear in P in the sense that

---

Research supported by NSF Grants DMS-8506638 and DMS-880305.

for any laws P and Q in $\tilde{P}$ and $0 \le \lambda \le 1$, we have

$$T(\lambda P + (1-\lambda)Q) = \lambda T(P) + (1-\lambda)T(Q).$$

A functional T is Fréchet differentiable, with respect to a norm $\|\cdot\|$ defined and finite on the set $\tilde{P}$ of laws, if T can be approximated, in the neighborhood of any $P \in \tilde{P}$, by a constant plus a linear functional: we have

$$T(Q) = T(P) + A(Q-P) + o(\|Q - P\|) \quad \text{as} \quad \|Q - P\| \to 0,$$

where A is a linear functional defined on the signed measures $Q - P$. In the theory of statistical functionals, A is not necessarily assumed continuous with respect to $\|\cdot\|$ (for which the space of signed measures $P - Q$ for P and Q in $\tilde{P}$ may not be complete). It will be assumed, as in most of the literature, that $A(\mu) \equiv \int f_P \, d\mu$ for some function $f_P = f_{P,T}$ integrable for each $Q \in \tilde{P}$. This can be proved under some conditions (§4 below). Since the integral of any constant with respect to $Q - P$ is 0, f_P is only determined up to an additive constant and we can choose f_P to make $\int f_P \, dP = 0$.

So the Fréchet derivative of T at P on $\tilde{P}$ is a linear form $\mu \mapsto \int f_P \, d\mu$ defined for suitable signed measures μ. Sometimes f_P itself will be called the derivative of T at P.

For distribution functions on $\mathbb{R}$, and their differences, the usual norm has been the supremum ("Kolmogorov") norm $\|G - F\| := \sup_t |(F - G)(t)|$. Some functionals of interest are differentiable for this norm (Boos and Serfling, 1980; Parr, 1985). But, for example, the square of the mean,

$$T(F) := \left(\int_{-\infty}^{\infty} x \, dF(x)\right)^2,$$

is not differentiable for the sup norm even when restricted to laws which have bounded (but not uniformly bounded) support. This can be seen where P is a point mass at 0, G_n is a point mass at n, and $Q_n =$

$(1 - \lambda_n)P + \lambda_n G_n$ where $\lambda_n = 1/n^p$ for $p = 1, 2$.

On $\mathbb{R}$, and of course on other spaces, it seems worth while to consider differentiability with respect to other norms. This paper will consider norms of the following type. Let $(X,\tilde{A})$ be a sample space and $\tilde{F}$ a collection of measurable functions on X. Let $\tilde{P}$ be a collection of laws on X such that $\int|f|dP < \infty$ for each $f \in \tilde{F}$ and $P \in \tilde{P}$. Then for P and Q in $\tilde{P}$ let

$$\|P - Q\|_{\tilde{F}} := \sup\{|\int f \, d(Q-P)|: f \in \tilde{F}\}.$$

The Kolmogorov (sup) norm for distribution functions is of this form, taking $\tilde{F}$ to be the collection of indicators $1_{]-\infty,t]}$, $t \in \mathbb{R}$. More generally, suppose that $\tilde{F}$ is a Donsker class for P (as defined in §2 below: a central limit theorem for empirical measures holds with respect to uniform convergence over $\tilde{F}$), that T is differentiable with respect to $\|\cdot\|_{\tilde{F}}$, and that the derivative functions f_P belong to $\tilde{F}$. Under these conditions limit theorems will be proved (in §2) on the asymptotic behavior of $T(P_n)$ for empirical measures P_n. For $T(P) = (\int x \, dP)^2$, T is differentiable in this sense if $\tilde{F}$ consists of the one function $f(x) \equiv x$, and the derivative of T(F) with respect to F at any $F = F_0$ can be written as a multiple of this same function.

Fréchet differentiability of a functional T at a point x with respect to a norm $\|\cdot\|$ can be formulated as follows: the derivative $D(x)(y)$ is linear in y, and $T(x + ty) - T(x) - tD(x)(y) = o(t)$ as $t \to 0$ uniformly on $\{y: \|y\| \le \delta\}$ for some $\delta > 0$. For the weaker "compact derivative" (cf. Esty et al., 1985), uniformity is assumed for y in (suitable) compact sets, possibly in a linear subspace of the original space. But a larger supply of norms $\|\cdot\|_{\tilde{F}}$ gives additional possibilities for Fréchet differentiability, possibly avoiding the need for compact differentiability. In Fréchet differentiability, the convergence of the remainder to 0 is simpler to characterize specifically and numerically than it is for compact differentiability, which seems to be worth while for a given norm only when Fréchet differentiability fails.

One possible reason that little attention had been given to differentiable functionals of multidimensional distribution functions, although a central limit theorem was available for them (Dudley, 1966, 1967), is that the asymptotic distribution of $n^{1/2}\sup|F_n - F|$ is not the same for all continuous distributions on $\mathbb{R}^d$ as it is in $\mathbb{R}$. One way to estimate such asymptotic distributions in general is to resample from the empirical measure P_n to get a randomized (bootstrap) empirical measure P_n^*. Recent results on bootstrap central limit theorems for classes of linear statistics $T(P_n) = \int f\, dP_n$ (Bickel and Freedman, 1981; Gaenssler, 1986; and especially the characterization theorems of Giné and Zinn, 1988) allow extensions to (classes of) non-linear functionals T (Theorem 4 below).

2. <u>Statements and proofs</u>. Let T be a functional defined on a set $\tilde{P}$ of laws on a measurable space $(X,\tilde{A})$. For a given $P \in \tilde{P}$, let X(1), X(2),..., be i.i.d. with distribution P. (Strictly speaking, the X(i) are coordinate functions on a countable Cartesian product of copies of X.) Let P_n be the nth empirical measure, $P_n := n^{-1}(\delta_{X(1)} + \ldots + \delta_{X(n)})$. $\tilde{P}$ will be called <u>finitely full</u> iff it contains every law Q with finite support (such as any P_n).

Let $\tilde{F}$ be a collection of measurable functions on X such that $\int f^2 dP < \infty$ for each $f \in \tilde{F}$ and $P \in \tilde{P}$. Let $\nu_n := n^{1/2}(P_n - P)$. For any signed measure μ and function f integrable for μ, let $\mu(f) := \int f\, d\mu$. For a function G defined on $\tilde{F}$, such as $G = \nu_n$, let $\|G\|_{\tilde{F}} := \sup\{|G(f)|: f \in \tilde{F}\}$. Then $\tilde{F}$ will be said to satisfy the <u>bounded central</u>

limit theorem for P iff $\|y_n\|_{\tilde{F}}$ is bounded in outer probability, that is, for every $\varepsilon > 0$ there is an $M < \infty$ such that for all n, $Pr^*(\|y_n\|_{\tilde{F}} > M) < \varepsilon$, where $Pr^*(A) := \inf\{Pr(B): A \subset B, B \text{ measurable}\}$ for any set A.

Usually, central limit theorems are stated in terms of convergence in law. For empirical processes, what currently seems to be the best definition is as follows. For an arbitrary function g on a probability space, let $E^*g := \inf\{Eg: g \text{ measurable}, g \geq f\}$. Likewise $E_*g := \sup\{Ef: f \text{ measurable}, f \leq g\}$. Let (S,d) be any metric space, in general non-separable. Let V_n, n = 0, 1,..., be a sequence of functions from possibly different probability spaces into S. Assume that V_0 takes values in a separable subspace and is measurable. Other V_n need not be measurable at all. Any function from a probability space into S will be called a random element. Then V_n is said to converge in law to V_0, or to its law P_0 on S, if for every bounded continuous function H on S, $E^*H(V_n) \to EH(V_0) = \int H(v)dP_0(v)$ as $n \to \infty$. This definition of convergence in law, due to Hoffmann-Jørgensen (1984), applies to empirical processes as follows. For any law P and random variables f and g in $\tilde{L}^2(P)$, let $cov_P(f,g) := \int fg\, dP - \int f\, dP \int g\, dP$. (Here $\tilde{L}^2(P)$ is the space of measurable functions square-integrable for P, as opposed to the Hilbert space $L^2(P)$ of equivalence classes.) Then the variance of f is $var_P(f) := cov_P(f,f)$. For any $\tilde{F} \subset \tilde{L}^2(P)$ there is a Gaussian stochastic process G_P indexed by $f \in \tilde{F}$, having mean 0 and such that for each f and g in $\tilde{F}$, $EG_P(f)G_P(g) = cov_P(f,g)$. So for the probability measure Pr for G_P, $cov_{Pr}(G_P(f),G_P(g)) \equiv cov_P(f,g)$. For any set $\tilde{T}$, $\ell^\infty(\tilde{T})$ denotes the set of all bounded real functions h on $\tilde{T}$, with the sup norm $\|h\|_{\tilde{T}} := \sup\{|h(t)|: t \in \tilde{T}\}$. Then $\tilde{F}$ is called a (functional) Donsker class (for P) if as $n \to \infty$, y_n converges in law to G_P in $\ell^\infty(\tilde{F})$ for (the metric defined by) $\|\cdot\|_{\tilde{F}}$.

A priori, ν_n need not even have values in $\ell^\infty(\tilde{F})$; it does if the envelope function $F_{\tilde{F}}(x) := \sup_{f \in \tilde{F}} |f(x)| < \infty$ is integrable for P. Also, the definitions of Donsker class and of convergence in law require that G_P can be taken to have a distribution in a separable subspace of $\ell^\infty(\tilde{F})$. The Hoffmann-Jørgensen definition has been proved equivalent in this case to others, specifically the "invariance principle" definition of Dudley and Philipp (1983), in Dudley (1985, p. 158). It is not hard to show that a functional Donsker class satisfies the bounded central limit theorem.

Given a functional T on a set $\tilde{P}$ and $P \in \tilde{P}$, T will be called differentiable at P for $\tilde{F}$ with derivative $f = f_{P,T}$ iff

$$T(Q) = T(P) + \int f \, d(Q - P) + o(\|Q - P\|_{\tilde{F}})$$

as $\|Q - P\|_{\tilde{F}} \to 0$, $Q \in \tilde{P}$. Thus $\int f \, dQ$ must be defined and finite at least for all Q in a $\|\cdot\|_{\tilde{F}}$ neighborhood of P in $\tilde{P}$. The linear functional $\mu \mapsto \int f \, d\mu$ for signed measures μ is the Fréchet derivative of T on $\tilde{P}$ at P with respect to $\|\cdot\|_{\tilde{F}}$.

If $f \in \tilde{F}$ or if $f = cg$ for some constant c and $g \in \tilde{F}$, then differentiability of T for $\tilde{F}$ at P implies continuity at P for $\|\cdot\|_{\tilde{F}}$, but not in general. Differentiability without continuity was perhaps understandably allowed if differentiability was only with respect to the sup norm for distribution functions. Now, if the class of functions $f_{Q,T}$, at least for Q in a neighborhood of P, is itself a Donsker class for P, a hypothesis that is quite often satisfied, then we can include these functions in $\tilde{F}$, since the union of any two Donsker classes is a Donsker class (Alexander, 1987, p. 183).

Here is a central limit theorem for a non-linear functional:

Theorem 1. Let $P \in \tilde{P}$, a convex, finitely full class of laws on a measurable space $(X,\tilde{A})$. Let $\tilde{F}$ satisfy the bounded central limit theorem for P. Let T be a functional on $\tilde{P}$, differentiable at P for $\tilde{F}$ with derivative $f \in \tilde{L}^2(P)$. Then as $n \to \infty$, $n^{1/2}(T(P_n) - T(P))$ converges in law to $N(0,\text{var}_P(f))$.

__Note.__ Although T is real-valued, differentiability of T at the one point P does not require T to be continuous at any other point or even measurable, so that $T(P_n)$ may not be measurable and the convergence in law is in the Hoffmann-Jørgensen sense.

__Proof.__ The differentiability implies

$$n^{1/2}(T(P_n) - T(P)) = \int f\, d\nu_n + n^{1/2}o(\|P_n - P\|_{\tilde{F}}),$$

which is $\int f\, d\nu_n + o_p(1)$ by the bounded central limit theorem. The law of $\int f\, d\nu_n$ converges to $N(0,\text{var}_P f)$ by the ordinary 1-dimensional central limit theorem, so the result follows (from the extended Slutsky's theorem, Theorem C in §5), Q.E.D.

Note that $n^{1/2}(T(P_n) - T(P))$ can be written as $T(\nu_n)$ for the "empirical process" signed measure ν_n if T is linear, $T(\mu) = \int f\, d\mu$, but not in general. So for non-linear T, the emphasis is more on the empirical measure P_n than on the process ν_n.

Next, let's consider classes of non-linear functionals. Given a class $\tilde{T}$ of real-valued functionals defined on a class $\tilde{P}$ of laws on $(X,\tilde{A})$, a distance will be defined on $\tilde{P}$ by $d_{\tilde{T}}(P,Q) :=$ $\sup\{|T(P) - T(Q)|: T \in \tilde{T}\}$. Clearly, $d_{\tilde{T}}$ is a pseudo-metric on $\tilde{P}$. Let $P \in \tilde{P}$. Then $\tilde{T}$ will be said to satisfy the __bounded central limit theorem__ for P iff $n^{1/2}d_{\tilde{T}}(P_n,P)$ is bounded in outer probability. For a set U of laws on $(X,\tilde{A})$, measurable function h on X and class $\tilde{F}$ of functions integrable for each law in U, let

$$\|h\|'_{\tilde{F},U} := \sup\{|\int h\,d(\mu - Q)|/\|\mu - Q\|_{\tilde{F}}: \mu, Q \in U, \|\mu - Q\|_{\tilde{F}} > 0\}.$$

Let $\tilde{P}$ be a set of laws on $(X,\tilde{A})$ and T a real functional on $\tilde{P}$. Let $\tilde{F}$ be a class of measurable functions on $(X,\tilde{A})$. Then T will be called C^1 <u>on</u> U <u>for</u> $\tilde{F}$ iff U is an open subset of $\tilde{P}$ for $\|\cdot\|_{\tilde{F}}$ such that for all $Q \in U$, T is differentiable at Q with derivative $f_Q \in \tilde{F}$, and such that f_Q is uniformly continuous in $Q \in U$ in the sense that $\|f_\mu - f_Q\|'_{\tilde{F},U} \to 0$ as $\|\mu - Q\|_{\tilde{F}} \to 0$ for μ and Q in U. A class $\tilde{T}$ of such functionals will be called <u>equi</u>-C^1 on U iff each $T \in \tilde{T}$ is C^1 on U and the uniform continuity of $Q \mapsto f_Q = f_{Q,T}$ for each T is also uniform in T for Q restricted to U, in other words these functions of Q are uniformly equicontinuous for $Q \in U$ and $T \in \tilde{T}$.

Here is a useful consequence of the equi-C^1 condition, a more or less standard fact, stated here in a form adapted to the current setup:

<u>Lemma 1</u>. If $\tilde{T}$ is an equi-C^1 class of functionals on $U \subset \tilde{P}$ for $\tilde{F}$, where $\tilde{P}$ is convex, then for any $P \in U$ there are neighborhoods V and W of P for $\|\cdot\|_{\tilde{F}}$ such that for Q in W, in the differentiability condition

$$T(Q + t(\mu - Q)) = T(Q) + t\int f_{Q,T}\,d(\mu - Q) + o(t) \quad \text{as} \quad t \to 0,$$

the $o(\cdot)$ error term holds uniformly in $T \in \tilde{T}$, $Q \in W$ and $\mu \in V$: in other words, given $\varepsilon > 0$, there is a $\delta > 0$ such that $|o(t)|/t < \varepsilon$ for $|t| < \delta$, where δ does not depend on T, Q or μ in the given sets.

<u>Proof</u>. For any $r > 0$ let $B(P,r) := \{\mu \in \tilde{P}: \|\mu - P\|_{\tilde{F}} < r\}$. Then for some $r > 0$, we have $B(P,r) \subset U$. Let $V := B(P,r)$ and $W := B(P,r/2)$. Then since $\tilde{P}$ is convex, both V and W are convex. For any $Q \in W$ and $\mu \in V$, consider the line segment $Q + t(\mu - Q)$, where $0 \le t \le 1$. The

uniformity of the $o(t)$ condition follows from the definition of equi-C^1 and a fact from analysis, for example Dieudonné (1960, (8.6.2) p. 156), Q.E.D.

Theorem 2. Let $\tilde{P}$ be a finitely full, convex class of laws on $(X,\tilde{A})$. Let $P \in U \subset \tilde{P}$. Let $\tilde{F}$ be a class of measurable functions on $(X,\tilde{A})$ satisfying the bounded central limit theorem for P. Let $\tilde{T}$ be a class of functionals on $\tilde{P}$ equi-C^1 on U for $\tilde{F}$ with $f_{P,T} \in \tilde{F}$ for each $T \in \tilde{T}$. Then $\tilde{T}$ satisfies the bounded central limit theorem for P.

Proof. In the proof of Theorem 1, the $o_p(1)$ term is uniform in $T \in \tilde{T}$ by Lemma 1. Also, $\sup\{|\int f_{P,T}\, d\gamma_n : T \in \tilde{T}|\}$ is bounded in outer probability since $f_{P,T} \in \tilde{F}$ and $\tilde{F}$ satisfies the bounded central limit theorem. Adding the uniform $o_p(1)$ term leaves the sequence bounded in outer probability, Q.E.D.

Let $\alpha_n(T) := n^{1/2}(T(P_n) - T(P))$. Then for a class $\tilde{T}$ of functionals as in Theorem 2, α_n is a sequence of stochastic processes indexed by $T \in \tilde{T}$. Say $\tilde{T}$ is a Mises-Donsker class for P iff these processes converge in law in $\ell^\infty(\tilde{T})$ as $n \to \infty$.

Theorem 3. If the hypotheses of Theorem 2 hold and $\tilde{F}$ is a functional Donsker class for P, then $\tilde{T}$ is a Mises-Donsker class for P.

Proof. In view of the proofs of Theorems 1 and 2, we need only the following fact, for $h(T) := f_{P,T}$:

Lemma 2. If A and B are any two sets, h is an arbitrary function from

A into B, and random elements $V_n \to V_0$ in law in $\ell^\infty(B)$, then $V_n \circ h \to V_0 \circ h$ in law in $\ell^\infty(A)$.

<u>Proof.</u> Lemma 2 is straightforward, so Theorem 3 also follows.

Next, let's consider the bootstrap (Efron 1979, 1981). Given a law P and one of its empirical measures P_n, the <u>bootstrap empirical measure</u> P_n^* is an empirical measure for P_n, so that $P_n^* = n^{-1}(\delta_{Y(1)} + \ldots + \delta_{Y(n)})$ where $Y(1),\ldots,Y(n)$ are i.i.d. with law P_n. Given a class $\tilde{F} \subset \tilde{L}^2(P)$, $\tilde{F}$ will be called an (<u>a.s.</u>) <u>bootstrap Donsker class</u> for P iff the processes $n^{1/2}\int f\, d(P_n^* - P_n)$ converge in law in $\ell^\infty(\tilde{F})$ to a limiting Gaussian process, for almost all sequences $X(1),\ldots,X(n),\ldots$ i.i.d. for P (used in forming P_n, and where P_n^* has its conditional distribution given P_n). A Mises-Donsker class $\tilde{T}$ of functionals for P in $\tilde{P}$ will be called an (<u>a.s.</u>) <u>bootstrap Mises-Donsker class</u> for P iff likewise, for almost all $X(1), X(2),\ldots,$ $n^{1/2}(T(P_n^*) - T(P_n))$ converges in law in $\ell^\infty(\tilde{T})$ to the same limit process as that for $n^{1/2}(T(P_n) - T(P))$.

A measurable space $(S,\tilde{B})$ is called <u>Suslin</u> iff there is a complete separable metric space Y with Borel σ-algebra and a measurable mapping of Y onto S. A useful measurability condition on a class $\tilde{F}$ of functions on a measurable space X is that it is <u>admissible Suslin</u>, namely, that X is Suslin and there is a σ-algebra $\tilde{B}$ of subsets of $\tilde{F}$ for which $(\tilde{F},\tilde{B})$ is Suslin and the mapping $(f,x) \mapsto f(x)$ is jointly measurable.

Recall $F_{\tilde{F}}(x) := \sup\{|f(x)|: f \in \tilde{F}\}$. Giné and Zinn (1988) proved that under some measurability conditions weaker than admissible Suslin, $\tilde{F}$ is a bootstrap Donsker class for P if and only if it is a Donsker class for P and $\int F_{\tilde{F}}^2\, dP < \infty$. Special cases proved earlier were Bickel and

Freedman (1981, Theorem 2.2) for bounded subsets of finite-dimensional subspaces of $\tilde{L}^2$, and Gaenssler (1986) for $\tilde{F} = \{1_C: C \in \tilde{C}\}$ where $\tilde{C}$ is a Vapnik-Cervonenkis class of sets (with suitable measurability), as is necessary for a class of sets to be a Donsker class for all P (Durst and Dudley, 1981; Dudley, 1984, p. 125).

Here is a theorem on the bootstrap:

Theorem 4. Let $\tilde{P}$ be a convex, finitely full set of laws on $(X,\tilde{A})$, $P \in \tilde{P}$ and $\tilde{F}$ an admissible Suslin functional Donsker class for P with $\int F_{\tilde{F}}^2\, dP < \infty$. Let U be a neighborhood of P in $\tilde{P}$ for $\|\cdot\|_{\tilde{F}}$. Let $\tilde{T}$ be an equi-C^1 class of functionals on U for $\tilde{F}$ such that the derivatives $f_{Q,T}$ for $Q \in U$ and $T \in \tilde{T}$ are all in $\tilde{F}$. Then $\tilde{T}$ is a bootstrap Mises-Donsker class for P.

Note. There is no assumption that $\tilde{F}$ is a Donsker class for any $Q \neq P$. The assumptions imply that $\tilde{F}$ is an a.s. bootstrap Donsker class for P (Giné and Zinn, 1988).

Proof. Take neighborhoods $W \subset V \subset U$ of P as in Lemma 1. Then since $\tilde{F}$ is a functional Donsker class, the inner probability Pr_* that $P_n \in W$ converges to 1 as $n \to \infty$, and since $\tilde{F}$ is a bootstrap Donsker class, so does the inner probability that $P_n^* \in W$. Then for $P_n \in W$ and $P_n^* \in W$, for each $T \in \tilde{T}$, we have

$$n^{1/2}(T(P_n^*) - T(P_n)) = T_n + U_n + V_n \quad \text{where}$$

$$T_n = n^{1/2}\int f_{P,T}\, d(P_n^* - P_n),\ U_n = n^{1/2}\int f_{P_n,T} - f_{P,T}\, d(P_n^* - P_n)$$

and $V_n = n^{1/2}o(\|P_n^* - P_n\|_{\tilde{F}})$; Lemma 1 implies that $V_n = o_p(1)$ uniformly in $T \in \tilde{T}$ as $n \to \infty$ since $\tilde{F}$ is a bootstrap Donsker class.

Next, $|U_n| \le n^{1/2}\|P_n^* - P_n\|_{\tilde{F}}\|f_{P_n,T} - f_{P,T}\|'_{\tilde{F},U}$. The latter factor goes to 0 in (outer) probability as $n \to \infty$ since $\|P_n - P\|_{\tilde{F}} \to 0$ in outer probability and by the equi-C^1 property for $T \in \tilde{T}$, while $n^{1/2}\|P_n^* - P_n\|_{\tilde{F}}$ is bounded in (outer) probability, again by the bootstrap central limit theorem for $\tilde{F}$.

Now $n^{1/2}\int f\, d(P_n^* - P_n)$ as a process on $\tilde{F}$ converges in law in $\ell^\infty(\tilde{F})$ to a G_P process. From Lemma 2, again for the function $h(T) := f_{P,T}$, we get convergence in law in $\ell^\infty(\tilde{T})$, and the terms are combined again by Slutsky's theorem (Theorem C in §5), proving Theorem 4.

In a special case of Theorems 1-4, $T(P)$ is a function of finitely many integrals (linear functionals),

$$T(P) = t(\int f_1 dP, \ldots, \int f_m dP)$$

for some functions $f_1, \ldots, f_m$ in $\tilde{L}^2(P)$ for each $P \in \tilde{P}$, and where t is a C^1 function of m real variables. Let $\tilde{F}$ be a set of linear combinations $\Sigma\, c_i f_i$ with c_i bounded, thus a bootstrap Donsker class (Bickel and Freedman, 1981). We can assume that each f_i is in $\tilde{F}$ and that $\tilde{F}$ is convex and symmetric. Then T is a C^1 function of P for $\tilde{F}$, with derivative

$$f = f_{P,T} = \Sigma_{1 \le i \le m}\, t_i(\{\int f_j dP\}_{1 \le j \le m}) f_i(\cdot)$$

where t_i is the partial derivative of t with respect to its ith argument, by a classical chain rule. Here $f = cg$ for some $g \in \tilde{F}$ and constant c, both depending on P.

Examples. (1) The variance: let $T(P) = \mathrm{var}_P(x)$, where x is the identity function on $\mathbb{R}$, and $\int x^2\, dP < \infty$ for all $P \in \tilde{P}$. Then we can take $m = 2$, $f_1(x) \equiv x^2$, $f_2(x) \equiv x$, and $t(u,v) := u - v^2$. Then $f_P(x) \equiv (2\int y\, dP(y))x$, and for the remainder,

$$|T(Q) - T(P) - \int f_P d(Q-P)| = (\int x\, d(Q-P))^2 \le \|Q-P\|_{\tilde{F}}^2 = o(\|Q-P\|_{\tilde{F}})$$

as $\|Q-P\|_{\tilde{F}} \to 0$, uniformly in P and Q. The constant $c = c(P)$ is bounded for sets of laws P such that $\int x \, dP$ remains bounded. For central limit theorems we will need $\int f_1^2 \, dPr = \int x^4 \, dPr(x) < \infty$, so let $\tilde{P}$ be the set of all laws P on $\mathbb{R}$ such that $\int x^4 \, dP < \infty$.

(2) The correlation: let $X = \mathbb{R}^2$ with coordinates Y and Z. Let $m = 5$ and let f_i be the functions Y, Z, Y^2, YZ, Z^2 for $i = 1,\ldots,5$ respectively. Then t is the function

$$t(y,z,u,v,w) = (v - yz)/((u - y^2)(w - z^2))^{1/2},$$

and the constants $c(P)$ are bounded uniformly on sets of laws P such that the means $\int Y \, dP$ and $\int Z \, dP$ are bounded while the variances of Y and Z are bounded away from 0.

3. Remarks. In the work of von Mises (1947), Filippova (1961) and others on functionals of distribution functions, there is considerable emphasis on higher order derivatives of T and (thus) on functionals of the form

$$T(F) = \int \ldots \int \phi(x_1,\ldots,x_k) dF(x_1) \ldots dF(x_k)$$

for kth order derivatives. (If $k \geq 2$, these functionals have some analogies and relations to U-statistics, see Mandelbaum and Taqqu (1984) and on the bootstrap, Bretagnolle (1983).) For the asymptotics of the empirical process, the leading term in $T(F_n) - T(F)$ is in terms of the non-zero derivative of T at F of lowest order. The results of this paper give a trivial (degenerate) limiting distribution when the first derivative is 0. A well-known example where this occurs is the chi-squared statistic for a decomposition of the line into disjoint intervals. When the first and possibly further derivatives are 0, it is natural to use a different normalization for the difference $T(F_n) - T(F)$ or, more generally, $T(P_n) - T(P)$, to get a non-trivial limiting distribution which for $k \geq 2$ is that of a non-linear function of normal variables and so in general not normal.

Now if P and Q are two laws in $\tilde{P}$, one can consider derivatives along the line from P to Q, in other words derivatives of the function $g(t) = T(P + t(Q - P))$, $0 \leq t \leq 1$. These are called Gâteaux derivatives of P. The Gâteaux derivative has been considered too "weak" (Esty et al., 1985, p. 110, citing Serfling, 1980, p. 216 and Huber, 1981, p. 40) unless there is some uniformity over different lines, as with Fréchet derivatives, and such uniformity is all the more needed in this paper.

On $\tilde{L}^2(P)$, a "centered" pseudo-metric is defined by $\rho_P(f,g) := (E_P(f-g)^2 - [E_P(f-g)]^2)^{1/2}$. In general, one of the equivalent conditions for a class $\tilde{F}$ to be a functional Donsker class (Dudley, 1985, p. 158) is that $\tilde{F}$ is totally bounded for ρ_P and satisfies the "uniform asymptotic equicontinuity" condition: for every $\varepsilon > 0$ there is an n_o and a $\delta > 0$ such that for all $n \geq n_o$,

$$Pr^*\{\sup\{|\int n^{1/2}(f-g)d(P_n-P)|: \rho_P(f,g) < \delta\} > \varepsilon\} < \varepsilon.$$

This plays the role of a "tightness" condition. It holds also for the bootstrap case, replacing $P_n - P$ with $P_n^* - P_n$ (Giné and Zinn, 1988, (2.16)). So in the definition of C^1 functional T, or equi-C^1 class of functionals $\tilde{T}$, the function $Q \mapsto f_{Q,T}$ can be assumed uniformly (equi)continuous from the $\|\cdot\|_{\tilde{F}}$ norm on $\tilde{P}$ to the ρ_P distance (rather than the $\|\cdot\|'_{\tilde{F},U}$ distance) and Theorem 4 still holds.

Generally, of course, the larger the Donsker class $\tilde{F}$ of functions, the larger the norm $\|\cdot\|_{\tilde{F}}$, and consequently the easier it is for the remainder in differentiation to be $o(\|Q-P\|_{\tilde{F}})$. On the other hand, for larger $\tilde{F}$, convergence in the central limit theorem would tend to be slower. In any case, there will be an error in the distribution of the first-derivative term, approximating the law of $n^{1/2}\int f_P \, d(P_n - P)$ by that of $G_P(f_P)$, uniformly over a possibly large collection of functions f_P, which may be larger than error terms given by higher-order derivatives of the functional T.

4. Linear functionals on signed measures as integrals. If a linear functional A on a space of signed measures on a set X is defined for point masses δ_x, let $f(x) := A(\delta_x)$. Then $A(\mu) = \int f \, d\mu$ for any signed measure μ which is a finite linear combination of point masses or a limit of such measures in any topology for which A is continuous. Often these may include all the signed measures Q − P for P and Q in $\tilde{P}$.

The derivative A of a statistical functional need only be defined on some signed measures $\mu = P - Q$ of net charge $\mu(X) = 0$ (and total variation at most 2). Still, if A is a linear functional defined on all signed measures $\delta_x - \delta_y$ for x and y in a set X, we can write $A(\delta_x - \delta_y) = f(x) - f(y)$ by setting $f(x) := A(\delta_x - \delta_u)$ for any fixed $u \in X$. (Likewise, more generally, the real functions g on $X \times X$ which are of the form $g(x,y) \equiv f(x) - f(y)$ for some f can be characterized as the functions satisfying

$$g(x,y) + g(y,z) + g(z,x) = 0 \text{ for all } x, y \text{ and } z \text{ in } X;$$

this is a special case of a fact well known in another field, "Alexander cohomology", see e.g. Hu, 1968, p. 67, Prop. 1.3, for $n = 1$.)

Now if $A(\delta_x - \delta_y) \equiv f(x) - f(y)$ for some real f on X, then $A(\mu) = \int f \, d\mu$ for any signed measure μ of net charge 0 which is a finite linear combination of point masses and so, again, for any limit of such signed measures in any topology for which A is continuous.

5. Convergence in law in non-separable metric spaces. This section is in the nature of an appendix. Some results from the theory of convergence of general (possibly non-measurable) random elements in a general (possibly non-separable) metric space (S,d) were used in some of the proofs above.

Here are the results, due to J. Hoffmann-Jørgensen (unpublished). The proofs parallel those for random variables with values in separable metric spaces (Dudley, 1989, Chapter 11).

For any set A in a topological space, the boundary is the intersection of the closures of A and its complement. A set A is called a <u>continuity set</u> for a measure μ if its boundary has μ-measure 0.

<u>Theorem A</u> (extended portmanteau theorem; Hoffmann-Jørgensen). For any random elements V_n with values in S, where V_0 is measurable and has separable range, the following are equivalent:

(a) $V_n \to V_0$ in law; specifically, for every bounded continuous real function h on S, $E^*h(V_n) \to Eh(V_0)$ as $n \to \infty$.

(b) $E_*g(V_n) \to Eg(V_0)$ for any bounded continuous g.

(c) For every open set $U \subset S$, $P(V_0 \in U) \leq \liminf_{n\to\infty} P_*(V_n \in U)$;

(d) For every closed set $F \subset S$, $P(V_0 \in F) \geq \limsup_{n\to\infty} P^*(V_n \in F)$;

(e) For every continuity set A of the law of V_0,

$$\lim_{n\to\infty} P^*(V_n \in A) = \lim_{n\to\infty} P_*(V_n \in A) = P(V_0 \in A).$$

<u>Proof</u>. Clearly (a) and (b) are equivalent (take $g = -h$), and (c) and (d) are equivalent by complementation. For the rest, the proof for S separable, given, for example, in Billingsley (1968, pp. 11-14) or Dudley (1989, p. 303), carries over straightforwardly, as Andersen and Dobrić (1987, Remark 2.13) point out.

Now some conditions for convergence in law will be developed of a more metric kind. For any metric space (S,d), let BL(S,d) be the space of all bounded Lipschitz real-valued functions f on S, with the Lipschitz seminorm $\|f\|_L := \sup\{|f(x)-f(y)|/d(x,y): x \neq y\}$, the sup norm

$\|f\|_s := \sup\{|f(x)|: x \in S\}$, and bounded Lipschitz norm $\|f\|_{BL} := \|f\|_L + \|f\|_s$. For any set $A \subset S$ and $\delta > 0$, we have $A^\delta := \{y: d(x,y) < \delta$ for some $x \in A\}$. For any functions V and Y from probability spaces into S, where Y has separable range and is measurable, let $d^*_{BL}(V,Y) := \sup\{|E^*f(V) - Ef(Y)|: \|f\|_{BL} \leq 1\}$, and $d^*_\rho(V,Y) := \inf\{\varepsilon > 0: P^*(V \in A) \leq P(Y \in A^\varepsilon) + \varepsilon$ for all Borel sets $A\}$.

<u>Theorem</u> <u>B</u>. For any random elements V_n with values in a metric space (S,d), where V_0 is measurable and has separable range, the following are equivalent:

(a) V_n converge in law to V_0;

(b) $d^*_{BL}(V_n,V_0) \to 0$ as $n \to \infty$;

(c) $d^*_\rho(V_n,V_0) \to 0$ as $n \to \infty$.

<u>Proof</u>. The proof for random variables in a separable metric space (e.g. Dudley, 1968, p. 1568, or 1989, Theorem 11.3.3) carries over with only minor changes. It does use the portmanteau theorem, Theorem A above.

The following result may seem obvious, but it will be proved since non-separability and non-measurability can make such statements less clear.

<u>Theorem</u> <u>C</u> (extended Slutsky's theorem). Let $(X,\|\cdot\|)$ be a normed linear space. Let Z_n be random elements of X with $Z_n \to Z_0$ in law as $n \to \infty$. Let W_n be random elements on the same probability space(s) such that $W_n \to 0$ in outer probability as $n \to \infty$. Then $Z_n + W_n \to Z_0$ in law.

<u>Proof</u>. For any f with $\|f\|_{BL} \leq 1$, $|f(Z_n + W_n) - f(W_n)| \leq \min(1,\|W_n\|)$ and so $\sup\{|E^*(f(Z_n + W_n)) - E^*f(Z_n)|: \|f\|_{BL} \leq 1\} \leq E^*\min(1,\|W_n\|)$, which

converges to 0 as $n \to \infty$. Then the equivalence of (a) and (b) in Theorem B gives the conclusion.

Acknowledgment. Many thanks to Jørgen Hoffmann-Jørgensen and Niels T. Andersen for conversations about the material in Section 5.

REFERENCES

Alexander, K. S. (1987). The central limit theorem for empirical processes on Vapnik-Červonenkis classes. Ann. Probability 15, 178-203.

Andersen, N. T., and V. Dobrić (1987). The central limit theorem for stochastic processes. Ann. Probability 15, 164-177.

Bickel, P. J., and D. A. Freedman (1981). Some asymptotic theory for the bootstrap. Ann. Statist. 9, 1196-1217.

Billingsley, P. (1968). Convergence of Probability Measures. Wiley, New York.

Boos, Dennis D., and R. J. Serfling (1980). A note on differentials and the CLT and LIL for statistical functions, with application to M-estimates. Ann. Statist. 8, 618-624.

Bretagnolle, J. (1983). Lois limites du Bootstrap de certaines fonctionelles. Ann. Inst. Henri Poincaré B 19, 281-296.

Dieudonné, J. (1960). Foundations of Modern Analysis. Academic Press, New York.

Dudley, R. M. (1966). Weak convergence of probabilities on nonseparable metric spaces and empirical measures on Euclidean spaces. Illinois J. Math. 10, 109-126.

__________ (1967). Measures on non-separable metric spaces. Ibid. 11 449-453.

Dudley, R. M. (1968). Distances of probability measures and random variables. Ann. Math. Statist. 39, 1563-1572.

________ (1984). A course on empirical processes. Ecole d'été de probabilités de St.-Flour, 1982. Lecture Notes in Math. (Springer) 1097, 1-142.

________ (1985). An extended Wichura theorem, definitions of Donsker class, and weighted empirical distributions. Probability in Banach Spaces V (Proc. Conf. Medford, 1984), Lecture Notes in Math. (Springer) 1153, 141-178.

________ (1987). Universal Donsker classes and metric entropy. Ann. Probability 15, 1306-1326.

________ (1989). Real Analysis and Probability. Brooks-Cole, Pacific Grove, Calif.

________ and Walter Philipp (1983). Invariance principles for sums of Banach space valued random elements and empirical processes. Z. Wahrscheinlichkeitsth. verw. Geb. 62, 509-552.

Durst, M., and R. M. Dudley (1981). Empirical processes, Vapnik-Chervonenkis classes and Poisson processes. Prob. Math. Statist 1 #2, 109-115.

Efron, B. (1979). Bootstrap methods: another look at the jackknife. Ann. Statist. 7, 1-26.

Efron, B. (1981). Nonparametric estimates of standard error: the jackknife, bootstrap and other methods. Biometrika 68, 589-599.

Esty, W., R. Gillette, M. Hamilton and D. Taylor (1985). Asymptotic distribution theory of statistical functionals: the compact derivative approach for robust estimators. Ann. Inst. Statist. Math. A 37, 109-129.

Filippova, A. A. (1961). Mises' theorem on the asymptotic behavior of functionals of empirical distribution functions and its statistical applications. Theor. Probability Appls. 7, 24-57.

Fortet, Robert (1958). Recent advances in probability theory. In Some Aspects of Analysis and Probability, pp. 171-240. Wiley, New York.

Fortet, R., and Edith Mourier (1954). Sur les fonctionelles de certaines fonctions aléatoires. Comptes Rendus Acad. Sci. Paris 238, 1557-9.

Gaenssler, Peter (1986). Bootstrapping empirical measures indexed by Vapnik-Červonenkis classes of sets. Probability Theory and Mathematical Statistics, ed. Yu. Prohorov et al., pp. 467-481. VNU Press, Netherlands.

Giné, Evarist, and Joel Zinn (1988). Bootstrapping general empirical measures. Preprint.

Hoffmann-Jørgensen, J. (1984). Personal communication.

_______ (unpublished). Stochastic Processes on Polish Spaces.

Hu, Sze-Tsen (1968). Cohomology Theory. Markham, Chicago.

Huber, Peter J. (1981). Robust Statistics. Wiley, N.Y.

Mandelbaum, Avi, and Murad S. Taqqu (1984). Invariance principle for symmetric statistics. Ann. Statist. 12, 483-496.

Misès, R. de [= von Mises, R.] (1936). Les lois de probabilite pour les fonctions statistiques. Ann. Inst. Henri Poincaré 6, 185-212.

Mises, R. v[on] (1947). On the asymptotic distribution of differentiable statistical functions. Ann. Math. Statist. 18, 309-348.

Parr, William C. (1985). Jackknifing differentiable statistical functionals. J. Roy. Statist. Soc. B 47, 56-66.

Pollard, David (1982). A central limit theorem for empirical processes. J. Austral. Math. Soc. Ser. A 33, 235-248.

Serfling, Robert J. (1980). Approximation Theorems of Mathematical Statistics. Wiley, N.Y.

SUR LA RÉGULARITÉ DE CERTAINES CLASSES DE FONCTIONS ALÉATOIRES,

Xavier Fernique,
Institut de Recherche Mathématique Avancée,
7, Rue René Descartes, F-67084, Strasbourg Cédex.

1. Introduction.

La régularité des trajectoires des variables aléatoires à partir de la régularité moyenne de leurs accroissements est analysée depuis longtemps ([9], [1]) quand ces variables aléatoires réelles sont définies sur **R**. Elle a été aussi étudiée de près depuis une quinzaine d'années lorsque les fonctions aléatoires sont définies sur $\mathbf{R}^k$ ou plus généralement un espace métrique T et sont à valeurs réelles ou même à valeurs dans un espace métrique E ([3], [6], [2], [5], [7], [12], [3]). L'étude de la compacité de leurs lois dans $\mathbb{C}(T ; E)$ a été poussée moins loin jusqu'ici (voir pourtant [8], [10]) alors qu'elle est au moins aussi importante en Statistique Mathématique. Nous avons publié récemment une Note ([4]) énonçant sans preuve détaillée des résultats dans ce domaine. Nous nous proposons ici de fournir les démonstrations. Postérieurement à la publication de la Note ci-dessus, M. Talagrand, par un schéma différent a d'ailleurs retrouvé ([11]) les mêmes résultats de compacité.

2. Régularité de certaines classes de pseudo-métriques aléatoires.

2.1 Notations : Nous notons (T, δ) un espace métrique ou pseudo-métrique ; pour tout t appartenant à T et tout u > 0, B(t, u) désigne la boule ouverte de centre t et de rayon u ; N(u) = $N_T(u)$ sera le nombre minimal de boules ouvertes de rayon u recouvrant T. Toutes les variables aléatoires utilisées seront notées pour la commodité sur le même espace (Ω, **A**, **P**) sans que cela ait quelque signification. Soit $D = \{D(\omega ; s, t), \omega \in \Omega, (s, t) \in T \times T\}$ une fonction aléatoire sur T × T ; nous disons que D est une *pseudo-métrique aléatoire* si pour tout triplet (s,t,u) d'éléments de T, on a :

2.1.1 $$\mathbf{P}\{ 0 = D(t,t) \leq D(s,t) = D(t,s) \leq D(t,u) + D(u,s) \} = 1.$$

Pour toute fonction X_0 positive, décroissante et intégrable sur]0, 1], nous notons $D(X_0)$ la classe des pseudo-métriques aléatoires séparables D sur T telles que :

2.1.2 $$\forall (s,t,u) \in T \times T \times \mathbf{R}^+,\ \mathbf{E}\{D(s,t)\, I_{D(s,t)\geq u}\} \leq \delta(s,t) \int_0^{\mathbf{P}\{D(s,t)\geq u\}} X_0(\omega)\, d\omega.$$

Dans ces conditions, on se propose de démontrer :

Théorème 2.2 : *On suppose que l'intégrale*

$$\iint_{0<\omega<1<N(u)} X_0\left(\frac{\omega}{N(u)}\right) d\omega \, du$$

est finie. Alors tout élément D de $D(X_0)$ a p.s. ses trajectoires continues sur $T \times T$ et vérifie :

2.2.1 $$\mathbf{E} \sup_{T\times T} D \leq 8 \iint_{0<\omega<1<N(u)} X_0\left(\frac{\omega}{N(u)}\right) d\omega \, du .$$

De plus sous la même hypothèse, pour tout $\varepsilon > 0$, il existe $\eta > 0$ tel que pour tout D appartenant à $D(X_0)$, on ait :

2.2.2 $$\mathbf{E} \sup_{\delta(s,t)\leq\eta} D(s,t) \leq \varepsilon.$$

La démonstration du théorème suivra un schéma voisin de celui employé dans [3] pour démontrer 2.2.1 et aussi dans [5] et [8] ; X_0 désignant une fonction positive, décroissante et intégrable sur [0, 1] fixée, on démontre d'abord pour tout élément D de $D(X_0)$ une majoration du type 2.2.1, on utilise ensuite des approximations des éléments de $D(X_0)$ par des espérances conditionnelles continues et convergeant uniformément. La difficulté ici sera la suivante : les tribus de conditionnement varieront suivant les éléments de $D(X_0)$; on devra pourtant contrôler uniformément la continuité des espérances conditionnelles et leur convergence uniforme. On utilisera trois lemmes qui analysent les propriétés des classes $D(X_0)$.

Lemme 2.3 ([3], lemme 1.3) : *(a) Pour toute v.a. f à valeurs dans [0, 1] et tout D appartenant à $D(X_0)$, on a :*

2.3.1 $$\forall (s,t) \in T \times T, \quad \mathbf{E}\{D(s,t)\, f\} \leq \delta(s,t) \int_0^{\mathbf{E}(f)} X_0(\omega)\, d\omega.$$

(b) Pour tout entier $n \geq 1$ et tout nombre $a \in [0, 1]$, on a :

2.3.2 $$\sup\left(\sum_{k=1}^{n} \int_0^{p_k} X_0(\omega)d\omega,\ (p_k) \in [0, 1]^n, \sum_{k=1}^{n} p_k = a\right) = n \int_0^{a/n} X_0(\omega)d\omega.$$

Démonstration du lemme 2.3 : (a) nous omettons les variables s et t ; notons pour commencer que sous les hypothèses (a) du lemme, pour tout nombre réel M et tout couple (λ, μ) de nombres positifs de somme 1, on a pour tout $\omega \in \Omega$:

$$(D(\omega) - M)\, f(\omega) \leq (D(\omega) - M)^+ \leq (D(\omega) - M)\{\lambda\, I_{D\geq M}(\omega) + \mu\, I_{D>M}(\omega)\} ;$$

on choisit alors M, λ, μ non aléatoires tels que :

$$\mathbf{P}\{D \geq M\} \leq \mathbf{E}\,(f) \leq \mathbf{P}\{D > M\}, \quad \mathbf{E}\,(f) = \lambda\, \mathbf{P}\{D \geq M\} + \mu\, \mathbf{P}\{D > M\} ;$$

en intégrant en ω, on obtient donc :

$$\mathbf{E}\{D\, f\} = M\, \mathbf{E}\,(f) + \mathbf{E}\,\{(D - M)\, f\} \leq \lambda\, \mathbf{E}\,\{D\, I_{D\geq M}\} + \mu\, \mathbf{E}\,\{D\, I_{D>M}\}.$$

Puisque D appartient à $D(X_0)$, ce dernier membre se majore par :

$$\delta\left(\lambda\int_0^{\mathbf{P}(D\geq M)} X_0(\omega)\,d\omega + \mu\int_0^{\mathbf{P}(D>M)} X_0(\omega)\,d\omega\right)$$

qui se majore lui-même par concavité par :

$$\delta\int_0^{\lambda\mathbf{P}(D\geq M)+\mu\mathbf{P}(D>M)} X_0(\omega)\,d\omega = \delta\int_0^{\mathbf{E}(f)} X_0(\omega)\,d\omega\ ,$$

c'est bien le résultat 2.3.1.

(b) La fonction indiquée est concave et symétrique, elle atteint donc son maximum au point p ayant toutes ses composantes p_k égales à a/n, c'est le résultat 2.3.2.

Lemme 2.4 : *(a) Soient D un élément de $D(X_0)$ et* ***B*** *une sous-tribu de* ***A****, alors toute version séparable D'de* ***E****{D /* ***B****} appartient aussi à $D(X_0)$. (b) Soit D un élément de $D(X_0)$ mesurable pour une sous-tribu* ***B*** *de* ***A*** *engendrée par n atomes, on a alors :*

2.4.1
$$\mathbf{E}\sup_{T\times T}\frac{D(s,t)}{\delta(s,t)} \leq n\int_0^{1/n} X_0(\omega)d\omega.$$

(ici et dans toute la suite, nous attribuons à la fraction 0/0 la valeur nulle)

Démonstration du lemme 2.4 : (a) On vérifie immédiatement que D' a la propriété 2.1.1, c'est donc une pseudo-métrique aléatoire ; de plus pour tout triplet (s, t, u) de $T \times T \times \mathbf{R}^+$, l'ensemble {D'(s, t) ≥ u} est un élément de **B** ; appliquant le lemme 2.3.(a), on en déduit :

$$\mathbf{E}\ \{D'(s,t)\ I_{D'(s,t)\geq u}\} = \mathbf{E}\ \{D(s,t)\ I_{D'(s,t)\geq u}\} \leq \delta(s,t)\int_0^{\mathbf{P}\{D'(s,t)\geq u\}} X_0(\omega)\,d\omega,$$

de sorte que D' est bien un élément de $D(X_0)$.

(b) Notons A l'ensemble des atomes non négligeables de **B**; pour tout atome a, on a :

$$\forall\ (\omega,s,t) \in a\times T\times T,\ \ D(\omega\ ;\ s,t) = \frac{1}{\mathbf{P}(a)}\int_a D(s,t)\ d\mathbf{P}\ ,$$

et donc d'après le lemme 2.3.(a) :

$$D(\omega\ ;\ s,t) \leq \left(\frac{1}{\mathbf{P}(a)}\int_0^{\mathbf{P}(a)} X_0(\omega')\ d\omega'\right)\delta(s,t)\ ;$$

on en déduit en intégrant successivement dans les différents atomes :

$$\mathbf{E} \sup_{T\times T} \frac{D(s,t)}{\delta(s,t)} \leq \sum_{a\in A} \int_0^{\mathbf{P}(a)} X_0(\omega)\, d\omega .$$

Le lemme 2.3.(b) fournit alors exactement le résultat.

Lemme 2.5 : *Supposons le diamètre $\delta(T)$ fini ; alors pour tout $\varepsilon > 0$, il existe un nombre $n = n(\varepsilon)$ tel que pour tout élément D de $D(X_0)$ et pour tout couple (s, t) d'éléments de T, il existe une sous-tribu $\boldsymbol{B} = \boldsymbol{B}_{D,s,t}$ de **A** engendrée par n atomes telle que pour toute tribu **B'** contenant **B**, on ait :*

2.5.1 $$\mathbf{E}\,|D(s,t) - \mathbf{E}\{D(s,t) \mid \mathbf{B}'\}| \leq \varepsilon.$$

Démonstration du lemme 2.5. : Pour tout élément D de $D(X_0)$ et tout couple (s, t) d'éléments de T, l'inégalité de Cebicev montre que, indépendamment de D, s, t, on a :

$$\forall\, u > 0,\ u\,\mathbf{P}\{D(s,t) \geq u\} \leq \delta(T) \int_0^1 X_0(\omega)\, d\omega ,$$

de sorte que les lemmes 2.3.(a) et 2.4.(a) impliquent :

2.5.2 $$\mathbf{E}\,\{D(s,t)\, I_{D(s,t)\geq u}\} \leq \delta(T) \int_0^{c/u} X_0(\omega)\, d\omega\ ,\quad c = \delta(T) \int_0^1 X_0(\omega)\, d\omega ,$$

et pour toute tribu **B'** :

$$\mathbf{E}\{\ \mathbf{E}\{D(s,t) \mid \mathbf{B}'\}\ I_{D(s,t)\geq u}\} \leq \delta(T) \int_0^{c/u} X_0(\omega)\, d\omega.$$

Dans la suite de la preuve du lemme, on fixe un nombre u dépendant de ε, indépendant de D, s, t tel que :

2.5.4 $$\delta(T) \int_0^{c/u} X_0(\omega)\, d\omega < \varepsilon/3.$$

Soient maintenant un élément D de $D(X_0)$ et (s, t) un couple d'éléments de T ; nous leur associons la tribu **B** engendrée par les $a_k = \{ D(s, t) \in [k\varepsilon/3, (k+1)\varepsilon/3[\}$, $k \in [0, 3u/\varepsilon]$ de sorte que le nombre d'atomes de **B** soit au plus $(2 + 3u/\varepsilon)$. Pour toute tribu **B'** contenant **B** et tout $k \in [0,3u/\varepsilon]$, on a :

2.5.5 $$|D(s,t) - \mathbf{E}\{D(s,t) \mid \mathbf{B}'\}|\ I_{a_k} = |D(s,t) - \mathbf{E}\{D(s,t)\, I_{a_k} \mid \mathbf{B}'\}| \leq (\varepsilon/3)\ I_{a_k}\ ;$$

On obtient alors 2.5.1 en regroupant 2.5.2, 2.5.3 et 2.5.5.

2.6 Nous démontrons maintenant le théorème 2.2 ; nous devrons noter avec soin le temps d'introduction des différentes variables. Les données initiales sont X_0 et (T, δ) vérifiant l'hypothèse intégrale du théorème 2.2 ; pour tout $u \leq \delta(T)/2$, N(u) est supérieur à 1 et on a donc :

$$\iint_{0<\omega<1<N(u)} X_0\!\left(\frac{\omega}{N(u)}\right) d\omega\, du \geq \int_0^{\delta(T)/2} \left(\int_0^1 X_0(\omega/2)\, d\omega \right) du \geq (1/8)\, \delta(T) \int_0^1 X_0(\omega) d\omega\ ;$$

si $\delta(T)$ est nul ou si X_0 est identiquement nul, la conclusion du théorème est triviale ; dans le cas contraire, l'inégalité ci-dessus montre que $\delta(T)$ et $\int_{[0,1]} X_0(\omega)\, d\omega$ sont finis ; ceci permet d'utiliser dans la suite de la preuve les lemmes 2.3, 2.4 et 2.5. Nous introduisons maintenant un nombre $\varepsilon > 0$ et nous fixons un nombre u tel que :

2.6.1 $$\int_0^u \Big[\int_0^1 X_0\Big(\frac{\omega}{N(v)}\Big) d\omega\Big] dv \leq \varepsilon/64 .$$

Pour tout entier n, nous notons S_n une partie de T de cardinal $N(u/2^n)$ telle que la famille $\{B(s,u/2^n),\ s \in S_n\}$ recouvre T et nous choisissons une application g_n de S_{n+1} dans S_n telle que pour tout s appartenant à S_{n+1}, $\delta(s,g_n(s))$ soit inférieur à $u/2^n$; nous fixons de plus un entier positif K et pour tout entier k appartenant à [1, K], nous notons f_k l'application de S_{K+1} dans S_k définie par la composition $g_k \circ \ldots \circ g_K$.

Nous fixons maintenant un élément D de $D(X_0)$; nous lui associons une application mesurable τ de Ω dans S_{K+1} vérifiant :

$$D(\tau, f_0(\tau)) = \sup_{t \in S_{K+1}} D(t, f_0(t)) ;$$

dans ces conditions, on a immédiatement :

$$\mathbf{E} \sup_{S_{K+1} \times S_{K+1}} | D(s,t) - D(f_0(s),f_0(t)) | \leq 2 \sum_{k=0}^{K} \sum_{t \in S_{k+1}} \mathbf{E} \{ D(t, g_k(t))\, I_{f_{k+1}(\tau)=t} \} .$$

Le lemme 2.3 permet de majorer le second membre ; chacun de ses termes se majore en effet à partir de 2.3.1 puisque D appartient à $D(X_0)$; les distances $\delta(t, g_k(t))$ se majorent indépendamment de t et on obtient :

$$\forall k \in [0, K], \sum_{t \in S_{k+1}} \mathbf{E} \{D(t, g_k(t))\, I_{f_{k+1}(\tau)=t}\} \leq \frac{u}{2^k} \sum_{t \in S_{k+1}} \int_0^{\mathbf{P}\{f_{k+1}(\tau)=t\}} X_0(\omega) d\omega ;$$

la somme des bornes supérieures des intégrales est égale à 1 de sorte que le lemme 2.3.(b), inégalité 2.3.2, montre que le second membre se majore à partir du cardinal de S_{k+1} ; en regroupant, on obtient :

$$\mathbf{E} \sup_{S_{K+1} \times S_{K+1}} | D(s,t) - D(f_0(s),f_0(t)) | \leq 2 \sum_{k=0}^{K} u\, 2^{-k} M \int_0^{1/M} X_0(\omega)\, d\omega, \quad M = N(u\, 2^{-k-1}),$$

et en utilisant l'évaluation intégrale de la somme et le choix de u (condition 2.6.1) :

2.6.2 $$\mathbf{E} \sup_{S_{K+1} \times S_{K+1}} | D(s,t) - D(f_0(s),f_0(t)) | \leq \varepsilon/8.$$

Cette majoration suffit pour établir la majoration 2.2.1 : choisir u égal au diamètre de T de sorte que $S_0 = f_0(T)$ ait un seul élément, faire tendre K vers l'infini et utiliser la séparabilité de D. Nous démontrons donc maintenant les autres affirmations du théorème à partir de cette même formule 2.6.2. Pour toute sous-tribu **B'** de **A** engendrée par un nombre fini d'atomes, le lemme 2.4.(a) permet d'appliquer en effet 2.6.2 à une version régulière de $\mathbf{E}\{D \mid \mathbf{B}'\}$; on en déduit, en posant $D' = |D - \mathbf{E}\{D \mid \mathbf{B}'\}|$, en utilisant l'inégalité triangulaire et la séparabilité ($K \to \infty$) :

2.6.3 $$\mathbf{E}\ \{\sup_{T\times T} D'\} \leq \mathbf{E}\ \{\sup_{S_0\times S_0} D'\} + \varepsilon/4.$$

On applique maintenant à D le lemme 2.5 (b) pour tout couple (s, t) d'éléments de S_0 ; ce lemme avec ses notations permet d'associer à D une tribu **B'** engendrée par des atomes dont le nombre est inférieur à $L(\varepsilon) = [n(\frac{\varepsilon}{2N^2(u)})]^{N^2(u)}$ telle que pour tous les couples (s, t) d'éléments de S_0, on ait simultanément :

2.6.4 $$\mathbf{E}\ \{D'(s,t)\} \leq \varepsilon/(2\ N^2(u))\ ;$$

en regroupant 2.6.3 et 2.6.4, on obtient donc :

2.6.5 $$\mathbf{E} \sup_{T\times T} |D(s,t) - \mathbf{E}\{D(s,t) \mid \mathbf{B}'\}| \leq 3\ \varepsilon/4\ ;$$

à partir de cette inégalité, le lemme 2.5 fournit alors pour tout $\eta > 0$:

$$\mathbf{E} \sup_{\delta(s,t)\leq\eta} D(s,t) \leq \frac{3\,\varepsilon}{4} + \eta\, L(\varepsilon) \int_0^{1/L(\varepsilon)} X_0(\omega)\, d\omega\ ,$$

et finalement le résultat 2.2.2 et la conclusion complète du théorème en choisissant :

$$\eta = \varepsilon\Big\{4\, L(\varepsilon) \int_0^{1/L(\varepsilon)} X_0(\omega) d\omega\Big\}^{-1}.$$

2.7. Les classes de pseudo-métriques aléatoires du type $D(X_0)$ ont été introduites ci-dessus parce que leurs propriétés sont bien adaptées au schéma des preuves utilisées ; à partir du théorème 2.2, on peut obtenir des énoncés ayant des formes plus classiques :

Corollaire 2.7 : *Soient (T, δ) un espace métrique ou pseudo-métrique et Φ une fonction $\boldsymbol{R}^+ \to \boldsymbol{R}^+$ convexe strictement croissante ; on suppose que :*

2.7.1 $$\lim_{x\to 0} \frac{\Phi(x)}{x} = 0\ ,\ \lim_{x\to\infty} \frac{\Phi(x)}{x} = \infty\ ,\quad \int_{N(u)>1} \Phi^{-1}(N(u))du < \infty.$$

on note $\Delta(\Phi)$ l'ensemble des pseudo-métriques aléatoires séparables sur (T, δ) telles que :

2.7.2 $$\forall\ (s,t) \in T \times T\ ,\ \mathbf{E}\{\Phi(\frac{D(s,t)}{\delta(s,t)})\} \leq 1.$$

Alors tout élément D de $\Delta(\Phi)$ a p.s. ses trajectoires continues sur $T \times T$ et vérifie :

2.7.3 $$\mathbf{E}\ \{\sup_{T\times T} D(s,t)\} \leq 8 \int_{N(u)>1} \Phi^{-1}(N(u))\ du.$$

De plus sous la même hypothèse, pour tout $\varepsilon > 0$, il existe $\eta > 0$ tel que pour tout D appartenant à $\Delta(\Phi)$, on ait :

2.7.4 $$\mathbf{E} \sup_{\delta(s,t)<\eta} D(s,t) < \varepsilon.$$

Démonstration du corollaire 2.7 : L'hypothèse 2.7.2 permet par convexité pour tout couple (s, t) d'éléments de T et tout élément A de **A**, de majorer $\mathbf{E}\{D(s,t)\, I_A\}$; on obtient :

2.7.5 $$\mathbf{E}\{D(s,t)\, I_A\} \leq \delta(s,t)\, \mathbf{P}(A)\, \Phi^{-1}\left(\frac{1}{\mathbf{P}(A)}\right);$$

la même hypothèse 2.7.2 montre que $\{p\, \Phi^{-1}(p)\}$ est positive croissante et concave sur]0, 1] et tend vers zéro avec p ; il existe donc une fonction positive et décroissante X_0 sur]0, 1] telle que :

$$\forall\, p \in]0, 1],\ \int_0^p X_0(\omega)\, d\omega = p\, \Phi^{-1}\left(\frac{1}{p}\right).$$

L'inégalité 2.7.5 signifie que D appartient à $D(X_0)$; par ailleurs la construction de X_0 montre que :

$$\iint_{0<\omega<1<N(u)} X_0(\omega)\, d\omega\, du = \int_{N(u)>1} \Phi^{-1}(N(u))\, du$$

l'application du théorème 2.2 est donc justifiée et fournit le corollaire.

2.8 Remarque : On peut réduire les hypothèses du corollaire 2.7 et conserver ses conclusions, hors l'inégalité 2.7.3. Supposons en effet donnée une application Φ convexe et croissante de $\mathbf{R}^+$ dans $\mathbf{R}^+$ telle que :

2.8.1.1 $$\lim_{x\to\infty} \frac{\Phi(x)}{x} = \infty,$$

alors pour x assez grand, $\Phi(x)$ est strictement croissante de sorte que pour t assez grand, $\Phi^{-1}(t)$ est bien défini et il est légitime de supposer que :

2.8.1.2 $$\int_0 \Phi^{-1}(N(u))du < \infty.$$

Définissons alors $\Delta(\Phi)$ comme dans le corollaire et supposons-le non vide; ceci impose que $\Phi(0)$ soit inférieur ou égal à 1. Nous distinguons deux éventualités : (a) Si $\Phi(0)$ est égal à 1, alors $\Delta(\Phi)$ est inclus dans $\{D : \forall\, s, t \in T,\ D(s,t) \leq \delta(s,t) \sup\{u : \Phi(u) = 1\}\}$ de sorte que les conclusions du corollaire sont immédiatement vérifiées, hors l'inégalité 2.7.3. (b) Si au contraire $\Phi(0)$ est strictement inférieur à 1, alors la fonction Ψ définie sur $\mathbf{R}^+$ par :

$$\Psi(x) = \Phi(x) - \Phi(0) - x\, \Phi'(0) + \beta \inf(x, x^2),\ \beta = \frac{1}{\Phi^{-1}(1)},$$

est positive, convexe, strictement croissante et vérifie les deux premières hypothèses 2.7.1 et même la troisième puisque pour x assez grand, $\Phi(x)$ est inférieur à $\Psi(x)$. Par ailleurs, pour tout élément D de $\Delta(\Phi)$, D' = D/2 vérifie pour tout couple (s, t) d'éléments de T et en omettant les variables s, t :

$$\mathbf{E}\, \Psi\left(\frac{D'}{\delta}\right) \leq \mathbf{E}\, \{(\Phi - \Phi(0))\left(\frac{D}{2\,\delta}\right)\} + \frac{\beta}{2} \mathbf{E}\, \{\frac{D}{\delta}\};$$

la convexité de $(\Phi - \Phi(0))$ pour le premier terme et l'inégalité de Jensen pour le second impliquent :

$$\mathbf{E}\,\Psi\left(\frac{D'}{\delta}\right) \le \frac{1}{2}\,\mathbf{E}\,\Phi\left(\frac{D}{\delta}\right) - \frac{1}{2}\Phi(0) + \frac{1}{2}\beta\,\Phi^{-1}(1) \le 1 ,$$

de sorte que D' est un élément de $\Delta(\Psi)$ et l'application du corollaire 2.7 à la fonction Ψ montre que les conclusions de ce corollaire sont aussi applicables, hors l'inégalité 2.7.3, à Φ et à $\Delta(\Phi)$.

3. Application à la régularité de classes de fonctions aléatoires.

3.1 Pour qu'une fonction aléatoire X séparable sur un espace métrique ou pseudo-métrique (T, δ) à valeurs dans un espace métrique séparable (E, Δ) ait p.s. des trajectoires continues, il faut et il suffit que la pseudo-métrique aléatoire séparable D définie par :

3.1.1 $$D_X(s, t) = \inf(1, |X(s) - X(t)|),$$

ait la même propriété sur (T, δ) ou sur (T, δ^α), $\alpha \in]0, 1]$ de sorte que δ^α soit effectivement une pseudo-métrique plus grande que δ au voisinage de zéro ; de même pour que les lois d'une classe $\mathbb{C}$ de fonctions aléatoires séparables sur (T, δ) à valeurs dans (E, Δ) soient relativement compactes dans $\mathbb{M}(\mathbb{C}(T, \delta)\,;\,(E, \Delta))$, il suffit que les deux conditions suivantes soient vérifiées :

3.1.2 Il existe un élément t de T tel que l'ensemble des lois $\{\mu_{X(t)}, X \in \mathbb{C}\}$ soit relativement compact dans $\mathbb{M}(E)$.

3.1.3 pour tout $\varepsilon > 0$, il existe $\eta > 0$ tel que pour tout élément X de $\mathbb{C}$, on ait :

$$\mathbf{E} \sup_{\delta(s,t)\le\eta} D_X(s,t) \le \varepsilon.$$

Ceci permet d'appliquer le théorème 2.2 et son corollaire 2.7 à de larges classes de fonctions aléatoires. Dans le cas de fonctions aléatoires réelles, on en déduit par exemple le résultat suivant annoncé dans [4] et qui a une forme particulièrement simple et utilisable :

Théorème 3.2 : *Soient T un espace topologique et F une fonction continue et bornée sur T ; soit de plus $\alpha \in]0, 1]$ et Φ une fonction $\mathbf{R}^+ \to \mathbf{R}^+$ croissante et convexe telle que :*

3.2.1 $$\int^{\infty} \frac{dx}{\Phi^\alpha(x)} < \infty.$$

On note $\mathbb{C}_\alpha(F, \Phi)$ l'ensemble des fonctions aléatoires X sur T séparables pour la pseudo-métrique δ définie par $\delta(s, t) = |F(s) - F(t)|^\alpha$ et telles que :

3.2.2 $$\forall\, s, t \in T, \quad \mathbf{E}\,\Phi\left[\frac{\inf\{1, |X(s) - X(t)|\}}{\delta(s, t)}\right] \le 1 ;$$

alors tout élément X de $\mathbb{C}_\alpha(F, \Phi)$ a p.s. ses trajectoires continues sur T. De plus, pour qu'un sous-ensemble C de $\mathbb{C}_\alpha(F, \Phi)$ ait des lois relativement compactes dans $\mathbb{M}(\mathbb{C}(T))$, il faut et il suffit qu'il existe un élément t de T tel que l'ensemble des lois $\{\mu_{X(t)}, X \in C\}$ soit relativement compact dans $\mathbb{M}(\mathbf{R})$.

L'intérêt d'un tel résultat, outre l'usage des moments tronqués, est qu'il n'introduit aucun paramètre lié à T à part la fonction F. C'est en fait un cas particulier du résultat plus général et moins explicite suivant :

Théorème 3.3 : *Soient (T, δ) un espace métrique ou pseudo-métrique et Φ une application croissante et convexe de $\mathbf{R}^+$ dans $\mathbf{R}^+$; on suppose que :*

3.3.1 $$\lim_{x \to \infty} \frac{\Phi(x)}{x} = \infty \ , \qquad \int_{N(u)>1} \Phi^{-1}(N(u))\, du < \infty \ ,$$

où $N(u)$ est le nombre minimal de boules ouvertes de rayon u recouvrant T.
On note $\Gamma(\Phi)$ l'ensemble des f.a. séparables sur (T, δ) telles que :

3.3.2 $$\forall\ (s, t) \in T \times T\ , \ \mathbf{E}\, \Phi \left[\frac{\inf(1, |X(s) - X(t)|)}{\delta(s, t)}\right] \leq 1.$$

Alors tout élément X de $\Gamma(\Phi)$ a p.s. ses trajectoires continues. De plus, pour qu'un sous-ensemble C de $\Gamma(\Phi)$ ait des lois relativement compactes dans $\mathbb{M}(\mathbb{C}(T))$, il faut et il suffit qu'il existe un élément t de T tel que l'ensemble des lois $\{\mu_{X(t)}, X \in C\}$ soit relativement compact dans $\mathbb{M}(\mathbf{R})$).

Remarque 3.4 : Dans l'un et l'autre de ces énoncés, les hypothèses impliquent que (T, δ) est un espace pseudo-métrique précompact ; $\mathbb{C}(T)$ est l'ensemble des fonctions uniformément continues sur (T, δ) muni de la topologie de la convergence uniforme ; c'est donc un espace polonais. $\mathbb{M}(\mathbb{C}(T))$ est l'ensemble des probabilités sur cet espace polonais muni de la topologie de la convergence étroite (w*-topologie).

3.5 **Démonstration du théorème 3.3 :** Supposons les hypothèses du théorème 3.3 réalisées, alors la remarque 2.8 permet d'appliquer à Φ les conclusions du corollaire 2.7, hors l'inégalité 2.7.3 ; les relations 1.3.2 et 2.7.2 montrent que pour tout élément X de $\Gamma(\Phi)$, la pseudo-métrique aléatoire D_X définie par 3.1.1 appartient à $\Delta(\Phi)$; les conclusions du corollaire 2.7 impliquent donc celles du théorème 3.3 qui est démontré.

3.6 **Démonstration du théorème 3.2** : pour justifier le théorème 3.2, on utilise sur T, la pseudo-métrique continue $\delta_\alpha(s,t) = |\ F(s) - F(t)\ |^\alpha$; le théorème 3.3 implique alors les conclusions du théorème 3.2 pourvu que l'intégrale $\int_{N(u)>1} \Phi^{-1}(N(u))du$ soit convergente. Nous calculons donc la fonction N associée à δ_α ; nous pouvons supposer que F est comprise entre 0 et 1 et dans ces conditions, les ensembles $\{F(s) \in [k\, u^{1/\alpha}, (k+1)\, u^{1/\alpha}[\}$ où k parcourt $[0, u^{-1/\alpha}[$ sont contenus dans des boules de rayon u qui recouvrent T, de sorte que N(u) est majoré par $u^{-1/\alpha}$. Pour que l'intégrale ci-dessus converge, il suffit donc que l'intégrale $\int_0 \Phi^{-1}(u^{-1/\alpha})\, du$ converge aussi ; or le changement de variables $\Phi(v) = u^{-1/\alpha}$ et une intégration par parties montrent que cette dernière condition est réalisée si et seulement si 3.2.1 est vérifiée. Le théorème 3.2 est donc établi.

Références.

[1] P. Billingsley, Convergence of Probability Measures, J.Wiley, New-York, 1968.

[2] R.M. Dudley, Metric entropy and the central limit theorem in $\mathbb{C}(S)$, Ann. Instit. Fourier, Grenoble, 24-2, 1974, 49-60.

[3] X. Fernique , Régularité de fonctions aléatoires non gaussiennes, Springer Lecture Notes in Math.,976, 1-74.

[4] X. Fernique, Sur la régularité de certaines classes de fonctions aléatoires, C.R.Acad.Sci. Paris, t.307, Série 1, 1988, 493-496.

[5] M.G. Hahn et J. Klass, Sample continuity of square integrable processes, Ann. of Prob., 5,1977, 361-370.

[6] I.A. Ibragimov, On Smoothness Conditions for Trajectories for random Functions, Theory of Prob. and Appl., 28, 2, 1983, 240-262.

[7] N. Kono, Sample path properties of stochastic processes, J. Math. of Kyoto Univ., 20-2, 1980, 295-313.

[8] G. Pisier, Conditions d'entropie assurant la continuité de certains processus et applications à l'analyse harmonique,Sém.d'Anal.Fonct.,1979-1980, Exposés 13-14, Paris, Ecole Polytechnique.

[9] E. Slutsky, Alcuno propozitioni sulla teoria delle funzioni aleatorie, Giorn.Inst. Italiano degli Attuari,8,1937,193-199.

[10] M. Talagrand, Sample boundedness of stochastic processes under incremental conditions, preprint.

[11] M. Talagrand, communication orale.

[12] M. Weber, Une méthode élémentaire pour l'étude de la régularité d'une large classe de fonctions aléatoires,C.R.Acad.Sci.Paris, 292, 1981, 599-602.

COMPARISON OF LOG-PROBABILITIES OF PARTIAL SUMS WITH THOSE OF POISSONIZED SUMS

Marjorie G. Hahn
Department of Mathematics, Tufts University
Medford, MA 02155 USA

Michael J. Klass
Departments of Mathematics and Statistics
University of California, Berkeley, CA 94720 USA

1. Introduction.

Let $Z, Z_1, Z_2, \ldots$ be i.i.d. random variables with values in a finite or infinite-dimensional space and partial sum $S_k = \sum_{j=1}^{k} Z_j$. Let N_k be an independent Poisson random variable with $EN_k = k$. Analysis of S_{N_k} is often easier than that of S_k. One reason is that the Poissonized sums $\{\sum_{j=1}^{N_k} Z_j I(Z_j \in A_i) : \ i \in F\}$, for some finite index set F, are independent if the A_i are disjoint. This facilitates the use of decomposition and recombination of the resulting probabilities. Consequently, it is advantageous to determine when sums of a Poissonized number of random variables provide good approximations for sums of a fixed number of random variables. The lack of appropriate comparisons for real-valued random variables, necessitates the consideration of this case first. The results will then pertain to the multi-dimensional case through consideration of one-dimensional projections.

In Hahn and Klass (1989), upper and lower bounds for the log-probability that a Poissonized sum reaches or exceeds a level y are obtained. These results are valid for *all* Z, y, and k. Under the assumptions of nonnegativity plus a probability constraint, Jain and Pruitt (1987) obtain the asymptotic log-probability of sums of a fixed number of i.i.d. random variables. Combining these two results allows for appropriate comparisons when the random variables are nonnegative. It is to be hoped that more general comparisons can ultimately be made.

Section 2 presents the general approximations from Hahn and Klass (1989). Section 3 presents the results of Jain and Pruitt (1987) and shows how to combine them with the results of Section 2 to compare the respective log-probabilities.

Previously, probabilities of Poissonized sums have been used to approximate probabilities of fixed sums in cases where the differences tend to zero (e.g. see Araujo and Giné (1980) or the accompanying laws theorem in Araujo, Giné, Mandrekar and Zinn (1981)). Such results give only the crudest information when the probabilities in question tend to zero. The use of log-probabilities permits comparison on these sets as well.

2. Main Approximations for Poissonized Sums.

From now on let $Z, Z_1, Z_2, \ldots$ be arbitrary i.i.d. real-valued random variables with $S_k = \sum_{j=1}^{k} Z_j$. Let N_k be an independent Poisson process with $EN_k = k$, $k = 1, 2\ldots$ or $k \in [0, \infty)$. Define a partitioning level, which is independent of the level y to be exceeded, by

$$v_k = v_k(Z) = \sup\{v : \quad kEZ^2 I(0 \le Z \le v) \ge v^2\}. \tag{2.1}$$

Define a rate function R for a random variable Z by

$$R_k(t) = R_{Z,k}(t) = R_{Z,k,y}(t) \equiv \begin{cases} kE\{(tZ-1)e^{tZ}+1\} & \text{if } 0 \le t < \infty \\ \infty & \text{if } t = \infty \text{ and } y > 0 \\ kP(Z < 0) & \text{if } t = \infty \text{ and } y = 0. \end{cases} \tag{2.2}$$

Let

$$t_y = t_y(Z) \equiv \begin{cases} \inf\{t \ge 0 : \quad kEZe^{tZ} \ge y\} & \text{if such } t \text{ exists} \\ \infty & \text{if no such } t \text{ exists} \end{cases} \tag{2.3}$$

If $t_y < \infty$, $y \le kEZe^{t_y Z}$, with equality $\iff$ $kEZ \le y$. Furthermore, when $t_y < \infty$, $t_y = 0 \iff y \le kEZ$.

The general theorem is obtained by combining bounds for two large classes of random variables for which a multiple of a power of the usual exponential upper bound is also a lower bound. The first class includes all $Z \le 0$.

Theorem 1 (Hahn and Klass (1989) Theorem 0.4). *Fix any $\varepsilon > 0$ and any Z, k, and y. Assume $P(Z \le v_k(Z)) = 1$. Then there exists $C_\varepsilon > 0$, independent of Z, k, and y, such that*

$$C_\varepsilon e^{-(1+\varepsilon)R_{Z,k}(t_y)} \le P(S_{N_k} \ge y) \le e^{-R_{Z,k}(t_y)}. \tag{2.4}$$

If $Z \ge 0$, then $P(\sum_{j=1}^{N_k} Z_j \ge y) = P(\sum_{j=1}^{N_k}(Z_j \wedge y) \ge y)$ for all $y \ge 0$. This motivates the form for the other class.

Theorem 2 (Hahn and Klass (1989) Theorem 0.6). *Take any Z, k, and y with $y \ge 0$. Then there exists $C > 0$, independent of Z, k, and y such that*

$$Ce^{-2R_{(Z\wedge y)I(Z>v_k),k}(\tilde{t}_y)} \le P\left(\sum_{j=1}^{N_k} Z_j I(Z_j > v_k) \ge y\right) \le e^{-R_{(Z\wedge y)I(Z>v_k),k}(\tilde{t}_y)} \tag{2.5}$$

where $\tilde{t}_y = t_y((Z \wedge y)I(Z > v_k))$.

The upper and lower coefficients of the rate functions in both Theorems 1 and 2 cannot be improved.

Conceptually, the idea for combining the above results is simple: Write an arbitrary random variable Z as $Z_j = Z_j^{'} + Z_j^{''}$ with

$$Z_j^{'} = Z_j I(Z_j \le v_k) \text{ and } Z_j^{''} = Z_j I(Z_j > v_k) \tag{2.6}$$

and the level y to be reached or exceeded as $y = y' + y''$. Exponential bounds give the appropriate exponential rate of decay for $P\left(\sum_{j=1}^{N_k} Z'_j \geq y'\right)$ and $P\left(\sum_{j=1}^{N_k}(Z''_j \wedge y'') \geq y''\right)$. The latter is sufficient since $P(\sum_{j=1}^{N_k} Z''_j \geq y'') = P(\sum_{j=1}^{N_k}(Z''_j \wedge y'') \geq y'')$ due to the fact that $Z'' \geq 0$. Finally, choose y' and y'' to make the two exponential bounds equal (if possible). This leads to the following general theorem:

Theorem 3 (Hahn and Klass (1989) Theorem 5.14). *Let Z be an arbitary random variable. Let*

$$x_y = \inf\{x : \ R_{Z',k}(t_x(Z')) \geq R_{Z'' \wedge (y-x),k}(t_{y-x}(Z'' \wedge (y-x)))\}. \tag{2.7}$$

Then given, $\varepsilon > 0$ there exist universal positive constants C_ε and C such that both

$$\begin{gathered} C_\varepsilon e^{-(3+\varepsilon)R_{Z',k}(t_{x_y}(Z'))} \leq P(S_{N_k} \geq y) \leq 2e^{-R_{Z',k}(t_{x_y}(Z'))} - e^{-2R_{Z',k}(t_{x_y}(Z'))} \\ \text{if either } x_y \neq 0 \text{ or } x_y = 0 \text{ and } R_{Z',k}(t_0(Z')) \geq R_{Z'' \wedge y,k}(t_y(Z'' \wedge y)), \end{gathered} \tag{2.8}$$

and

$$\begin{gathered} Ce^{-3R_{Z'' \wedge y,k}(t_y(Z'' \wedge y))} \leq P(S_{N_k} \geq y) \leq e^{-R_{Z'' \wedge y,k}(t_y(Z'' \wedge y))} \\ \text{if } x_y = 0 \text{ and } R_{Z',k}(t_0(Z')) < R_{Z'' \wedge y,k}(t_y(Z'' \wedge y)). \end{gathered} \tag{2.9}$$

The next corollary of Theorem 3 provides sufficient conditions for deducing the correct asymptotic exponential order of the upper tail.

Corollary 4 (Hahn and Klass (1989) Corollary 5.21). *Let Z be an arbitrary random variable. Let v_k be as in (2.1) and x_{y_k} be as in (2.7). If y_k is such that*

$$\lim_{k \to \infty} R_{Z',k}(t_{y_k}(Z')) = \infty \tag{2.10}$$

and

$$\lim_{k \to \infty} \frac{R_{Z',k}(t_{y_k}(Z'))}{R_{Z',k}(t_{x_{y_k}}(Z'))} = 1 \tag{2.11}$$

then

$$P(S_{N_k} \geq y_k) = e^{-(1+o(1))R_{Z',k}(t_{y_k}(Z'))} \quad \text{as } k \to \infty. \tag{2.12}$$

3. Comparison of Fixed Sums and Poissonized Sums.

Corollary 4 and known results about sums of a fixed number of i.i.d. random variables can be used to determine conditions under which fixed sums and Poissonized sums have the same asymptotic behavior. Alternatively, this comparison can be viewed as a recipe for approximating probabilities of fixed sums by probabilities of the corresponding Poissonized sums.

Throughout this section we invoke the following assumptions and definitions:

Assumptions and Definitions:

Let $Z, Z_1, Z_1, \ldots$ be i.i.d. nonnegative and nonconstant random variables. Let b be a nonnegative real number with $x_* \equiv \operatorname{essinf} Z \leq b \leq EZ$. Define

$$t(b) = \inf\{t \geq 0: \quad EZe^{-tZ} \leq b\}. \tag{3.1}$$

Thus,

$$t(b) = \begin{cases} \infty & \text{if } b = x_* \\ 0 & \text{if } b \geq EZ \end{cases}$$

and otherwise $t(b)$ is the unique positive real such that $EZe^{-t(b)Z} = b$. Also, define

$$\overline{R}(t) = \begin{cases} E(1-(1+tZ)e^{-tZ}) & \text{if } t < \infty \\ P(Z > 0) & \text{if } t = \infty, \end{cases} \tag{3.2}$$

In particular, $k\overline{R}(t(b)) = R_{-Z,k}(t(b))$. Finally, define

$$Q(t) = -\log\left(Ee^{-tZ}e^{\frac{tEZe^{-tZ}}{Ee^{-tZ}}}\right). \tag{3.3}$$

If $x_* < b \leq EZ$, let $t^*(b)$ denote the unique real number satisfying

$$\frac{EZe^{-t^*(b)Z}}{Ee^{-t^*(b)Z}} = b. \tag{3.4}$$

If $b = x_*$, let $t^*(b) = \infty$ and $Q(\infty) = -\log P(Z = x_*)$. Observe that for all $x_* \leq b \leq EZ$, $t^*(b) \geq t(b)$. Moreover, $\lim_{t\to\infty} Q(t) = Q(\infty)$.

Our main comparative result is

Theorem 5. *Let Z be a nonnegative and nonconstant random variable. Let b_k be constants such that $x_* \equiv \operatorname{essinf} Z \leq b_k \leq EZ$ and let $t_k = t(b_k)$. Suppose*

$$k\overline{R}(t_k) \to \infty. \tag{3.5}$$

Then

$$\liminf_{k\to\infty} \frac{\log P(S_k \leq kb_k)}{\log P(S_{N_k} \leq kb_k)} \geq 1; \tag{3.6}$$

$$\limsup_{k\to\infty} \frac{\log P(S_k \leq kb_k)}{\log P(S_{N_k} \leq kb_k)} < \infty \iff \liminf_{k\to\infty} P(Z \leq b_k) > 0; \tag{3.7}$$

and

$$\lim_{k\to\infty} \frac{\log P(S_k \leq kb_k)}{\log P(S_{N_k} \leq kb_k)} = \begin{cases} 1 & \iff EZ^2 = \infty \text{ and } b_k \to EZ \\ \frac{EZ^2}{VarZ} & \text{if } EZ^2 < \infty \text{ and } b_k \to EZ. \end{cases} \tag{3.8}$$

The proof of Theorem 5 will rely heavily on Corollary 4 and the following reformulation of several results of Jain and Pruitt:

Theorem 6 (Jain and Pruitt (1987)). *Let Z be a nonnegative and nonconstant random variable with b_k as above. Then*

$$P(S_k \leq kb_k) \leq e^{-kQ(t_k^*)} \tag{3.9}$$

and

$$P(S_k \leq kb_k) \to 0 \quad \Longleftrightarrow \quad kQ(t_k^*) \to \infty. \tag{3.10}$$

If

$$kQ(t_k^*) \to \infty \tag{3.11}$$

and if

$$\exists \; c > 0 \; \text{ such that } \; P(Z \leq b_k) \geq c \; \forall \; k \text{ large} \tag{3.12}$$

then

$$\lim_{k\to\infty} \frac{-\log P(S_k \leq kb_k)}{kQ(t_k^*)} = 1. \tag{3.13}$$

Remark 1. The rate functions $\overline{R}$ and Q, of (3.2) and (3.3) respectively, are both continuous on $(0, \infty)$, strictly increasing, and are equal to 0 at 0. $\overline{R}(\infty) = \infty$ while $Q(\infty) = -\log P(Z = \text{essinf } Z)$ which is infinite if and only if $P(Z = \text{essinf } Z) = 0$.

Remark 2. In analogy with (3.10), notice that from Theorem 3, it is immediate that whenever $Z \geq 0$,

$$k\overline{R}(t_k) \to \infty \quad \Longleftrightarrow \quad P(S_{N_k} \leq kb_k) \to 0. \tag{3.14}$$

Remark 3. The following are equivalent:

(i) $t_k \to 0$;
(ii) $t_k^* \to 0$;
(iii) $b_k \to EZ$;
(iv) $Q(t_k^*) \to 0$;
(v) $\overline{R}(t_k) \to 0$;

Clearly, (i) and (ii) are each equivalent to (iii) while (iv) is equivalent to (ii) and (v) is equivalent to (i).

Furthermore, each of the above implies

(vi) $\liminf_{k\to\infty} P(Z \leq b_k) > 0$.

The latter fact is an immediate consequence of (iii) and $P(Z < EZ) > 0$ since Z is nonconstant.

The Proof of Theorem 5 requires some comparisons between the rate functions for fixed and Poissonized sums. The first lemma will be used when any of the following equivalent conditions holds: $t_k \to 0$, $t_k^* \to 0$ or $b_k \to EZ$.

Lemma 7. *If $EZ^2 = \infty$, then $\lim_{t\to 0} \frac{Q(t)}{\overline{R}(t)} = 1$.* (3.15)

If $EZ^2 < \infty$, then

$$\lim_{t\to 0} \frac{Q(t)}{\frac{1}{2}t^2 VarZ} = 1 \tag{3.16}$$

and

$$\lim_{t\to 0} \frac{\overline{R}(t)}{\frac{1}{2}t^2 EZ^2} = 1. \tag{3.17}$$

Proof. All are easy. (3.15) and (3.16) can be found in Lemma 2.2 of Jain and Pruitt (1987). For (3.17), notice that since $EZ^2 < \infty$,

$$\begin{aligned}\lim_{t\to 0} \frac{\overline{R}(t)}{t^2} &= \lim_{t\to 0} \frac{E(1-(1+tZ)e^{-tZ})}{t^2} \\ &= \lim_{t\to 0} \frac{E(tZ^2e^{-tZ})}{2t} \quad \text{by L'Hopital's Rule} \\ &= \frac{1}{2}EZ^2. \quad \blacksquare\end{aligned}$$

The next lemma will be used when $b_k \not\to EZ$.

Lemma 8.
(i.) For any $x_* < b < EZ$, $\frac{Q(t^*(b))}{\overline{R}(t(b))} > 1$.
(ii.) $\liminf_{b\searrow x_*} \frac{Q(t^*(b))}{\overline{R}(t(b))} > 1$.
(iii.) If $\lim_{k\to\infty} \frac{Q(t^*(b_k))}{\overline{R}(t(b_k))} = 1$, *then* $b_k \to EZ$.

Proof. Take any $x_* < b < EZ$,

$$\begin{aligned}\frac{Q(t^*(b))}{\overline{R}(t(b))} &= \frac{-\log Ee^{t_b^*(b-Z)}}{\log Ee^{t_b(b-Z)}} \frac{\log Ee^{t_b(b-Z)}}{\overline{R}(t(b))} \\ &= \frac{-\log Ee^{t_b^*(b-Z)}}{\log Ee^{t_b(b-Z)}} \left(\frac{bt_b + \log Ee^{-Zt_b}}{\overline{R}(t(b))}\right) \\ &> \frac{-\log Ee^{t_b^*(b-Z)}}{\log Ee^{t_b(b-Z)}} \left(\frac{bt_b - E(1-e^{-Zt_b})}{\overline{R}(t(b))}\right) \\ &= \frac{\log Ee^{t_b^*(b-Z)}}{\log Ee^{t_b(b-Z)}} \\ &> 1,\end{aligned}$$

since $Ee^{t_b^*(b-Z)} = \inf_{w>0} Ee^{w(b-Z)}$. This verifies (i).

For (ii), suppose first that $x_* > 0$. Then $t(x_*) < \infty$.

$$\begin{aligned}\liminf_{b\searrow x_*} \frac{Q(t^*(b))}{\overline{R}(t(b))} &\geq \liminf_{b\searrow x_*} \frac{\log Ee^{t_b^*(b-Z)}}{\log Ee^{t_b(b-Z)}} \\ &= \frac{\log P(Z = x_*)}{\log\left(Ee^{t(x_*)(x_*-Z)}\right)} \\ &> 1,\end{aligned}$$

since $Ee^{t(x_*)(x_*-Z)} > P(Z = x_*)$.

Next, assume $x_* = 0$. Then

$$\begin{aligned}\liminf_{b \searrow x_*} \frac{Q(t^*(b))}{\overline{R}(t(b))} &= \frac{Q(\infty)}{\overline{R}(\infty)} \\ &= \frac{-\log P(Z=0)}{P(Z>0)} \\ &\geq \inf_{0<\lambda<1} \frac{-\log \lambda}{1-\lambda}.\end{aligned}$$

Let

$$g(\lambda) = \frac{-\log \lambda}{1-\lambda}$$

and observe that $\lim_{\lambda \searrow 0} g(\lambda) = \infty$ and $\lim_{\lambda \uparrow 1} g(\lambda) = 1$. To minimize $g(\cdot)$ notice that

$$g'(\lambda) = \frac{-\frac{(1-\lambda)}{\lambda} - \log \lambda}{(1-\lambda)^2}$$

so that

$$g'(\lambda^*) = 0 \iff -\log \lambda^* = \frac{1-\lambda^*}{\lambda^*}.$$

But since $\log \frac{1}{\lambda^*} < \frac{1}{\lambda^*} - 1$ for $\lambda^* \neq 1$, we must have $\lambda^* = 1$. Hence, $g(\lambda) > 1 \iff 0 \leq \lambda < 1$. Since Z is nonconstant, $P(Z = 0) < 1$. Thus, (ii) holds.

(iii) is now an immediate consequence of (i) and (ii). ∎

The preliminaries for Theorem 5 are now complete.

Proof of Theorem 5.

By taking subsequences if necessary, it may be supposed that $b_k \to b_\infty$, $t_k \to t_\infty$ and $t_k^* \to t_\infty^*$, where $x_* \leq b_\infty \leq \infty$, $0 \leq t_\infty \leq \infty$, and $0 \leq t_\infty^* \leq \infty$.

Case 1. $P(Z = x_*) > 0$ or $b_\infty > x_*$. Then

$$\liminf_{k\to\infty} P(Z \leq b_k) \geq P(Z < b_\infty) \vee P(Z = x_*) > 0$$

so that

$$\frac{\log P(S_k \leq kb_k)}{\log P(S_{N_k} \leq kb_k)} \sim \frac{-kQ(t_k^*)}{-k\overline{R}(t_k)}.$$

Now if $b_\infty < EZ$ then $\overline{R}(t_k) \to \overline{R}(t_\infty) > 0$ so that

$$\lim_{k\to\infty} \frac{\log P(S_k \leq kb_k)}{\log P(S_{N_k} \leq kb_k)} = \frac{Q(t_\infty^*)}{\overline{R}(t_\infty)}. \tag{3.18}$$

This limit is finite since if $x_* < b < EZ$ then

$$Q(t^*(b)) < \infty,$$

and if $b_\infty = x_*$ then

$$\lim_{b \searrow x_*} Q\left(t^*(b)\right) = -\log P(Z = x_*)$$

(which is finite in this case because $P(Z = x_*) > 0$). Moreover, Lemma 8 shows that the limit in (3.18) exceeds 1 in this case.

If $b_\infty = EZ$, it is shown in Lemma 9 below that

$$\lim_{k\to\infty} \frac{\log P(S_k \le kb_k)}{\log P\left(S_{N_k} \le kb_k\right)} = \begin{cases} 1 & \text{if } EZ^2 = \infty \\ \frac{EZ^2}{VarZ} & \text{if } EZ^2 < \infty. \end{cases} \tag{3.19}$$

Case 2. $b_\infty = x_*$ and $P(Z = x_*) = 0$.

Fix any $\varepsilon > 0$.

$$\begin{aligned} \liminf_{k\to\infty} \frac{\log P(S_k \le kb_k)}{\log P\left(S_{N_k} \le kb_k\right)} &\ge \liminf_{k\to\infty} \frac{\log P\left(S_k \le k(x_* + \varepsilon)\right)}{-k\overline{R}(t_k)} \\ &= \frac{Q\left(t^*(x_* + \varepsilon)\right)}{\overline{R}(t(x_*))} \ge Q\left(t^*(x_* + \varepsilon)\right). \end{aligned}$$

Since $\varepsilon > 0$ is arbitrary,

$$\begin{aligned} \liminf_{k\to\infty} \frac{\log P(S_k \le kb_k)}{\log P\left(S_{N_k} \le kb_k\right)} &\ge \lim_{b \searrow x_*} Q(t^*(b)) \\ &= -\log P(Z = x_*) = \infty. \end{aligned}$$

Hence all of Theorem 5 has been verified modulo the proof of (3.19). ∎

Lemma 9. *If* $b_\infty = EZ$,

$$\lim_{k\to\infty} \frac{\log P(S_k \le kb_k)}{\log P\left(S_{N_k} \le kb_k\right)} = \begin{cases} 1 & \textit{if } EZ^2 = \infty \\ \frac{EZ^2}{VarZ} & \textit{if } EZ^2 < \infty. \end{cases}$$

Proof. When $b_k \to EZ$ and $k\overline{R}(t_k) \to \infty$, note that both $t_k \to 0$ and $t_k^* \to 0$ by Remark 3. Since $k\overline{R}(t_k) \to \infty$, $t_k > 0$ for all k large and consequently $t_k^* > 0$ for such k as well. Hence for all k sufficiently large,

$$EZe^{-t_k Z} = \frac{EZe^{-t_k^* Z}}{Ee^{-t_k^* Z}} (= b_k). \tag{3.20}$$

Case 1. $0 < \text{Var } Z < \infty$. Expanding both sides of (3.20) in terms of t_k or t_k^*,

$$\begin{aligned} EZ - (1 + o(1))t_k EZ^2 &= \frac{EZ - (1 + o(1))t_k^* EZ^2}{1 - (1 + o(1))t_k^* EZ} \\ &= EZ - (1 + o(1))t_k^* \text{Var} Z. \end{aligned}$$

Consequently,

$$t_k \sim t_k^* \frac{\mathrm{Var}Z}{EZ^2}. \tag{3.21}$$

Therefore, since $t_k^* \to 0$,

$$\begin{aligned}
\frac{\log P(S_k \le kb_k)}{\log P(S_{N_k} \le kb_k)} &\sim \frac{Q(t_k^*)}{\overline{\overline{R}}(t_k)} \text{ by Corollary 4 and Theorem 6} \\
&\sim \frac{Q(t_k^*)}{\overline{\overline{R}}\left(t_k^* \frac{\mathrm{Var}Z}{EZ^2}\right)} \text{ by (3.21)} \\
&\sim \frac{Q(t_k^*)}{\left(\frac{\mathrm{Var}Z}{EZ^2}\right)^2 \overline{\overline{R}}(t_k^*)} \text{ by (3.16) and (3.17)} \\
&\sim \frac{EZ^2}{\mathrm{Var}Z} \text{ again by (3.16) and (3.17).}
\end{aligned}$$

This completes the verification if $0 < \mathrm{Var}Z < \infty$.

Case 2. $EZ^2 = \infty$.

Since hypothesis (3.5) implies (3.11) and $t_k \to 0$ is equivalent to $t_k^* \to 0$,

$$\begin{aligned}
\lim_{k\to\infty} \frac{\log P(S_k \le kb_k)}{\log P(S_{N_k} \le kb_k)} &= \lim_{k\to\infty} \frac{Q(t_k^*)}{\overline{\overline{R}}(t_k)} \text{ by Corollary 4 and Theorem 6} \\
&= \lim_{k\to\infty} \frac{Q(t_k^*)}{\overline{\overline{R}}(t_k^*)} \frac{\overline{\overline{R}}(t_k^*)}{\overline{\overline{R}}(t_k)} \\
&= \lim_{k\to\infty} \frac{\overline{\overline{R}}(t_k^*)}{\overline{\overline{R}}(t_k)} \text{ by (3.15) since } t_k^* \to 0 \\
&\to 1 \text{ by Lemma 10 below.}
\end{aligned}$$

Therefore, the proof is complete modulo the next lemma. ∎

Lemma 10. *Let $Z \ge 0$ with $EZ^2 = \infty$. Then*

$$\lim_{t\searrow 0} \frac{\left(EZe^{-tZ}\right)^2}{EZ^2 e^{-tZ}} = 0. \tag{3.22}$$

If $0 < t_k < t_k^*$ satisfies

$$EZe^{-t_k Z} = \frac{EZe^{-t_k^* Z}}{Ee^{-t_k^* Z}} \tag{3.23}$$

and if

$$t_k^* \to 0, \tag{3.24}$$

then

$$\lim_{k\to\infty} \frac{\overline{\overline{R}}(t_k)}{\overline{\overline{R}}(t_k^*)} = 1. \tag{3.25}$$

Proof. (3.22): For every $0 < c < \infty$,

$$L \equiv \lim_{t \searrow 0} \frac{EZ^2 e^{-tZ}}{(EZe^{-tZ})^2}$$

$$= \lim_{t \searrow 0} \frac{EZ^2 e^{-tZ}}{(EZe^{-tZ} I(Z > c))^2} \quad \text{since} \quad EZ^2 = \infty.$$

Hence there exists $c_t \to \infty$ *such that*

$$L = \lim_{t \searrow 0} \frac{EZ^2 e^{-tZ}}{\left(EZe^{-\frac{tZ}{2}}(e^{-\frac{tZ}{2}} I(Z > c_t))\right)^2}$$

$$\geq \lim_{t \searrow 0} \frac{EZ^2 e^{-tZ}}{(EZ^2 e^{-tZ})(Ee^{-tZ} I(Z > c_t))} \quad \text{by the Cauchy – Schwarz inequality}$$

$$= \infty.$$

(3.25): First notice that since $v^{-2}(1 - e^{-v}(1+v)) \downarrow$, we have

$$(1 - \frac{2}{e})(v^2 \wedge 1) \leq 1 - e^{-v}(1+v) \quad \forall\, v \geq 0. \tag{3.26}$$

Now

$$0 \leq 1 - \frac{\overline{R}(t_k)}{\overline{R}(t_k^*)} \quad \text{since } \overline{R}(\cdot) \text{ increases and } t_k^* > t_k$$

$$= \frac{\overline{R}(t_k^*) - \overline{R}(t_k)}{\overline{R}(t_k^*)} = \frac{\int_{t_k}^{t_k^*} \overline{R}'(t)\, dt}{\overline{R}(t_k^*)} = \frac{\int_{t_k}^{t_k^*} EtZ^2 e^{-tZ}\, dt}{\overline{R}(t_k^*)}$$

$$\leq \frac{t_k^* \int_{t_k}^{t_k^*} EZ^2 e^{-tZ}\, dt}{\overline{R}(t_k^*)} = t_k^* \frac{(EZe^{-t_k Z} - EZe^{-t_k^* Z})}{\overline{R}(t_k^*)}$$

$$= \frac{t_k^* EZe^{-t_k^* Z} E(1 - e^{-t_k^* Z})}{Ee^{-t_k^* Z} \overline{R}(t_k^*)} \quad \text{by inserting (3.23)}$$

$$\leq \frac{t_k^* EZe^{-t_k^* Z} E(t_k^* Z \wedge 1)}{Ee^{-t_k^* Z} \overline{R}(t_k^*)}$$

$$\leq (1 - \frac{2}{e})^{-1} \frac{t_k^* EZe^{-t_k^* Z} E(t_k^* Z \wedge 1)}{(1 - o(1)) E\{(t_k^* Z)^2 \wedge 1\}} \quad \text{by (3.26).}$$

Observe that

$$\lim_{k \to \infty} \frac{t_k^* EZe^{-t_k^* Z} Et_k^* Z I(t_k^* Z < 1)}{E\{(t_k^* Z)^2 \wedge 1\}} \leq \lim_{k \to \infty} \frac{(t_k^* EZe^{-t_k^* Z})^2}{E(t_k^* Z)^2 e^{-t_k^* Z}},$$

which is zero by (3.22). As for the other part of the preceeding term,

$$\lim_{k\to\infty} \frac{Et_k^* Ze^{-t_k^* Z} P(t_k^* Z \geq 1)}{E\{(t_k^* Z)^2 \wedge 1\}} \leq \lim_{k\to\infty} Et_k^* Ze^{-t_k^* Z} = 0$$

by the dominated convergence theorem. Therefore (3.25) holds. ∎

References

Araujo, A. and E. Giné (1980). *The central limit theorem for real and Banach valued random variables.* Wiley, New York.

Araujo, A., E. Giné, V. Mandrekar and J. Zinn (1981). On the accompanying laws theorem in Banach spaces. *Ann. Probab.* **9**, 202-210.

Jain, N. C. and W. E. Pruitt (1987). Lower tail probability estimates for subordinators and nondecreasing random walks. *Ann. Probab.* **15**, 75-101.

Hahn, M. G. and M. J. Klass (1989). Log-probability bounds for Poissonized sums formed from arbitrary random variables. Preprint.

A LAW OF LARGE NUMBERS FOR RANDOM VECTORS HAVING LARGE NORMS

Bernard HEINKEL
Département de Mathématique
7, rue René Descartes
67084 STRASBOURG Cédex

During the past decade the knowledge on the law of large numbers for random vectors has made a very substantial progress. In particular Prohorov's theorem, which is the "nearly optimal form" of the strong law of large numbers (SLLN) in the scalar case, has found a completely satisfactory analogue for Banach space valued random variables in a recent result of Ledoux and Talagrand, which was the endpoint of a whole series of works on Prohorov's theorem in Banach spaces ([9], [4,5,6], [2]). The statement of this infinite dimensional version of Prohorov's theorem the necessity part of which is due to Alt ([2], Théorème 4) and the sufficiency part of which is due to Ledoux and Talagrand ([10], Theorem 13) is as follows :

THEOREM 0.1. — *Let (X_k) be a sequence of r.v. taking their values in a real separable Banach space $(B, \| \ \|)$ which are independant and centered. One supposes that there exists a positive constant M such that :*

$$\forall\, k, \quad \|X_k\| \leq M(k/L_2 k) \quad a.s.\,,$$

where L_2 denotes the iterated logarithm function :

$$\forall\, x > 0, \quad L_2 x = Log\,(Log\ \sup(x, e^e))\,.$$

The sequence (X_k) satisfies the SLLN-i.e. $S_n/n = (1/n)(X_1 + \cdots + X_n)$ converges a.s. to 0 - if and only if the following properties hold :

a) *$S_n/n \to 0$ in probability;*

b) *$\forall\, \varepsilon > 0, \quad \sum_{n\geq 1} \exp(-\varepsilon/\sigma_n^2) < +\infty$,*

where for every n :

$$\sigma_n^2 = 2^{-2n} \sup\Big(\sum_{k\in I(n)} Ef^2(X_k), \quad \|f\|_{B'} \leq 1\Big)\,,$$

and : $I(n) = (2^n + 1 \dots 2^{n+1})$.

For being convinced that this result is -in some sense- nearly optimal, consider (X_k) a sequence of r.v. taking their values in a real separable

Banach space $(B, \| \ \|)$, being independent and, for simplicity, symmetrically distributed. Let's suppose that (X_k) satisfies the SLLN. It is well known that this implies that (X_k/k) converges a.s. to 0, and so, by the Borel-Cantelli lemma, it is possible to find a sequence of positive numbers (α_k), converging to 0 and such that :

$$\sum_{k \geq 1} P(\|X_k\| > k\alpha_k) < +\infty. \tag{0.1}$$

If one defines for every k :

$$U_k = X_k I_{(\|X_k\| \leq k\alpha_k)},$$

one notes that the SLLN for the sequence (X_k) reduces to the SLLN for the sequence of truncated r.v. (U_k). If one now splits each of the r.v. U_k into two r.v. in the following way :

$$U_k = Y_k + Z_k,$$

with :

$$Y_k = U_k I_{(\|U_k\| \leq k/L_2 k)} \ ,$$

and :

$$Z_k = U_k - Y_k \ ,$$

one notices that the almost sure asymptotic behavior of the sequence $S_n(Y) = Y_1 + \cdots + Y_n$ is completely cleared up by Theorem 0.1; so the only thing remaining to be done for checking if (X_k) satisfies the SLLN is to study the asymptotic behavior of the sequence $(S_n(Z))$.

In the scalar setting, Nagaev's theorem [13] allows, at least in theory, to study completely this asymptotic behavior. Unfortunately, in practice, Negaev's result is very difficult to apply, except for very simple r.v. (Z_k); this remark makes probabilists and statisticians very angry, because they would like to have at their disposal a simple criterion -as for instance condition (b) in Theorem 0.1- allowing to recognize immediately if (Z_k) satisfies the SLLN of not. Thinking of such a criterion immediately leads to the natural idea of expressing in one way or another that $P(Z_k \neq 0)$ is small when k is large. In [9] (Theorem 1) for instance the following condition- which seems at first glance rather mysterious- is used :

$$\exists\, p \in [1,2],\ \exists\, s > 0 : \sum_{n \geq 1} \Big(2^{-np} \sum_{k \in I(n)} E\|X_k\|^p\Big)^s < +\infty \ ; \tag{0.2}$$

in [11] (Theorem 3.1) the following more general assumption is made :

$$(0.2)' \qquad \exists\, p > 0,\ \exists\, s > 0 : \sum_{n \geq 1} (2^{-np} \sum_{k \in I(n)} E\|X_k\|^p)^s < +\infty \ .$$

Under (0.2) -with $p = 2$- the following series of course converges :

$$\sum_{n \geq 1} \Big\{ (\mathrm{Log}(n+1))^{-2} \sum_{k \in I(n)} P(\|X_k\| > k/L_2 k) \Big\}^s < +\infty \ ,$$

and this shows that at least for a large class of indices, $P(Z_k \neq 0)$ is small.

Theorem 2.2 which we will prove in Section 2 will help us to better understand the nature of assumption (0.2) -with $p = 2$- and to appreciate its degree of optimality.

For finding a natural, simple to check, and efficient condition for saying that $P(Z_k \neq 0)$ is small, we go back again to Prohorov's work. The result we will take as a model is less famous than the finite dimensional version of Theorem 0.1., but it is nevertheless very sharp :

THEOREM 0.2 ([17] THEOREM 2). — *Let (X_k) be a sequence of real-valued r.v., which are independent and centered. One supposes that there exists a function :*

$$f : N \to \mathbb{R}^+$$

such that :

1) $f(n) 2^{-n} \mathrm{Log}\, n \underset{n \to +\infty}{\longrightarrow} +\infty$;
2) $f(n) 2^{-n} \underset{n \to +\infty}{\longrightarrow} 0$;
3) $\forall\, k \in I(n),\ |X_k| \leq f(n)$ *a.s.* .

Under these hypotheses the following condition is sufficient for implying that the SLLN holds for the sequence (X_k) :

$$(0.3) \qquad \forall\, \varepsilon > 0,\ \sum_{n \geq 1} \exp\Big\{ -\frac{\varepsilon 2^n}{f(n)} \mathrm{Arc\ sh} \Big(\frac{\varepsilon f(n) 2^n}{\sum_{k \in I(n)} E(X_k^2)} \Big) \Big\} < +\infty \ .$$

Remark : Prohorov has given an example showing that assumption (0.3) is optimal ([16] Theorem 6.1) : Lets consider (X_k) a sequence of real-valued r.v., which are independent, each of them taking only three values $-b_k, 0, b_k$, with the following restrictions :

$$\begin{array}{lll} \lim_{k\to+\infty} & b_k/k & = 0\,; \\ \forall\, k, & P(X_k = b_k) & = P(X_k = -b_k) = p_k/2, \\ & P(X_k = 0) & = 1 - p_k\,; \end{array}$$

furthermore there exist two positive constants C_1 and C_2 such that :

$$\begin{array}{lll} \forall\, n, & \inf_{k\in I(n)} b_k / \sup_{k\in I(n)} b_k & \geq C_1\,, \\ & \inf_{k\in I(n)} p_k / \sup_{k\in I(n)} p_k & \geq C_2\,; \end{array}$$

roughly speaking these last two conditions mean that the r.v. X_k, when k belongs to $I(n)$, are nearly identically distributed.

In this special case, Prohorov showed that condition (0.3) is necessary for the SLLN; by taking into account the information given by this example, he stated the following :

THEOREM 0.3 ([17] THEOREM 2). — *Let f be a function $N \to \mathbb{R}^+$, such that :*

1) $f(n)2^{-n} \operatorname{Log} n \underset{n\to+\infty}{\longrightarrow} +\infty$,

2) $f(n)2^{-n} \underset{n\to+\infty}{\longrightarrow} 0$.

Let (δ_n) be a sequence of positive numbers.

In order that every sequence of independent, centered, real-valued r.v. (X_k) such that :

3) $\forall\, k \in I(n), \quad |X_k| \leq f(n)$ *a.s.* ;

4) $\forall\, n, \quad \sum_{k\in I(n)} E(X_k^2) = \delta_n$;

satisfies the SLLN, it is necessary and sufficient that the following holds :

$$\forall\, \varepsilon > 0, \quad \sum_{n\geq 1} \exp\Big\{-\frac{\varepsilon\, 2^n}{f(n)} \operatorname{Arc\,sh}\Big(\frac{\varepsilon\, 2^n f(n)}{\delta_n}\Big)\Big\} < +\infty.$$

For scalar valued X_k our sequence (Z_k) enters the scope of Theorem 0.2; condition (0.3) ensures that $\sum_{k\in I(n)} P(Z_k \neq 0)$ is small, in a way which is both efficient and simple to check.

The purpose of this paper is to extend Theorem 0.2 to Banach space valued r.v. As a corollary of our result and of the above Theorem 0.1, we will obtain a good sufficient condition for the SLLN, for vector valued r.v. with large norms.

Prohorov's proof of Theorem 0.2 involves a nice exponential inequality, which is proved by tricky computations that don't go through to infinite dimension. It is obvious that the proof of an infinite dimensional version of Theorem 0.2 will also involve exponential inequalities; what kind of such inequalities? One knows that the classical "vectorial" exponential inequalities (Yurinskii [18], de Acosta [1]) are efficient for r.v. which are small in norm; but here -contrary to Theorem 0.1- we are interested in situations where the r.v. $\|X_k\|/k$ can be relatively large. So we need exponential inequalities from a different kind, the so-called "weak l_p" inequalities ([7], [8]). Furthermore, we need also a new exponential inequality of that type. This brings us to start this work by a short section on " weak l_p" exponential inequalities.

§1. NON-INCREASING REARRANGEMENTS OF SEQUENCES OF R.V. AND EXPONENTIAL INEQUALITIES.

Let's first recall some facts on weak l_p spaces.

For every $0 < p < +\infty$ one defines :

(1.1)

$$l_{p,\infty} = \Big\{(a_n) \in \mathbb{R}^N : \|(a_n)\|_{p,\infty} = (\sup_{t>0}(t^p \mathrm{card}(n : |a_n| > t))^{1/p} < +\infty\Big\} .$$

That space $l_{p,\infty}$ is called the weak l_p space; if $p > 1$ the functional $\| \ \|_{p,\infty}$ is equivalent to a norm and $l_{p,\infty}$ equipped with that norm is a Banach space. In the sequel we will call the quantity $\|(a_n)\|_{p,\infty}$ the "weak l_p norm" of the sequence (a_n), and this for any value of p. It is obvious that a sequence (a_n) belonging to $l_{p,\infty}$ converges to 0; so the non-increasing rearrangement (a_n^*) of $(|a_n|)$ can be defined without any problem; from that remark it is easy to check that the following holds :

$$\|(a_n)\|_{p,\infty} = \sup_n(n^{1/p}a_n^*) \tag{1.2}$$

Weak l_p spaces were involved in exponential inequalities for the first time in a theorem of Pisier, Rodin and Semyonov [15] stating that for every $p \in]1, 2[$ there exists a positive constant $k(p)$ such that for every finite sequence of real numbers $(a_1, \dots, a_n)$ and for a sequence of independent Rademacher r.v. $(\varepsilon_1, \dots, \varepsilon_n)$ one has :

$$E \ \exp\Big\{\Big(\frac{|\sum_{1\le k\le n} a_k\varepsilon_k|}{k(p)\|(a_k)\|_{p,\infty}}\Big)^{p/(p-1)}\Big\} \le 2. \tag{1.3}$$

In [7] and [8] a whole family of exponential inequalities of the same spirit as (1.3) is studied; these inequalities apply to r.v. which are much more general than weighted Rademacher r.v. . The main result in [7] and [8] is the following :

THEOREM 1.1. — *Let $p \in]1, +\infty[$ and let q be its conjugate : $(1/p) + (1/q) = 1$. Consider $X_1, \dots, X_n$ r.v. taking their values in a real* *separable Banach space* $(B, \| \ \|)$, *which are independent, symmetrically distributed and strongly square integrable. Define :*

$$\begin{aligned} S &= \textstyle\sum_{1\leq k\leq n} X_k \\ \Lambda^2 &= \textstyle\sum_{1\leq k\leq n} E\|X_k\|^2 . \end{aligned}$$

Then : $\forall\, t > 0,\ \forall\, c > 0,$

(1.4)

$$P(\|S\| > t) \leq P(\|\ \|X_k\|\ \|_{p,\infty} > c) + 2\ \exp\Big\{ \frac{-\dfrac{t^2}{8(q+1)^2} + \dfrac{t}{4(q+1)} E\|S\|}{\Lambda^2 + \dfrac{c^q(q+1)^{q-2}}{2} t^{\frac{p-2}{p-1}}} \Big\} .$$

If one takes no care of the correcting term $P(\|\ \|X_k\|\ \|_{p,\infty} > c)$, one notices that Theorem 1.1 provides a bound for $P(\|S\| > t)$ which is much sharper for large values of t than the usual exponential bounds : the bound in (1.4) behaves like $\exp(-t^q)$ when Yurinskii's bound [18] behaves like $\exp(-t)$ and de Acosta's [1] like $\exp(-t \text{ Log } t)$! For applying Theorem 1.1 to concrete situations it is of course useful to have a sharp and handy bound for $P(\|\ \|X_k\|\ \|_{p,\infty} > c)$. Such a bound is provided by a well known result of Marcus and Pisier [12] :

PROPOSITION 1.2. — *Let* (Z_n) *be a sequence of independent, positive r.v. . Then one has for every* $0 < p < +\infty$ *and all* $c > 0$ *:*

$$(1.5) \qquad c^p P(\|(Z_n)\|_{p,\infty} > c) \leq 2e \sup_{t>0}(t^p \sum_{n\geq 1} P(Z_n > t)) .$$

Proposition 1.2 will not be adapted to our needs; we will use a more precise bound -in the case $p = 2$- which applies when the r.v. Z_n are small a.s. . This result due to Andersen, Giné, Ossiander and Zinn [3] (Lemma 2.16) is as follows :

PROPOSITION 1.3. — *Let* $(Z_1, \ldots, Z_n)$ *be a sequence of independent, positive r.v. such that there exists a constant* $b > 0$, *with :*

$$\forall\, k = 1, \ldots, n \qquad Z_k \leq b \quad a.s. .$$

If one defines : $K = \sup_{t>0}(t^2 \sum_{1\leq k\leq n} P(Z_k > t))$, *then one has :*

$$(1.6)\ \forall\, c > eK, \quad P\Big(\|(Z_k)\|^2_{2,\infty} > c\Big) \leq \frac{1}{(1 - \frac{eK}{c})} \exp\Big\{ -\frac{c}{b^2} \text{Log}(\frac{c}{eK}) \Big\} .$$

The exponential bound of Theorem 1.1 will be one of the tools that we will need in Section 2. The other main tool will be a weak-l^p version of the inequality of de Acosta [1]. Let's state and prove this new exponential inequality :

THEOREM 1.4. — *Let $p \in]1,+\infty[$ be given and let q be its conjugate : $(1/p)+(1/q)=1$. Consider $(X_1,\dots,X_n)$ a sequence of r.v. taking their values in a real separable Banach space $(B,\|\ \|)$, which are independent, symmetrically distributed and strongly square integrable. One defines :*

$$\begin{aligned} S &= \textstyle\sum_{1\le k\le n} X_k \\ \Lambda^2 &= \textstyle\sum_{1\le k\le n} E\|X_k\|^2. \end{aligned}$$

Then one has :

(1.7)
$\forall\, t>0,\ \forall\, c>0,$
$P(\|S\|-E\|S\|>t) \le P(\|\ \|X_k\|\ \|_{p,\infty}>c)$
$$+2\ exp\Big(\frac{t^q}{2(c(q+1))^q}\Big\{1-(1+\frac{2\Lambda^2(q+1)^{(\frac{p-2}{p-1})}}{c^q t^{(\frac{p-2}{p-1})}})\mathrm{Log}(1+\frac{c^q t^{(\frac{p-2}{p-1})}}{2\Lambda^2(q+1)^{(\frac{p-2}{p-1})}})\Big\}\Big)$$

Proof : Let's suppose that the r.v. X_k are defined on a probability space $(\Omega,\mathcal{F},P)$; let $(\varepsilon_1,\dots,\varepsilon_n)$ be a sequence of independent Rademacher r.v., defined on another probability space $(\Omega',\mathcal{F}',P')$.

Fix $c>0$ and define the following event :

$$A=(\omega\in\Omega : \|\ \|X_k\|(\omega)\|\ \|_{p,\infty}\le c).$$

By symmetry one has immediately for all $t>0$:

(1.8)
$$\begin{aligned} P(\|S\|>t+E\|S\|) &= P\otimes P'(\|\textstyle\sum_{1\le k\le n}\varepsilon_k X_k\|>t+E\|S\|) \\ &\le P(A^c)+P\otimes P'(\|\textstyle\sum_{1\le k\le n}\varepsilon_k X_k\| I_A>t+E\|S\|)\ . \end{aligned}$$

Now, write t as : $t=(q+1)cu$.

If $u\ge n^{1/q}$, the second term in the righthandside in inequality (1.8) clearly vanishes -apply (1.2)- and inequality (1.7) is therefore obvious. From now we suppose that $u<n^{1/q}$. Of course there exists a measurable function Θ :

$$\Theta : (\Omega\times(1,2,\dots,n),\ \mathcal{F}\otimes\mathcal{P}(1,2,\dots,n)) \longrightarrow ((1,2,\dots,n),\ \mathcal{P}(1,2,\dots,n))\,,$$

such that for every ω, $(\|X_{\Theta(\omega,k)}(\omega)\|)_{k\le n}$ is a non-increasing rearrangement of the sequence $(\|X_j(\omega)\|)_{j\le n}$.

Fix an $\omega \in A$; by the definition of the weak $-l_p$ norm one gets :

$$(1.9) \qquad \forall\, k = l, \dots, n \quad \|X_{\Theta(\omega,k)}(\omega)\| \leq c\, k^{-1/p}\, .$$

By defining now :

$$a(\omega) = P'(\|\sum_{1 \leq k \leq n} \varepsilon_k X_k(\omega)\| > (q+1)c\; u + E\|S\|)\, ,$$

it immediately follows from (1.9) that :

$$(1.10) \qquad a(\omega) \leq P'\Big(\|\sum_{k \geq [u^q]+1} \varepsilon_{\theta(\omega,k)} X_{\theta(\omega,k)}(\omega)\| > cu + E\|S\|\Big)\, ,$$

where [ ] stands for the integer part of a real number. By (1.2) and Lévy's inequality one deduces the following inequality from (1.10) :

$$(1.11) \; a(\omega) \leq 2P'\Big(\|\sum_{1 \leq k \leq n} \varepsilon_k X_k(\omega) I_{(\|X_k\| \leq cu^{-1/(p-1)})}(\omega)\| > cu + E\|S\|\Big)$$

By applying now Fubini's theorem, one obtains :

$$(1.12) \qquad \begin{aligned} &P(\|S\| > (q+1)cu + E\|S\|) \leq P(A^c) \\ &\quad +2P(\|\textstyle\sum_{1 \leq k \leq n} X_k I_{(\|X_k\| \leq cu^{-1/(p-1)})}\| > cu + E\|S\|\; .) \end{aligned}$$

For bounding the righthandside of this inequality we will now use a classical exponential inequality : the "vectorial" version of Bennett's inequality, due to de Acosta [1] (Theorem 2.1). For the reader's convenience we recall the statement of de Acosta's result :

LEMMA 1.5. — *Let $\eta_1, \dots, \eta_n$ be r.v. taking their values in a real separable Banach space $(B, \|\;\|)$, which are independent; denote by η their sum. Suppose furthermore that these r.v. are bounded by a positive constant M :*

$$\forall\, k = 1, \dots, n \quad \|\eta_k\| \leq M \quad a.s.\, .$$

If one notes $b = \sum_{1 \leq k \leq n} E\|\eta_k\|^2$, one has for all $t > 0$:

$$P(\|\eta\| - E\|\eta\| > t) \leq \exp\{(t/2M)(1 - (1 + (2b)/(Mt))\mathrm{Log}(1 + (Mt)/(2b))\}\, .$$

Observe that by symmetry and independence :

$$E\|\sum_{1\le k\le n} X_k I_{(\|X_k\|\le cu^{-1/(p-1)})}\| \le E\|S\| .$$

By putting now :

$$S' = \sum_{1\le k\le n} X_k I_{(\|X_k\|\le cu^{-1/(p-1)})} ,$$

and :

$$b' = \sum_{1\le k\le n} E\|X_k I_{(\|X_k\|\le cu^{-1/(p-1)})}\|^2$$

it follows from Lemma 1.5 that :
(1.13)

$$P(\|S'\| > cu+E\|S\|) \le \exp\Big(\frac{u^q}{2}\Big\{1-(1+\frac{2u^{(\frac{2-p}{p-1})}b'}{c^2})\mathrm{Log}(1+\frac{c^2}{2u^{(\frac{2-p}{p-1})}b'})\Big\}\Big) .$$

If one denotes by v the quantity $v = 2b'/c^2u^{(p-2)/(p-1)}$, one notices that the expression involved in the exponential in the righthandside in inequality (1.13) is :

$$u^q(1-(1+v)\mathrm{Log}(1+1/v))/2 .$$

It is easy to see that this function of $v > 0$ is increasing. So by replacing b' by Λ^2 in (1.13) one obtains an inequality which is rougher than (1.13), but simpler to apply.

This simplified inequality will be labelled (1.14).

Finally one chooses $u = t/c(q+1)$ and $c =$ "c" in (1.14); this choice provides the announced exponential inequality.

Remark : If one would keep the quantity b' until the end of the proof of Theorem 1.4. and replace $E\|S\|$ by $E\|S'\|$ one would obtain a more precise bound than (1.7), needing no integrability assumption at all on the X_k, because only truncated r.v. deduced from the X_k are involved in the proof. Such a result would be of the same spirit as some inequalities of Nagaev [14].

Now we can state and prove our SLLN.

§2. A SLLN FOR R.V. TAKING VALUES WHICH ARE LARGE IN NORM.

In this Section we will first prove a SLLN similar to Theorem 0.2, the hypotheses involving absolute values of scalar r.v. being replaced by assumptions on the norms of random vectors. Afterwards we will deduce a SLLN as a consequence of this result under "mixed" hypotheses -both in terms of weak and strong moments-; in Section 3, we will show that this latter result is the best possible of its kind.

Let's begin with the analogue of Theorem 0.2 :

THEOREM 2.1. — *Let (X_k) be a sequence of r.v. taking their values in a real separable Banach space $(B, \| \ \|)$, which are independent and centered. Suppose that there exists a sequence (a_n) of positive numbers such that :*

i) $\lim_{n\to+\infty} a_n 2^{-n} \mathrm{Log}\, n = +\infty$;

ii) $\lim_{n\to+\infty} a_n 2^{-n} = 0$;

iii) $\forall\, k \in I(n), \quad \|X_k\| \leq a_n \quad a.s.$,

where : $I(n) = (2^n + 1, \ldots, 2^{n+1})$.

Finally we suppose that the following holds :

a) $S_n/n \to 0$ *in probability.*

b)

$$(2.2) \qquad \forall\, \varepsilon > 0, \quad \sum_{n\geq 1} \exp\{-(\varepsilon 2^n/a_n)\mathrm{Arc\ sh}(\varepsilon a_n/2^n\Lambda^2(n))\} < +\infty ,$$

where :

$$\Lambda^2(n) = 2^{-2n} \sum_{k\in I(n)} E\|X_k\|^2 .$$

Then :

$$S_n/n \longrightarrow 0 \quad a.s.$$

Proof : It is well known that it suffices to consider the symmetric case. So let's suppose that the r.v. X_k are symmetrically distributed. By independence and symmetry it is of course sufficient to show that the following property is true :

$$(2.3) \qquad \forall\, t > 0, \quad \sum_{n\geq 1} P(T_n > t) < +\infty ,$$

where :

$$T_n = 2^{-n} \| \sum_{k \in I(n)} X_k \| \; . \tag{2.4}$$

Let $t > 0$ and n be fixed; we will bound $P(T_n > t)$ in an efficient way be separately considering two situations : $a_n / 2^n \Lambda^2(n) \leq 1$ and $a_n / 2^n \Lambda^2(n) > 1$.

First case :

$$a_n / 2^n \Lambda^2(n) \leq 1 \; . \tag{2.5}$$

Notice first that the (X_k) satisfying the weak law of large numbers, being symmetrically distributed and obeying the boundedness assumption (2.1), a classical result [9] gives :

$$\lim_{n \to +\infty} E(T_n) = 0 \; .$$

For bounding $P(T_n > t)$ we will use Theorem 1.1. Define :

$$g_n = \| (2^{-n} \| X_k \|)_{k \in I(n)} \|_{2,\infty} \; ,$$

and choose in Theorem 1.1, $p = 2$ and $c = (2e\Lambda^2(n))^{1/2}$. For n large enough, (1.4) gives :

$$P(T_n > t) \leq P(g_n > (2e\Lambda^2(n))^{1/2}) + 2 \exp(-t^2 / 144(1+e)\Lambda^2(n)) \; .$$

As

$$K = \sup_{s > 0} (s^2 \sum_{k \in I(n)} P(2^{-n} \| X_k \| > s)) \leq \Lambda^2(n) \; ,$$

one obtains by an easy application of Proposition 1.3 :

$$P(g_n^2 > 2c\Lambda^2(n)) \leq 2 \; \exp(-e\Lambda^2(n) 2^{2n+1} \mathrm{Log} \; 2/a_n^2) \; .$$

From (2.5) follows that :

$$P(g_n^2 > 2e\Lambda^2(n)) \leq 2 \; \exp(-e 2^{n+1} \mathrm{Log} \; \; 2/a_n) \; ;$$

so finally, for $\varepsilon > 0$ small enough :

$$P(g_n^2 > 2e\Lambda^2(n)) \leq 2 \; \exp(-(\varepsilon 2^n / a_n) \mathrm{Arc \; sh}(\varepsilon a_n / 2^n \Lambda^2(n))) \; . \tag{2.6}$$

Now one notices that if n satisfies (2.5) one has for every $\delta > 0$ small enough :

$$\exp(-2\delta^2/\Lambda^2(n)) \le \exp(-(\delta 2^n/a_n)\text{Arc sh}(\delta a_n/2^n\Lambda^2(n))) \ . \tag{2.7}$$

Putting (2.6) and (2.7) together, one finally sees that there exists an integer $n(t)$ and $\varepsilon(t) > 0$ such that :

(2.8)
$$\forall\, n \ge n(t), \quad P(T_n > t) \le 4\exp(-(\varepsilon(t)2^n/a_n)\text{Arc sh}(\varepsilon(t)a_n/2^n\Lambda^2(n))) \ .$$

Second case :

$$a_n/2^n\Lambda^2(n) > 1. \tag{2.9}$$

For bounding $P(T_n > t)$, this time we will use Theorem 1.4 with $p = 2$ and $c = 4(a_n 2^{-n})^{1/2}$.

As $\lim_{n\to+\infty} E(T_n) = 0$, one gets, by applying Theorem 1.4 and Proposition 1.3, that for every n large enough :

(2.10)

$$\begin{aligned} P(T_n > t) &\le P(T_n - E(T_n) > t/2) \\ &\le (1/(1-e/16))\exp(-(2^{n+4}/a_n)\text{Log}(16a_n/2^n\Lambda^2(n))) \\ &\quad +2\ \exp\{(2^n t^2/1152a_n)(1-(1+2^n\Lambda^2(n)/8a_n)\text{Log}(1+8a_n/2^n\Lambda^2(n)))\} \end{aligned}$$

From (2.10) follows that there exists $C_1 > 0$ such that :

$$\begin{aligned} P(T_n) &\le (4/3)\exp\{-(2^{n+4}/a_n)\text{Log}(16a_n/2^n\Lambda^2(n))\} \\ &\quad +2\ \exp\{-(C_1 2^n t^2/a_n)\text{Log}(1+8a_n/2^n\Lambda^2(n))\} \ . \end{aligned} \tag{2.11}$$

So there exists $C_2 > 0$ such that for n large enough, and t small enough :

$$P(T_n > t) \le 4\ \exp\{-(C_2 2^n t^2/a_n)\text{Arc sh}(C_2 t^2 a_n/2^n\Lambda^2(n))\} \ . \tag{2.12}$$

Finally, by putting (2.8) and (2.12) together one sees that :

$$\forall\, t > 0, \quad \sum_{n\ge 1} P(T_n > t) < +\infty \ .$$

and this concludes the proof of Theorem 2.1.

In the recent papers on the SLLN in infinite dimension -and also in the papers on the central limit theorem and the law of the iterated logarithm- it appears that the natural analogues of the variances of the scalar case aren't the strong second order moments of the r.v., but their weak variances. So it is natural to look if it is possible to improve Theorem 2.1 by introducing weak integrability assumptions instead of strong ones. The best possible statement which can be obtained is the following :

THEOREM 2.2. — *Let (X_k) be a sequence of r.v. taking their values in a real separable Banach space $(B, \| \ \|)$, which are centered and independent. Suppose that the following assumptions hold :*

1) *There exists a sequence (a_n) of positive numbers such that :*

i) $\lim_{n\to+\infty} a_n 2^{-n} \text{Log } n = +\infty$;

ii) $\lim_{n\to+\infty} a_n 2^{-n} = 0$;

iii)

$$\forall\, n, \forall\, k \in I(n), \|X_k\| \leq a_n \quad a.s.$$

where : $I(n) = (2^n + 1, \ldots, 2^{n+1})$.

2) *$S_n/n \to 0$ in probability.*

3)

$$\forall\, \varepsilon > 0, \sum_{n\geq 1} \exp(-\varepsilon/\sigma_n^2) < +\infty \tag{2.13}$$

where :

$$\forall\, n, \quad \sigma_n^2 = 2^{-2n} \sup\Big(\sum_{k\in I(n)} Ef^2(X_k), \|f\|_{B'} \leq 1 \Big).$$

4) *There exists $M > 0$ such that, if one defines :*

$$\forall\, n, \forall\, k \in I(n), \quad Y_k = X_k I_{(\|X_k\| > M\ 2^n/Ln)}$$

-where $L\, x = \text{Log } \sup(x, e)$- and

$$\Delta^2(n) = 2^{-2n} \sum_{k\in I(n)} E\|Y_k\|^2 ,$$

then one has :

$$\forall\, \varepsilon > 0, \sum_{n\geq 1} \exp(-(\varepsilon 2^n/a_n)\text{Arc sh}(\varepsilon a_n/2^n \Delta^2(n))) < +\infty . \tag{2.14}$$

Under these assumptions (X_k) *satisfies the SLLN.*

Proof : By hypothesis (2) the sequence (X_k) satisfies the weak law of large numbers; so if (X'_k) denotes an independent copy of (X_k), it is well known that (X_k) satisfies the SLLN if and only if $(X_k - X'_k)$ does [9]. It is clear that the sequence $(X_k - X'_k)$ fulfills hypotheses (1), (2) and (3) above (if one replaces a_n by $2a_n$). What about hypothesis (4)?

For checking that this hypothesis also is fulfilled -of course for $M' = 2M$- we define for every $k \in I(n)$:

$$Y'_k = (X_k - X'_k) I_{(\|X_k - X'_k\| > M2^{n+1}/Ln)}.$$

One notices immediately that :
$\forall\, t \geq M2^{n+1}/Ln$,

$$\begin{aligned} P(\|Y'_k\|) > t) &\leq 2P(\|Y'_k\| > t\ ;\ \|X_k\| \leq M2^n/Ln\ ; \|X'_k\| > M2^n/Ln) \\ &\quad + P(\|Y'_k\| > t\ ; \|X_k\| > M2^n/Ln\ ; \|X'_k\| > M2^n/Ln) \\ &\leq 2P(\|Y_k\| > t/2) + 2P(\|Y_k\| > t/2) = 4P(\|Y_k\| > t/2). \end{aligned}$$

Therefore :

$$2^{-2n} \sum_{k \in I(n)} E\|Y'_k\|^2 \leq 16\Delta^2(n). \tag{2.15}$$

and the sequence (Y'_k) satisfies hypothesis (4), the constant M being replaced by $2M$. So the proof of Theorem 2.2 reduces to show that the SLLN holds for a sequence of independent r.v. (X_k), which are symmetrically distributed and which fulfill hypotheses (1) to (4).

Let such a sequence (X_k) be given; define for every n and every $k \in I(n)$:

$$\begin{aligned} Z_k &= X_k I_{(\|X_k\| \leq M2^n/Ln)}\ , \\ Y_k &= X_k - Z_k\ . \end{aligned}$$

The symmetry of the X_k and the assumptions (1) iii) and (2) imply :

$$\lim_{n \to +\infty} E\| \sum_{1 \leq k \leq n} X_k\|/n = 0\ ,$$

and also :

$$\lim_{n \to +\infty} E\| \sum_{1 \leq k \leq n} Z_k\|/n = 0\ .$$

So it appears that the sequence (Z_k) fulfills the hypotheses of Theorem 0.1; therefore (Z_k) satisfies the SLLN.

One also has :

$$\lim_{n\to+\infty} E\|\sum_{1\le k\le n} Y_k\|/n = 0 ;$$

so sequence (Y_k) fulfills the assumptions of Theorem 2.1 and therefore also satisfies the SLLN.

This observation concludes the proof of Theorem 2.2.

The statement of Theorem 2.2 naturally raises several remarks :

1. — Let's first make some comments on hypothesis (3). By analogy with Theorem 0.2 one would like to have the following hypothesis instead of hypothesis (3) :

$$(2.16) \qquad \forall\, \varepsilon > 0, \quad \sum_{n\ge 1} \exp\{-(\varepsilon 2^n/a_n)\mathrm{Arc\ sh}(\varepsilon a_n/2^n\sigma_n^2)\} < +\infty .$$

It is obvious that (2.16) implies condition (2.13) in hypothesis (3), because for all $x > 0$, Arc sh $x < x$. So hypothesis (3) is the good one. The reader interested in minimal hypotheses would also like to replace X_k by $X_k I_{(\|X_k\|\le M2^n/Ln)}$ in the definition of σ_n^2. Let's show that this idea doesn't bring more generality.

Let n be a fixed integer; for every $k \in I(n)$ one denotes :

$$\begin{aligned} Z_k &= X_k I_{(\|X_k\|\le M2^n/ln)} , \\ Y_k &= X_k - Z_k . \end{aligned}$$

Define :

$$\sigma_n^2(Z) = \sup(2^{-2n} \sum_{k\in I(n)} Ef^2(Z_k) ; \|f\|_{B'} \le 1) ,$$

and similarly $\sigma_n^2(Y)$.

The follwing inequalities are obvious :

$$\sigma_n^2 \le 2\sigma_n^2(Z) + 2\sigma_n^2(Y) \le 2\sigma_n^2(Z) + 2\Delta^2(n) ,$$

and :

$$(2.17) \qquad 1/\sigma_n^2 \ge 1/(4\sup(\sigma_n^2(Z), \Delta^2(n))) .$$

So if one requires hypothesis (3) to hold only for $\sigma_n^2(Z)$, then by applying (2.17) and hypothesis (4) one obtains assumption (3) for σ_n^2

(the easy computations being again based on the elementary inequality $\forall\ x > 0$, Arc sh $x < x$). So hypothesis (3) can definitely be accepted as the best possible of its kind.

2). — Now we will try to shed some light on the nature of hypothesis (4) and of condition (0.2) mentioned in the introduction. Let (X_k) be a sequence of B-valued r.v., which are independent, symmetrically distributed, and which fulfill the classical necessary conditions for the SLLN :

a) $S_n/n \to 0$ in probability;

b) $X_k/k \to 0$ a.s.

As noticed in the introduction, this latter property allows to construct a sequence (α_k) of positive numbers, converging to 0, and such that :

$$\sum_{k\geq 1} P(\|X_k\| > k\alpha_k) < +\infty \ ;$$

by defining then :

$$\eta_k = X_k I_{(\|X_k\|\leq k\alpha_k)} \ ,$$

is is obvious that (X_k) satisfies the SLLN if and only if (η_k) does. This will be the case if for instance (η_k) fulfills the hypotheses of Theorem 2.2. With the notations of Theorem 2.2 applied to the sequence (η_k) one has obviously :

$$\exp\{-(\varepsilon 2^n/a_n)\text{Arc sh}(\varepsilon a_n/2^n\Delta^2(n))\} \leq (2^n\Delta^2(n)/\varepsilon a_n)^{\varepsilon 2^n/a_n} \ ;$$

the quantity at the righthandside in this inequality very much recalls the one involved in (0.2).

So the Theorems 0.2 and 2.2 make the condition (0.2) less mysterious.

3). — In [11] (Theorem 3.1) it is shown that if a sequence (X_k) satisfies (0,2)' and (2.13) and is such that (X_k/k) converges a.s. to 0, then the SLLN holds for (X_k) if and only if the weak law holds. So the following question arises very naturally from the statement of Theorem 2.2 : " Is it possible to weaken hypothesis (4) of Theorem 2.2, in replacing $\Delta^2(n)$ by :

$$\gamma_n^2(Y) = 2^{-2n}\sup(\sum_{k\in I(n)} Ef^2(Y_k), \quad \|f\|_{B'} \leq 1)?''$$

The example we are going to study now will bring a negative answer to this question.

§3. AN EXAMPLE.

Let $(\Omega, \mathcal{F}, P)$ and $(\Omega', \mathcal{F}', P')$ be two probability spaces. For every integer n, $(\varepsilon_1, \dots, \varepsilon_{2^n})$ will denote a sequence of independent Rademacher r.v. defined on $(\Omega, \mathcal{F}, P)$ and $(\eta_1, \dots, \eta_{2^n})$ will be a sequence of i.i.d. r.v. defined on $(\Omega', \mathcal{F}', P')$, their common law being the following one :

$$\begin{aligned} P'(\eta_1 = 1) &= 2^{-n} ; \\ P'(\eta_1 = 0) &= 1 - 2^{-n} . \end{aligned}$$

We will construct a sequence $(Y_1, \dots, Y_{2^n})$ of ℓ_2-valued r.v.; the Banach space ℓ_2 will be equipped with its usual norm, and (e_j) will denote its canonical basis.

Let's begin by dividing the set of integers $(1, \dots, 2^n)$ into adjacent intervals, all of which having $[(L_2 n)^4]$ elements, except perhaps the last one; let $B_1, \dots, B_s$ be these "intervals". For every $i = 1, \dots, s$ and all $k \in B_i$ one defined :

$$Y_k = (2^n / L_2 n) \varepsilon_k \eta_k e_i .$$

Now we will write down some properties of this sequence $(Y_k)_{k \le 2^n}$, properties which will be needed later.

Property 1. $\forall\, k = 1, \dots, 2^n$, $\|Y_k\|_2 \le 2^n / L_2 n$.
If one defines : $a_n = 2^n / L_2 n$, one has obviously :

$$a_n 2^{-n} \mathrm{Log}\, n \underset{n \to +\infty}{\longrightarrow} +\infty ; \quad a_n 2^{-n} \underset{n \to +\infty}{\longrightarrow} 0 .$$

Property 2.

$$\Lambda^2(n) = 2^{-2n} \sum_{1 \le k \le 2^n} E\|Y_k\|^2 = (L_2 n)^{-2} \underset{n \to +\infty}{\longrightarrow} 0 .$$

From this Property 2 the following property follows immediately :

Property 3. There exists no positive number ε for which

$$\exp\{-(\varepsilon 2^n / a_n) \mathrm{Arc\ sh}(\varepsilon a_n / 2^n \Lambda^2(n))\}$$

is the general term of a convergent series.

Property 4. If one defines :

$$\sigma_n^2 = 2^{-2n} \sup\Big(\sum_{1 \le k \le 2^n} E f^2(Y_k) ; \ \|f\|_2 \le 1 \Big) ,$$

one observes that : $\sigma_n^2 \le 2^{-n} (L_2 n)^2$.

So it is obvious that for every positive ε, $\exp\{-(\varepsilon 2^n/a_n)\text{Arc sh}(\varepsilon a_n/2^n\sigma_n^2)\}$ is the general term of a convergent series.

Property 5. The series whose general term is

$$u_n = P \otimes P'(2^{-n}\| \sum_{1\le k\le 2^n} Y_k\| > 1)$$

diverges.

Let k be a fixed integer, $k \le s$, and let $A(k)$ be the following element of $\mathcal{F}'$:

$A(k) = \{\omega' : \exists\ j_1,\dots,j_k \in (1,\dots,s)$ such that for every $r = 1,\dots,k$ there exists one -and only one- element $v_r \in B_{j_r}$ with $\eta_{v_r}(\omega') = 1$, and for

$$j \notin (j_1,\dots,j_k),\ \text{for all } v \in B_j,\ \text{ one has : } \eta_v(\omega') = 0\}\ .$$

For making the notation more convenient we put : $a = [(L_2 n)^4]$.
First we will compute the probability of $A(k)$:

$$\begin{aligned} P'(A(k)) &\ge \tbinom{s}{k}(1-2^{-n})^{a(s-k)}2^{-nk}(1-2^{-n})^{(a-1)k}a^{k-1} \\ &\ge s(s-1)\dots(s-k+1)(1-2^{-n})^{as-k}2^{-nk}a^{k-1}/k! \end{aligned}$$

From the obvious inequalities : $2^n/a \le s \le (2^n/a)+1$, it follows :

$$P'(A(k)) \ge (1-1/s)\dots(1-(k-1)/s)(1-2^{-n})^{2^n+a-k}/k!a$$

By choosing now $k = a$, one gets :

$$P'(A(a)) \ge C_1 \exp(-2^{-n+1}a^3)/(a+1)!\ ;$$

by applying Stirling's formula it follows that when n is large enough :

$$P'(A(a)) \ge C_2(L_2 n)^{-5(L_2 n)^4}\ . \tag{3.1}$$

Let now ω' be a fixed element of $A(a)$ and let $j_1(\omega'),\dots,j_a(\omega')$ be the indexes of the intervals B_i in which there exists one $r(\omega')$ -and only one- such that : $\eta_{r(\omega')} = 1$. Denote by $r_1(\omega'),\dots,r_a(\omega')$ these remarkable indexes.

For every $\omega \in \Omega$ one defines now a special element $\xi(\omega,\omega')$ of the closed unit ball of l_2 :

$$\xi(\omega,\omega') = (\xi_j(\omega,\omega'))\ ,$$

with

$$\forall \quad j \neq j_1(\omega'), \ldots, j_a(\omega') \qquad \xi_j(\omega, \omega') = 0 \ ,$$
$$\forall\, j = j_i(\omega') \qquad \xi_j(\omega, \omega') = a^{-1/2} \varepsilon_{r_i(\omega')}(\omega) \ .$$

Then :

$$(3.2) \qquad 2^{-n} \| \sum_{1 \leq j \leq 2^n} Y_j(\omega, \omega') \|_2 \geq (1/L_2 n) \sum_{1 \leq j \leq a} (L_2 n)^{-2} > L_2 n/2 \ .$$

From the relations (3.1) and (3.2) one deduces immediately :

$$(3.3) \qquad \forall\, n, \quad u_n \geq P'(A(a)) \geq C_2 (L_2 n)^{-5(L_2 n)^4} \ .$$

Property 5 then easily follows from (3.3).

Let now $(X_k)_{k \geq 2}$ be a sequence of l_2-valued r.v., which are independent, and such that for every integer n and for all $k \in I(n)$, X_k has the same law as Y_{k-2^n}. The space l_2 being of type 2, this sequence (X_k) satisfies the weak law of large numbers, by the above Property 2.

By Properties 1 and 4 -and again by the elementary inequality $\forall\, x > 0$, Arc sh $x \leq x$- the sequence (X_k) fulfills the hypotheses 1 to 3 of Theorem 2.2. But -by Property 3- hypothesis (4) of Theorem 2.2 doesn't hold.

From Property 5 it follows immediately that (X_k) doesn't satisfy the SLLN. But -by Property 4- the "weakened hypothesis (4)" holds.

From this example one sees that it is impossible to improve the statement of Theorem 2.2 by replacing in its hypothesis (4) the strong second order moments by weak variances.

REFERENCES

[1] De Acosta, A. : *Strong exponential integrability of sums of independent B-valued random vectors.* Prob. and Math. Stat. 1 (1980), 133-150.

[2] Alt, J.C. : *La loi des grands nombres de Prohorov dans les espaces de type p.* Ann. Inst. Henri Poincaré 23 (1987), 561-574.

[3] Andersen, N.T., Giné, E., Ossiander, M., Zinn, J. : *The central limit theorem and the law of the iterated logarithm for empirical processes under local conditions.* Probab. Th. Rel. Fields 77 (1988), 271-305.

[4] Heinkel, B. : *Une extension de la loi des grands nombres de Prohorov.* Z. Wahrscheinlichkeitstheorie verw. Gebiete 67 (1984), 349-362.

[5] Heinkel, B. : *The non i.i.d. strong law of large numbers in 2-uniformly smooth Banach spaces.* Probability theory on vector spaces 3-Lublin 1983-. Lecture Notes in Math 1080, Springer (1984), 90-118.

[6] Heinkel, B. *An application of a martingale inequality of Dubins and Freedman to the law of large numbers in Banach spaces.* Actes des Journées SMF de Calcul des Probabilités dans les espaces de Banach, Strasbourg 1985, Lecture Notes in Math. 1193, Springer 1986, 29-43.

[7] Heinkel, B. : *Some exponential inequalities with applications to the central limit theorem in* $C[0,1]$. To appear in the Proceedings of the Conference "Probability in Banach spaces 6" Sandbjerg, Denmark 1986.

[8] Heinkel, B. : *Rearrangements of sequences of random variables and exponential inequalities* (1987). To appear in Probability and Mathematical Statistics.

[9] Kuelbs, J. and Zinn, J. : *Some stability results for vector valued random variables.* Ann. Probability 7 (1979), 75-84.

[10] Ledoux, M. and Talagrand, M. : *Comparison theorems, random geometry and some limit theorems for empirical processes* (1986). To appear in Ann. Probability.

[11] Ledoux, M. and Talagrand, M. : *Some applications of isoperimetric methods to strong limit theorems* (1987). Preprint

[12] Marcus, M.B. and Pisier, G. : *Characterizations of almost surely continuous p-stable random Fourier series and strongly stationary processes.*

Acta Math. 152 (1984), 245-301.

[13] NAGAEV, S.V. : *On necessary and sufficient conditions for the strong law of large numbers.* Theor. Prob. Appl. 17 (1972), 573-581.

[14] NAGAEV, S.V. : *Large deviations of sums of independent random variables.* Ann. Probability 7 (1979), 745-789.

[15] PISIER, G. : *De nouvelles caractérisations des ensembles de Sidon.* Mathematical Analysis and Applications. Adv. in Math. Suppl. Stud. 7B (1981) 686-725.

[16] PROHOROV, YU. V. : *Strong stability of sums and infinitely divisible distributions.* Theor. Prob. Appl. 3 (1958), 141-152.

[17] PROHOROV, YU. V. *Some remarks on the strong law of large numbers.* Theor. Prob. Appl. 4 (1959), 204-208.

[18] YURINSKII, V.V. : *Exponential bounds for large deviations.* Theor. Prob. Appl. 19 (1974), 154-155.

Uniform Convergence of Martingales

J. HOFFMANN-JØRGENSEN

Introduction

The uniform law of large numbers, or equivalently the law of large numbers in Banach spaces, has been studied intensively in the past two decades, e.g. see [2] and [4]. Suppose that $\xi_1, \xi_2, \ldots$ are independent identically distributed random variables taking values in a measurable space $(S, \mathcal{S})$, and let $f: S \times T \to \mathbb{R}$ be a given function, where T is a given set. Suppose that $m(t) = Ef(\xi_1, t)$ exists for all $t \in T$, then in [2] and [4] you may find a series of necessary and sufficient conditions for the following form of the uniform law of large numbers:

$$\frac{1}{n}\sum_{j=1}^{n} f(\xi_j, t) \to m(t) \quad \text{uniformly on } T \text{ a.s.}$$

Now for each $t \in T$, we know that the sequence

$$X_n(t) = \frac{1}{n}\sum_{j=1}^{n} f(\xi_j, t) \qquad (n = 1, 2, \ldots)$$

is a reversed martingale. So it is natural to ask for necessary conditions for a collection of martingales or reversed martingales to converge uniformly. And in Theorem 2.5 below you will find such conditions for uniform convergence of martingales. The condition is surprisingly weak compared to the necessary and sufficient conditions for the uniform law of large numbers.

However the arguments below only work for ordinary (forwards) martingales, and I have not been able to find any sensible necessary condition for uniform convergence of reversed martingales. Usually reversed martingales behave much more regularly than ordinary martingales (the optimal sampling theorem and the martingale convergence theorem holds for all reversed martingales with no extra conditions added). But it seems that we we here have a case, where ordinary martingales behave more regularly than reversed martingales. The main point in the proof of Theorem 2.4 and Theorem 2.5 is Lemma V-2-9 in Neveu's excellent and inspiring book [3], and the reader easily checks that this lemma fails for reversed martingales.

2. The Martingale Convergence Theorem

Throughout this paper we let $(\Omega, \mathcal{F}, P)$ denote a complete probability space, and $\mathcal{F}_1 \subseteq \mathcal{F}_2 \subseteq \cdots \subseteq \mathcal{F}$ denotes a given increasing sequence of sub-σ-algebras of $\mathcal{F}$.

Let T be a set, then $\mathbf{R}^T$ resp. $B(T)$ denotes the set of all *real valued* resp. *bounded real valued* functions on T. If $S \subseteq T$ and $f \in \mathbf{R}^T$, then we put

$$M_S(f) = \sup_{s \in S} f(s), \qquad \|f\|_S = \sup_{s \in S} |f(s)|.$$

Note that $\|f-g\|_T$ defines a metric on $\mathbf{R}^T$ (if you are worried about infinite metrics, you should just replace $\|f-g\|_T$ by Arc $tg\|f-g\|_T$, which is an equivalent bounded metric), and we let $(\mathbf{R}^T, \|\cdot\|_T)$ denote this *metric space,* and $\mathcal{B}$ denotes *the Borel σ-algebra* on $(\mathbf{R}^T, \|\cdot\|_T)$.

If T is a topological space, we let $\mathcal{C}(T)$ resp. $C(T)$ denote the set of all *continuous* resp. *bounded continuous* functions from T into $\mathbf{R}$. And $\mathcal{U}sc(T)$ resp. $\mathcal{L}sc(T)$ denotes the set of all *upper* resp. *lower semicontinuous* functions from T into $\mathbf{R}$.

Let T be a topological space and let $f \in \mathbf{R}^T$. If $t \in T$ and U is a neighbourhood of t we define *the lower, upper* and *absolute oscillation of f at t over U* by:

$$W_U^+(t,f) = \sup_{s \in U}(f(t) - f(s)), \qquad W_U^-(t,f) = \sup_{s \in U}(f(s) - f(t))$$

$$W_U(t,f) = \sup_{s \in U} |f(t) - f(s)| = W_U^+(t,f) \vee W_U^-(t,f)$$

And we define *the lower, upper* and *absolute jump* of f at t by:

$$\partial^+(t,f) = \inf_{U \in \mathcal{N}_t} W_U^+(t,f), \qquad \partial^-(t,f) = \inf_{U \in \mathcal{N}_t} W_U^-(t,f)$$

$$\partial(t,f) = \inf_{U \in \mathcal{N}_t} W_U(t,f) = \partial^+(t,f) \vee \partial^-(t,f)$$

where $\mathcal{N}_t$ is the set of all neighbourhoods of t. Clearly we have

(2.1) $\qquad f$ is lower semicontinuous at $t \iff \partial^+(t,f) = 0$

$$(2.2) \qquad f \text{ is upper semicontinuous at } t \iff \partial^-(t,f) = 0$$

$$(2.3) \qquad f \text{ is continuous at } t \iff \partial(t,f) = 0$$

If ρ is a pseudo-metric on T, we let $\mathcal{C}_u(T,\rho)$ resp. $C_u(T,\rho)$ denote the set of all *uniformly ρ-continuous* resp. *bounded uniformly ρ-continuous* functions from T into $\mathbb{R}$.

Let $X = \{X(t,\omega) \mid t \in T, \omega \in \Omega\}$ be a real valued stochastic process, then we shall consider X as a map from Ω into $\mathbb{R}^T$ by $X(\omega) = X(\cdot,\omega)$. And we say that X is *Bochner measurable,* if there exists a separable subset $\mathcal{S}$ of $(\mathbb{R}^T, \|\cdot\|_T)$, such that $X(\omega) \in \mathcal{S}$ for a.a. $\omega \in \Omega$. Lemma 2.1 below shows, that this definition coincides with the usual definition of Bochner measurability, see e.g. [1, p. 42–44].

Recall that if $\{X_n \mid n \geq 1\}$ is a sequence of integrable random variables, then we say that $\{X_n, \mathcal{F}_n \mid n \geq 1\}$ is a *submartingale, supermartingale* or *martingale,* if X_n is integrable and $\mathcal{F}_n$-measurable for all $n \geq 1$, and $X_n \leq E(X_{n+1} \mid \mathcal{F}_n)$, $X_n \geq E(X_{n+1} \mid \mathcal{F}_n)$ or $X_n = E(X_{n+1} \mid \mathcal{F}_n)$ resp.

Let $\{Z_n\}$ be a sequence of real valued *random elements,* i.e., Z_n is an arbitrary function from Ω into $\mathbb{R}$. Then we write $Z_n \to 0$ a.s. if there exists a P-null set such that $Z_n(\omega) \to 0$ for all $\omega \in \Omega \setminus N$, and we write $Z_n \to 0$ a.s.*, if there exists random variables Z_n^*, such that $Z_n^* \to 0$ a.s. and $|Z_n| \leq Z_n^* \; \forall n \geq 1$. Similarly we write:

$$Z_n \to 0 \quad \text{in} \quad pr_* \iff \lim_{n\to\infty} P_*(|Z_n| > \varepsilon) = 0 \qquad \forall \varepsilon > 0$$

$$Z_n \to 0 \quad \text{in} \quad pr^* \iff \lim_{n\to\infty} P^*(|Z_n| > \epsilon) = 0 \qquad \forall \varepsilon > 0$$

where P_* and P^* denote the inner and outer P-measures. Clearly we have

$$(2.4) \qquad Z_n \to 0 \quad \text{a.s.} \Rightarrow Z_n \to 0 \quad \text{in } pr_*$$

$$(2.5) \qquad Z_n \to 0 \quad \text{a.s.}^* \Rightarrow Z_n \to 0 \quad \text{in } pr^*$$

$$(2.6) \qquad Z_n \to 0 \quad \text{a.s.}^* \Rightarrow Z_n \to 0 \quad \text{a.s.}$$

$$(2.7) \qquad Z_n \to 0 \quad \text{in } pr^* \Rightarrow Z_n \to 0 \quad \text{in } pr_*$$

But no other implication holds in general.

Recall that a topological space T is *separable,* if there exists a countable dense subset. And T is *hereditarily separable,* if every subspace of T is separable in its induced topology. It is well known that we have

(2.8) A continuous image of a separable (hereditarily separable space) is is separable (hereditarily separable)

(2.9) Every separable pseudo metric space is hereditarily separable

(2.10) The real line with its left (or right) Sorgenfrey topology (i.e. $]x-\varepsilon, x]$ resp. $[x, x+\varepsilon[$ are neighbourhoods of x) is hereditarily separable.

However $\mathbb{R}^2$ with the product left Sorgenfrey topology is separable, but not hereditarily separable.

Lemma 2.1. *Let $\mathcal{S}$ be a $\|\cdot\|_T$-separable subset of $\mathbb{R}^T$, and let $X = \{X(t) \,|\, t \in T\}$ be a stochastic process, such that $X(\cdot,\omega) \in \mathcal{S}$ for a.e. ω. Then there exists a separable pseudo-metric δ on T, such that $\mathcal{S} \subseteq \mathcal{C}_u(T,\delta)$, and $\omega \curvearrowright X(\omega)$ is a measurable map from $(\Omega, \mathcal{F})$ into $(\mathbb{R}^T, \mathcal{B})$, such that $X(\omega) \in \mathcal{C}_u(T,\delta)$ for a.a. ω.*

Moreover if τ is a hereditarily separable topology on T, then there exist a topology η on T satisfying

(2.1.1) $\quad \eta$ *is finer than* τ

(2.1.2) $\quad \eta$ *is finer than the δ-topology*

(2.1.3) $\quad (T,\eta)$ *is hereditarily separable*

In particular we have that $S \subseteq \mathcal{C}(T,\eta)$ and so $X(\omega) \in \mathcal{C}(T,\eta)$ for a.e. ω.

Proof. Let $\{\varphi_k\}$ be a countable $\|\cdot\|_T$-dense subset of $\mathcal{S}$, and put

$$\delta(s,t) = \sum_{k=1}^{\infty} 2^{-k} \text{Arc } tg|\varphi_k(s) - \varphi_k(t)|$$

Then δ is a separable pseudo-metric on T, such that $\varphi_k \in \mathcal{C}_u(T,\delta)$, since

$$|\varphi_k(s) - \varphi_k(t)| \le tg(2^k \delta(s,t))$$

for all $s,t \in T$ with $\delta(s,t) < \pi 2^{-k-1}$. Hence $\mathcal{S} \subseteq \mathcal{C}_u(T,\delta)$ since $\mathcal{C}_u(T,\delta)$ is a $\|\cdot\|_T$-closed subset of $\mathbb{R}^T$. And so $X(\omega) \in \mathcal{C}_u(T,\delta)$ for all $\omega \in \Omega_0$, for some $\Omega_0 \in \mathcal{F}$ with $P(\Omega_0) = 1$. Now let D be a countable δ-dense subset of T. Then we have

$$\|X(\omega) - f\|_T = \|X(\omega) - f\|_D \quad \forall \omega \in \Omega_0 \qquad \forall f \in \mathcal{S}$$

and so $\omega \curvearrowright \|X(\omega) - f\|_T$ is measurable for all $f \in \mathcal{S}$, and since $X(\omega) \in \mathcal{S}$ for a.a. ω, then by $\|\cdot\|_T$-separability of $\mathcal{S}$, it follows easily that $\omega \curvearrowright X(\omega)$ is measurable from $(\Omega, \mathcal{F})$ into $(\mathbb{R}^T, \mathcal{B})$.

Now let τ be a hereditarily separable topology on T and let $\mathcal{G}$ be a countable base for the δ-topology, which is stable under formation of finite unions and intersections. Then $\mathcal{V} = \{(U \cap G \,|\, G \in \mathcal{G}, U$ is τ-open$\}$ is a base for a topological η, which is finer than both τ and δ. Now let T_0 be a subset of T, then for every $G \in \mathcal{G}$, there exists a countable set $D_G \subseteq T_0 \cap G$, such that D_G is τ-dense in $T_0 \cap G$. Now put

$$D = \bigcup_{G \in \mathcal{G}} D_G$$

Then D is countable, since $\mathcal{G}$ and D_G are so, and $D \subseteq T_0$. Now let $V \in \mathcal{V}$ such that $V \cap T_0 \neq \emptyset$. Then $V = U \cap G$, for some $G \in \mathcal{G}$ and some τ-open set U. Since $U \cap G \cap T_0$ is a non-empty relatively open subset of $G \cap T_0$, we have that $D_G \cap U \cap G \cap T_0 \neq \emptyset$, since D_G is τ-dense in $T_0 \cap G$. Hence $D \cap V \cap T_0 \neq \emptyset$, and so D is a countable η-dense subset of T_0. Thus η is hereditarily separable and the lemma is proved. QED

Lemma 2.2. *Let $\{X_n(t), \mathcal{F}_n \mid n \geq 1\}$ be a submartingale for all $t \in T$, and let $\{X(t) \mid t \in T\}$ be a Bochner measurable stochastic process satisfying*

$$\text{(2.2.1)} \qquad \sup_n EM_S(X_n^+) < \infty \qquad \forall S \text{ countable } \subseteq T$$

$$\text{(2.2.2)} \qquad M_S(X^-) < \infty \text{ a.s.} \qquad \forall S \text{ countable } \subseteq T$$

$$\text{(2.2.3)} \qquad X_n(t) \to X(t) \text{ a.s.} \qquad \forall t \in T$$

Let $\Phi: \mathbf{R}^T \to [0, \infty]$ be a function, let D be a countable set and let $\mathcal{C} \subseteq \mathbf{R}^T$, such that

$$\text{(2.2.4)} \qquad \|(f-g)^+\|_T \leq \Phi(f) + \|(f-g)^+\|_D \qquad \forall f \in \mathbf{R}^T \quad \forall g \in \mathcal{C}$$

Then there exists a sequence $\{V_n\}$ of random variables, such that $V_n \to 0$, a.s. and

$$\text{(2.2.5)} \qquad \|(X_n(\omega) - X(\omega))^+\|_T \leq \Phi(X_n(\omega)) + V_n(\omega) \qquad \forall \omega \in \Omega_0$$

where $\Omega_0 = \{\omega \mid X(\omega) \in \mathcal{C}\}$.

Proof. By Lemma 2.1 there exists a separable pseudo-metric ρ on T and a P-nullset N_0, such that $X(\omega) \in \mathcal{C}_u(T, \rho)$ for all $\omega \notin N_0$. Let E be a countable ρ-dense subset of T containing D. Then by ρ-continuity of $X(\cdot, \omega)$ we have

$$\text{(i)} \qquad M_T(X^+(\omega)) = M_E(X^+(\omega)), \qquad M_T(X^-(\omega)) = M_E(X^-(\omega))$$

for all $\omega \notin N_0$. Moreover by Lemma V-2-9 in [3] we have

$$M_E(X_n^+(\omega)) \to M_E(X^+) = M_T(X^+) \quad \text{a.s.}$$

and so by (2.2.1) and Fatou's lemma we deduce that $EM_T(X^+) < \infty$, and thus $M_T(X^+) < \infty$ a.s. Hence by (2.2.2) and (i) there exists a P-nullset $N_1 \supseteq N_0$, such that $X(\omega) \in C_u(T,\rho)$ for all $\omega \notin N_1$. Moreover by Bochner measurability of X, there exist a $\|\cdot\|_T$-separable set $\mathcal{S} \subseteq C_u(T,\rho)$, and a P-nullset $N_2 \supseteq N_1$, such that

$$\text{(ii)} \qquad X(\omega) \in \mathcal{S} \subseteq C_u(T,\rho) \qquad \forall \omega \in \Omega \setminus N_2$$

If $f \in \mathcal{S}$ and $\omega \in \Omega$, we define

$$L_n(f,\omega) = M_E((X_n(\omega) - f)^+), \qquad L(f,\omega) = M_E((X(\omega) - f)^+)$$

$$V_n(\omega) = M_E((X_n(\omega) - X(\omega))^+)$$

Since E is countable we have that $L_n(f,\cdot)$, $L(f,\cdot)$ and V_n are measurable for all $n \geq 1$ and all $f \in \mathcal{S}$, and since $x \curvearrowright x^+$ is subadditive we have

$$\text{(iii)} \qquad L_n(f,\omega) \leq L_n(g,\omega) + \|f-g\|_T \qquad \forall \omega \in \Omega, \quad \forall f,g, \in \mathcal{S}$$

$$\text{(iv)} \qquad L(f,\omega) \leq L_n(g,\omega) + \|f-g\|_T \qquad \forall \omega \in \Omega, \quad \forall f,g \in \mathcal{S}$$

Let $f \in \mathcal{S}$, then $\{(X_n(t) - f(t))^+, \mathcal{F}_n \,|\, n \geq 1\}$ is a submartingale for all $t \in T$, such that

$$(X_n(t) - f(t))^+ \to (X(t) - f(t))^+ \text{ a.s.} \qquad \forall t \in T$$

$$\sup_{t\in E}(X_n(t) - f(t))^+ = L_n(f) \leq M_E(X_n^+) + \|f\|_T$$

and $\|f\|_T < \infty$ since $f \in \mathcal{S} \subseteq B(T)$. Hence by (2.2.1) and Lemma V-2-9 in [3] we know that $L_n(f) \to L(f)$ a.s. for all $f \in \mathcal{S}$. Now let $\mathcal{S}_0$ be a countable $\|\cdot\|_T$-separable dense subset of $\mathcal{S}$, then there exists a P-nullset $N_3 \supseteq N_2$, such that $L_n(f,\omega) \to L(f,\omega)$ for all $f \in \mathcal{S}_0$ and all $\omega \notin N_3$. But then by (iii), (iv) and $\|\cdot\|_T$-density of $\mathcal{S}_0$ in we have

$$\text{(v)} \qquad L_n(f,\omega) \to L(f,\omega) \qquad \forall f \in \mathcal{S} \quad \forall \omega \in \Omega \setminus N_3$$

Let $\omega \in \Omega \setminus N_3$, then $f(t) = X(t,\omega)$ belongs to $\mathcal{S}$ by (ii), and so by (v) we have

$$V_n(\omega) = L_n(X(\omega),\omega) \to L(X(\omega),\omega)$$

But evidently we have that $L(X(\omega),\omega) = 0$, and so we have

$$\text{(vi)} \qquad V_n(\omega) \to 0 \qquad \forall \omega \in \Omega \setminus N_3$$

Now suppose that $\omega \in \Omega_0$, then $X(\omega) \in \mathcal{C}$, and so by (2.2.4) we get

$$\begin{aligned}\|(X_n(\omega) - X(\omega))^+\|_T &\leq \Phi(X_n(\omega)) + \|(X_n(\omega) - X(\omega))^+\|_D \\ &\leq \Phi(X_n(\omega)) + V_n(\omega)\end{aligned}$$

since $D \subseteq E$. Thus the lemma is proved. QED

Lemma 2.3. *Let $\{X_n(t), \mathcal{F}_n \mid n \geq 1\}$ be a martingale for all $t \in T$, and let $\{X(t) \mid t \in T\}$ be a Bochner measurable stochastic process satisfying*

(2.3.1) $\qquad \sup_n E\|X_n\|_S < \infty \qquad \forall S$ *countable* $\subseteq T$

(2.3.2) $\qquad X_n(t) \to X(t)$ *a.s.* $\qquad \forall t \in T$

Let $\Phi: \mathbb{R}^T \to [0, \infty]$ be a function, let $D \subseteq T$ be a countable set and let $\mathcal{C} \subseteq \mathbb{R}^T$, such that

(2.3.3) $\qquad \|f - g\|_T \leq \Phi(f) + \|f - g\|_D \qquad \forall f \in \mathbb{R}^T \quad \forall g \in \mathcal{C}$

There exists a sequence of random variables $\{V_n\}$, such that $V_n \to 0$ a.s. and

(2.3.4) $\qquad \|X_n(\omega) - X(\omega)\|_T \leq \Phi(X_n(\omega)) + V_n(\omega) \quad \forall \omega \in \Omega_0$

where $\Omega_0 = \{\omega \mid X(\omega) \in \mathcal{C}\}$.

Proof. By (2.3.1) and Lemma V-2-9 in [3], we have that $\|X_n\|_S \to \|X\|_S$ a.s. for all countable sets $S \subseteq T$. So by (2.3.1) and Fatou's lemma we have that $E\|X\|_S < \infty$ and thus $\|X\|_S < \infty$ a.s. for all countable sets $S \subseteq T$.

The rest of the proof is exactly as the proof of Lemma 2.2, if you just replace $(f - g)^+$, $(X_n - f)^+$ etc. by $|f - g|$, $|X_n - f|$ etc.

Theorem 2.4. *Let $\{X_n(t), \mathcal{F}_n \mid n \geq 1\}$ be a submartingale for all $t \in T$, where T is a separable topological space, and let $\{X(t) \mid t \in T\}$ be a Bochner measurable stochastic process satisfying*

(2.4.1) $\qquad \sup_n EM_S(X_n^+) < \infty \qquad \forall S$ *countable* $\subseteq T$

(2.4.2) $\qquad M_S(X^-) < \infty$ *a.s.* $\qquad \forall S$ *countable* $\subseteq T$

(2.4.3) $\qquad X_n(t) \to X(t)$ *a.s.* $\qquad \forall t \in T$

(2.4.4) $\qquad X(\cdot, \omega) \in \mathcal{U}sc(T)$ $\qquad$ *for a.a.* ω

Then $X(\cdot,\omega) \in B(T)$ for a.a. ω, and there exists a sequence $\{V_n\}$ of random variables, such that

$$V_n \to 0 \ a.s. \tag{2.4.5}$$

$$\|(X_n(\omega) - X(\omega))^+\|_T \le \Delta_n^+(\omega) + V_n(\omega) \qquad \forall \omega \in \Omega \tag{2.4.6}$$

where $\Delta_n^+(\omega) = \sup\{\partial^+(t, X_n(\omega)) \mid t \in T\}$. Moreover if T is hereditarily separable, then (2.4.5) and (2.4.6) hold even if we drop the assumption (2.4.4).

In particular by (2.4.5) and (2.4.6) we have that, if $\{\Delta_n^+\}$ tends to 0 in pr_*, or in pr^*, or a.s., or a.s., then so does $\{\|(X_n - X)^+\|_T\}$.*

Remarks. (1) By (2.4.1) and the usual martingale convergence theorem, we have that $\overline{X}(t) = \lim X_n(t)$ exists a.s. for all $t \in T$, and (2.4.3) just states that X is a particular version of $\lim X_n(t)$.

(2) The exceptional nullsets in (2.4.2) and (2.4.3) may depend on S and t resp.

Proof. Let $\Phi(f) = \sup\{\partial^+(t, f) \mid t \in T\}$ for all $f \in \mathbb{R}^T$, and let D be a countable dense subset of T. Let $f, g \in \mathbb{R}^T$ and $t \in T$ be given, if U is any given neighbourhood of t, then by denseness of D there exists $s \in D \cap U$, and so

$$\begin{aligned} f(t) - g(t) &= (f(t) - f(s)) + (g(s) - g(t)) + (f(s) - g(s)) \\ &\le W_U^+(t, f) + W_U^-(t, g) + \|(f - g)^+\|_D \end{aligned}$$

Hence taking the limit over $U \in \mathcal{N}_t$ we find

$$(f(t) - g(t))^+ \le \partial^+(t, f) + \partial^-(t, g) + \|(f - g)^+\|_D \quad \forall t \in T$$

so taking sup over t we get

$$\|(f - g)^+\|_T \le \Phi(f) + \|(f - g)^+\|_D \qquad \forall f \in \mathbb{R}^T \quad \forall g \in \mathcal{U}sc(T) \tag{i}$$

since $\partial^-(t, g) = 0$ for all $g \in \mathcal{U}sc(T)$. Hence putting $\mathcal{C} = \mathcal{U}sc(T)$, we see that (2.2.4) holds, and so by (2.4.4) and Lemma 2.2 we deduce that there exists a sequence of random variables $\{V_n\}$ satisfying (2.4.5) and (2.4.6).

Now let us drop condition (2.4.4), but let us assume that T is hereditarily separable, if τ denotes the topology on T, then by Lemma 2.1 there exist a finer topology η on T, such that $X(\omega) \in \mathcal{C}(T, \eta)$ for a.a. ω, and such that (T, η) is hereditarily separable. Since η is finer than τ, we have that

$$\partial_\eta^+(t, f) \le \partial_\tau^+(t, f) \qquad \forall t \in T \quad \forall t \in \mathbb{R}^T$$

where ∂_η^+ and ∂_τ^+ are the lower jump functionals for the topologies η and τ resp. Hence by the first part of the theorem we see that there exists random variables $\{V_n\}$ satisfying (2.4.5) and (2.4.6). QED

Theorem 2.5. *Let $\{X_n(t), \mathcal{F}_n \mid n \geq 1\}$ be a martingale for all $t \in T$, where T is separable topological space, and let $\{X(t) \mid t \in T\}$ be a Bochner measurable stochastic process satisfying*

(2.5.1) $$\sup_n E\|X_n\|_S < \infty \qquad \forall S \text{ countable } \subseteq T$$

(2.5.2) $$X_n(t) \to X(t) \text{ a.s.} \qquad \forall t \in T$$

(2.5.3) $$X(\cdot, \omega) \in C(T) \qquad \text{for a.a. } \omega$$

Then $E\|X\|_T < \infty$ and there exists a sequence $\{V_n\}$ of random variables satisfying

(2.5.4) $$V_n \to 0 \quad a.s.$$

(2.5.5) $$\|X_n(\omega) - X(\omega)\|_T \leq \Delta_n(\omega) + V_n(\omega) \qquad \forall \omega \in \Omega$$

where $\Delta_n(\omega) = \sup\{\partial(t, X_n(\omega)) \mid t \in T\}$. Moreover if T is hereditarily separable, then (2.5.4) and (2.5.5) hold even if we drop the assumption (2.5.3).

In particular by (2.5.4) and (2.5.5) we have, that if $\{\Delta_n\}$ tends to 0 in pr^*, or in pr_*, or a.s., or a.s.*, then so does $\{\|X_n - X\|_T\}$.

Remark. The most severe condition in Theorem 2.5 is the Bochner measurability of the limiting process X. Suppose that $\mathcal{F}_n = \sigma(F_{jn} \mid j \in \mathbb{N})$, where $\{F_{jn} \mid j \leq 1\}$ is a disjoint partition of Ω, for all $n \geq 1$, then

$$X_n(t, \omega) = \sum_{j=1}^{\infty} c_{jn}(t) 1_{F_{jn}}(\omega) \qquad \forall (t, \omega) \in T \times \Omega$$

for some $c_{jn} \in \mathbb{R}^T$. Hence if $\|X_n - X\|_T \to 0$ a.s. we see that $X(\omega)$ belongs to the $\|\cdot\|_T$-closure of the countable set $\{c_{jn} \mid j, n \in \mathbb{N}\}$, and so X is necessarily Bochner measurable. This shows that Bochner measurability of the limiting process X is indispensable, even in the most simple cases.

Proof. Using $\Phi(f) = \sup\{\partial(t, f) \mid t \in T\}$ and Lemma 2.3 the proof goes in exactly the same way as the proof of Theorem 2.4. QED

Example 2.6. Let $\{\varepsilon_{jk} \mid j,k \in \mathbb{N}\}$ be independent random variables, such that

$$P(\varepsilon_{jk} = 0) = p_j, \qquad P\left(\varepsilon_{jk} = \frac{1}{1-p_j}\right) = 1 - p_j$$

where $0 < p_j < 1$ for all $j \geq 1$. Put $T = \mathbb{N}$ and

$$X_n(t,\omega) = \prod_{j=1}^{n} \varepsilon_{jt}(\omega), \qquad X(t,\omega) = \prod_{j=1}^{\infty} \varepsilon_{jt}(\omega)$$

$$\mathcal{F}_n = \sigma(\varepsilon_{jt} \mid 1 \leq j \leq n, t \in T)$$

Then $\{X_n(t), \mathcal{F}_n \mid n \geq 1\}$ is a martingale such that $X_n(t) \to X(t)$ a.s. for all $t \in T$. Now put

$$q = \prod_{j=1}^{\infty}(1 - p_j), \qquad q_n = \prod_{j=1}^{n}(1 - p_j)$$

Then $0 \leq q < 1$. And $q > 0$, if and only if $\Sigma p_j < \infty$. Moreover since $\{X_n(t) \mid t = 1,2,\ldots\}$ are independent identically distributed we have

$$\|X_n\|_T = \frac{1}{q_n} \text{ a.s.} \qquad \forall n \tag{2.6.1}$$

And similarly we find

$$\|X_n - X\|_T = \max\left\{\frac{1}{q_n}, \frac{1}{q} - \frac{1}{q_n}\right\} \text{ a.s.} \quad \text{if} \quad q > 0 \tag{2.6.2}$$

$$\|X_n - X\|_T = \frac{1}{q_n} \quad \text{if} \quad q = 0 \tag{2.6.3}$$

Moreover by the Borel–Cantelli lemmas it follows easily that we have

$$X(t) = 0 \text{ a.s.} \quad \text{if} \quad q = 0 \tag{2.6.4}$$

$$P(X(t) = 0) = 1 - q, \qquad P\left(X(t) = \frac{1}{q}\right) = q \quad \text{if} \quad q > 0 \tag{2.6.5}$$

If $q > 0$ then $X(1), X(2), \ldots$ are independent, identically distributed, non-degenerated random variables, and so it is well known that $\{X(t) \mid t \in$

$\mathbb{N}\}$ is not Bochner measurable. But when $q > 0$, then by (2.6.1) we see that condition (2.5.1) holds.

If $q = 0$, then $X(t) = 0$ a.s. for all t, and so evidently we have that $\{X(t) \,|\, t \in \mathbb{N}\}$ is Bochner measurable. But when $q = 0$, then by (2.6.1) we see that condition (2.5.1) fails.

Now by (2.6.2) and (2.6.3) we see that $\|X_n - X\|_T \to \frac{1}{q} \neq 0$ a.s., in either case. Hence neither (2.5.1) nor Bochner measurability can be suppressed in Theorem 2.5. QED

Example 2.7. Let $\Omega = [0,1]$ and let P be Lebesgue measure on $[0,1]$. Put $T = [0,1]$ and let $\{\eta_n\}$ be any sequence of non-negative random elements. Now put

$$X_n(t,\omega) = \begin{cases} 0 & \text{if} \quad t \neq \omega \\ \eta_n(\omega) & \text{if} \quad t = \omega \end{cases}$$

$$X(t,\omega) = 0$$

Then $X_n(t) = 0 = X(t)$ a.s., and evidently $\{X_n(t)\}$ is a martingale for all $t \in T$. Let $[0,1]$ have its usual Euclidean topology, then it is easily checked that we have

$$\|X_n(\omega) - X(\omega)\|_T = \eta_n(\omega) = \Delta_n(\omega) \qquad \forall \omega \in \Omega \tag{2.7.1}$$

where Δ_n is defined as in Theorem 2.5. Hence the inequality (2.5.5) cannot be improved in general, and $\|X_n - X\|_T$ can be any given sequence of non-negative random elements.

REFERENCES

[1] A. Badrikian, *Prolégomène au calcul des probabilités dans les Banach,* Ecole d'Eté de Probabilités de Saint-Flour V-1975 (ed. P.-L. Hennequin), LNS 539, Springer-Verlag (1976), 1–167.

[2] J. Hoffmann-Jørgensen, *Necessary and sufficient conditions for the uniform law of large numbers,* Probability in Banach spaces V, Proceeding, Medford 1984 (ed. A. Beck et al.), LNS 1153, Springer-Verlag (1985), 258–272.

[3] J. Neveu, *Discrete parameter martingales,* North-Holland Publ. Co. and American Elsevier Publ. Co., 1975.

[4] M. Talagrand, *The Glivenko-Cantelli problem,* Ann. Prob., **15** (1987), 837–870.

Mathematics Institute
Aarhus University
Ny Munkegade
Denmark

Continuity in ℓ^p of certain Ornstein–Uhlenbeck Processes

MICHAEL. B. MARCUS

The City College of CUNY

Consider $\mathbf{X}(t) = \{X_k(t)\}_{k=1}^{\infty}$, $t \in [0,T]$, as a vector of independent Ornstein–Uhlenbeck processes, i.e., mean zero Gaussian processes defined by

$$EX_k(t)X_k(s) = \frac{a_k}{\lambda_k}\exp(-\lambda_k|t-s|) \tag{1}$$

If

$$\sum_{k=1}^{\infty}\left|\frac{a_k}{\lambda_k}\right|^p < \infty \tag{2}$$

then for each fixed t, $\mathbf{X}(t) \in \ell^p$ a.s. (i.e., $\sum_{k=1}^{\infty}|X_k(t)|^p < \infty$ a.s.). The value of $T > 0$ is unimportant as long as $[0,T]$ is a bounded interval. In all that follows we will assume that $T > 0$ is fixed. The constants in Theorem 1 below depend on T. Note also that we allow the case $p = \infty$ with the usual interpretation. In [**I**], motivated by a question raised in the study of infinite dimensional stochastic differential equations, we were concerned with obtaining general conditions for the continuity of $\mathbf{X}(t)$ in ℓ^2. Our approach used a Corollary of Talagrand's [**T**] necessary and sufficient condition for continuity of Gaussian processes, due to Fernique [**F**], which gives necessary and sufficient conditions for continuity of Banach space valued Gaussian processes with stationary increments. In order to use this result we needed a generalization of an interesting inequality of Boas which was first used in the study of Gaussian processes by the author and L. Shepp [**MS**] twenty years ago. The generalization we obtained, Lemma 5 in [**I**], gave us what we needed for that paper but it was only sharp for $p = 2$. The purpose of this paper is to extend (1.15) of Lemma 5 in [**I**] to all $1 \le p \le \infty$. Then, as an extension of some of the results in [**I**], we can give equivalent lower and upper bounds for $E\sup_t\|\mathbf{X}(t)\|_p$ in certain cases. These bounds imply the almost sure continuity of these sequences in ℓ^p.

An interesting consequence of the extension of Lemma 5 in [**I**], which is given as Lemma 1 in this paper, is that it enables us, in Theorem 1, to relate $E\sup_t\|\mathbf{X}(t)\|_p$ to a function of $\{\frac{a_k}{\lambda_k}\}_{k=1}^{\infty}$ which is used to define membership of $\{\frac{a_k}{\lambda_k}\}_{k=1}^{\infty}$ in the sequence space $\ell_{q,p}$.

This research was supported in part by a grant from the National Science Foundation.

LEMMA 1. *Let $\{c_k\}_{k=1}^{\infty}$ be non-negative numbers such that c_k is non-increasing in k. Let $1 \le p \le \infty$ and $1/p + 1/q = 1$. Let $\beta > 0$. Then there exist constants $0 < c_\beta \le C_\beta < \infty$ such that*

(3)
$$c_\beta \left(\sum_{k=1}^{\infty} \frac{(k^{\beta/2} c_k)^p}{k^{p/2 \wedge 1}} \right)^{1/p} \le \sup_{\|\{b_k\}\|_q \le 1} \sum_{j=1}^{\infty} j^{\beta/2-1} \left(\sum_{k=j}^{\infty} b_k^2 c_k^2 \right)^{1/2} \le c_\beta \left(\sum_{k=1}^{\infty} \frac{(k^{\beta/2} c_k)^p}{k^{p/2 \wedge 1}} \right)^{1/p}$$

except that, when $p < 2$ and $\beta < 1$, we require, in addition to the above, that $k^{(1-\beta)(1/p-1/q)/2} c_k$ is non-increasing in k.

PROOF: Set

$$A_\beta = \sum_{j=1}^{\infty} j^{\beta/2-1} \left(\sum_{k=j}^{\infty} b_k^2 c_k^2 \right)^{1/2}$$

and

$$\mathcal{A}_{\beta,q} = \sup_{\|\{b_k\}\|_q \le 1} A_\beta$$

We will first obtain the lower bound in (3) when $p \ge 2$. We have

(4)
$$A_\beta \ge c_\beta \sum_{n=0}^{\infty} 2^{n\beta/2} \left(\sum_{k=2^n}^{\infty} b_k^2 c_k^2 \right)^{1/2} \ge c_\beta \sum_{n=0}^{\infty} 2^{n\beta/2} b_{2^n} c_{2^n}$$

(In this and all that follows, c_β and C_β are constants depending on β such that $0 < c_\beta, C_\beta < \infty$ but they are not necessarily the same at each occurence.) Recall that $\sup_{\|\{\alpha_k\}\|_q \le 1} \sum \alpha_k \beta_k = (\sum \beta_k^p)^{1/p}$. Using this in (4) we get

$$\mathcal{A}_{\beta,q} \ge c_\beta \sup_{\|\{b_k\}\|_q \le 1} \sum_{n=0}^{\infty} 2^{n\beta/2} b_{2^n} c_{2^n} \ge c_\beta \left(\sum_{n=0}^{\infty} 2^{np\beta/2} c_{2^n}^p \right)^{1/p} \ge c_\beta \left(\sum_{k=1}^{\infty} k^{p\beta/2-1} c_n^p \right)^{1/p}$$

where we make the usual interpretation when $p = \infty$. This completes the proof of the lower bound in (3) when $p \ge 2$.

We now obtain the lower bound when $p < 2$. Assume that

(5)
$$B = \sum_{k=1}^{\infty} k^{(\beta-1)p/2} c_k^p < \infty$$

and set

(6)
$$b_k^q = k^{(\beta-1)p/2} c_k^p / B \qquad \forall k \ge 1$$

Then, with $\{b_k\}_{k=1}^{\infty}$ taking the values in (6), we have

$$
\begin{aligned}
(7) \qquad A_\beta &= B^{-1/q}\sum_{j=1}^{\infty} j^{\beta/2-1}\left(\sum_{k=j}^{\infty} k^{(\beta-1)p/q}c_k^{2p}\right)^{1/2} \\
&= B^{-1/q}\sum_{j=1}^{\infty}\left(j^{\beta-2}c_{2j-1}^{2p}\sum_{k=j}^{2j-1} k^{(\beta-1)p/q}\right)^{1/2} \\
&\geq c_\beta B^{-1/q}\sum_{j=1}^{\infty} j^{(\beta-1)p/2}c_{2j-1}^{p}
\end{aligned}
$$

where, at the last step, we take $k = j$ or $k = 2j$ depending on whether $\beta \geq 1$ or $\beta < 1$. Extrapolating and substituting for B in the final line of (6) gives us the left-hand-side of (3) when $p < 2$ and $B < \infty$. However, by restricting attention to $\{c_k\}_{k=1}^{n}$ and then passing to the limit, it is clear that this also holds when $B = \infty$.

We proceed to obtain the upper bound in (3) when $p \geq 2$. Note that by homogeneity we may assume that $c_1 \leq 1$. Define $m(n) = \#\{k : c_k^q > 2^{-n}\}$ and define

$$
(8) \qquad F(n) = \sum_{j=m(n-1)+1}^{m(n)} j^{\beta/2-1}
$$

and

$$
G(h) = \sum_{k=m(h-1)+1}^{m(h)} b_k^q
$$

We have

$$
\begin{aligned}
(9) \qquad A_\beta &\leq \sum_{j=1}^{\infty} j^{\beta/2-1}\left(\sum_{k=j}^{\infty} b_k^q c_k^q\right)^{1/q} \\
&\leq \sum_{n=1}^{\infty} F(n)\left(\sum_{k=m(n-1)+1}^{\infty} b_k^q c_k^q\right)^{1/q} \\
&\leq \sum_{n=1}^{\infty} F(n)\left(\sum_{h=n}^{\infty} 2^{1-h}G(h)\right)^{1/q} \\
&= \sum_{n=1}^{\infty} \frac{F(n)}{2^{n/q}}\left(\sum_{h=n}^{\infty} 2^{n+1-h}G(h)\right)^{1/q} \\
&\leq \left(\sum_{n=1}^{\infty} \frac{F^p(n)}{2^{np/q}}\right)^{1/p}\left(\sum_{n=1}^{\infty}\sum_{h=n}^{\infty} 2^{n+1-h}G(h)\right)^{1/q} \\
&\equiv I_1 \cdot I_2
\end{aligned}
$$

where, obviously, at the next to the last step we used Hölder's Inequality. It is easy to see, by changing the order of summation, that

(10)
$$I_2 \leq \left(4 \sum_{k=1}^{\infty} b_k^q\right)^{1/q}$$

Furthermore, since $F(n) \leq C_\beta(m^{\beta/2}(n) - m^{\beta/2}(n-1))$ and $p \geq 2$ we also have

(11)
$$\begin{aligned} I_1 &\leq C_\beta \left(\sum_{n=1}^{\infty} \frac{m^{p\beta/2}(n) - m^{p\beta/2}(n-1)}{2^{np/q}}\right)^{1/p} \\ &\leq C_\beta \left(\sum_{k=1}^{\infty} k^{p\beta/2-1} c_k^p\right)^{1/p} \end{aligned}$$

Using (9), (10) and (11) we get the upper bound in (3) when $p \geq 2$.

Lastly, we obtain the upper bound in (3) when $p < 2$. Let $\delta = (\beta - 1)(1 - p/q)$. In what follows we will assume that $k^{-\delta} c_k^p$ is non–increasing. This is obviously the case when $\beta \geq 1$ and when $\beta < 1$ it is also true because of the additional hypothesis imposed in this case. The proof is similar to the proof of the upper bound when $p \geq 2$. As in that case we can assume that $c_1 \leq 1$. We define $\overline{m}(n) = \#\{k : k^{-\delta} c_k^p > 2^{-n}\}$. Set

$$J(h) = \sum_{k=m(h-1)+1}^{m(h)} b_k^2 k^\delta c_k^{2-p}$$

Then

(12)
$$\begin{aligned} A_\beta &= \sum_{n=1}^{\infty} F(n) \left(\sum_{h=n}^{\infty} \sum_{k=\overline{m}(h-1)+1}^{\overline{m}(h)} b_k^2 c_k^2\right)^{1/2} \\ &\leq \sum_{n=1}^{\infty} \frac{F(n)}{2^{n/2}} \left(\sum_{h=n}^{\infty} 2^{n+1-h} J(h)\right)^{1/2} \\ &\leq \left(\sum_{n=1}^{\infty} \frac{F^2(n)}{2^n}\right)^{1/2} \left(\sum_{n=1}^{\infty} \sum_{h=n}^{\infty} 2^{n+1-h} J(h)\right)^{1/2} \\ &\equiv I_1 \cdot I_2 \end{aligned}$$

where, obviously, at the last stage we use the Schwartz Inequality. It is easy to see, as in (10), by a change of order of summation and Hölder's Inequality, that

(13)
$$\begin{aligned} I_2 &\leq \left(4 \sum_{k=1}^{\infty} b_k^2 k^\delta c_k^{2-p}\right)^{1/2} \\ &\leq 2 \left(\sum_{k=1}^{\infty} b_k^q\right)^{1/q} \left(\sum_{k=1}^{\infty} k^{(\beta-1)p/2}\right)^{(q-2)/q} \end{aligned}$$

Also, as in (11)

$$I_1 \le C_\beta \left(\sum_{n=1}^{\infty} \frac{\overline{m}^\beta(n) - \overline{m}^\beta(n-1)}{2^n} \right)^{1/2} \tag{14}$$

$$\le C_\beta \left(\sum_{k=1}^{\infty} k^{\beta-1-\delta} c_k^p \right)^{1/2}$$

Using (12), (13) and (14) and taking into account the value of δ we get the upper bound in (3) when $p < 2$. This completes the proof of Lemma 1.

For a non-increasing sequence $\{c_k\}_{k=1}^{\infty}$ we set

$$|\{c_k\}|_{q,p} = \left(\sum_{k=1}^{\infty} \frac{(k^{1/q} c_k)^p}{k} \right)^{1/p} \tag{15}$$

for $1 \le p < \infty$ and $0 < q < \infty$ and

$$|\{c_k\}|_{q,\infty} = \sup_k k^{1/q} c_k \tag{16}$$

for $0 < q < \infty$. Also for $1 \le q = p \le \infty$ we set

$$|\{c_k\}|_{q,p} = \|\{c_n\}\|_p$$

where $\|\{c_k\}\|_p$ is the usual norm in ℓ^p. The function $|\{c_k\}|_{q,p}$ defines membership of $\{c_k\}$ in $\ell_{q,p}$, (In general $\{c_k\}$ is replaced in (15) and (16) by $\{c_k^*\}$, the non-decreasing rearrangement of $\{c_k\}$. So defined $|\{c_k\}|_{q,p}$ is a norm when $1 \le p < q$ and is equivalent to a norm when $q > 1$).

It is curious that $|\{c_k\}|_{q,p}$ is related to questions of continuity and boundedness in ℓ^p of certain sequences of independent Ornstein–Uhlenbeck processes. We define

$$\mathbf{X}_\beta(t) = \{c_k X_k(2^{k^\beta} t)\}_{k=1}^{\infty}$$

where $\{X_k\}_{k=1}^{\infty}$ is defined in (1) and $\{c_k\}_{k=1}^{\infty}$ is a non-increasing sequence of non-negative numbers.

THEOREM 1. *Let $\{c_k\}_{k=1}^{\infty}$ satisfy all the conditions of Lemma 1 including the additional condition imposed when $p < 2$ and $\beta < 1$. Then $\forall\ 1 \le p \le \infty$ and $0 < \beta < \infty$ there exist constants $0 < d_{\beta,p}, D_{\beta,p} < \infty$ such that*

(i) *If $0 < \beta \le 2/p \wedge 1$*

$$d_{\beta,p} \|\{c_k\}\|_p \le E \sup_t \|\mathbf{X}_\beta(t)\|_p \le D_{\beta,p} \|\{c_k\}\|_p$$

(ii) *If* $2/p < \beta \le 1$

$$d_{\beta,p}\left|\{c_k\}\right|_{2/\beta,p} \le E\sup_t \|\mathbf{X}_\beta(t)\|_p \le D_{\beta,p}\left|\{c_k\}\right|_{2/\beta,p}$$

(iii) *If* $\beta \ge 1$

$$d_{\beta,p}\left|\{c_k\}\right|_{r,p} \le E\sup_t \|\mathbf{X}_\beta(t)\|_p \le D_{\beta,p}\left|\{c_k\}\right|_{r,p}$$

where $r = \dfrac{2p}{p(\beta-1)+2} \wedge \dfrac{2}{\beta}$.

Note that (i), (ii) and (iii) include all cases for $1 \le p \le \infty$ and $0 < \beta < \infty$. Also $\dfrac{2p}{p(\beta-1)+2} \le \dfrac{2}{\beta}$ if and only if $p \le 2$. Note also that the upper bounds in Theorem 1 are valid for all processes of the form

$$\mathbf{X}(t) = \{c_k X_k(\lambda_k t)\}_{k=1}^{\infty} \tag{17}$$

where $\lambda_k \le 2^{k^\beta}$ and similarly the lower bounds are valid for processes of the form of (17) where $\lambda_k \ge 2^{k^\beta}$.

PROOF: All the statements above follow from Lemma 1 and the fact that $\forall\ 1 \le p \le \infty$, $0 < \beta < \infty$ there exist constants $0 < d'_{\beta,p}$, $D'_{\beta,p} < \infty$ such that

$$\begin{aligned} d'_{\beta,p}\left[E\|\mathbf{X}_\beta(0)\|_p + \left(\sum_{k=1}^{\infty}\frac{(k^{\beta/2}c_k)^p}{k}\right)^{1/p}\right] &\le E\sup_t \|\mathbf{X}_\beta(t)\|_p \\ &\le D'_{\beta,p}\left[E\|\mathbf{X}_\beta(0)\|_p + \left(\sum_{k=1}^{\infty}\frac{(k^{\beta/2}c_k)^p}{k}\right)^{1/p}\right] \end{aligned} \tag{18}$$

To obtain (18) we use Theorem 4 of [I] along with (1.47) and (1.48) of [I]. (We also use the bounds for $E\|\{X_k(0)\}\|_p$ given after (1.48) of [I] for $1 < p < \infty$. When $p = \infty$ we are either in case (ii) or (iii) and the term $E\|\{c_k\}\|_\infty$ is dominated by $\left|\{c_k\}\right|_{2/\beta,\infty}$). We see that all the cases $1 \le p \le \infty$ and $0 < \beta < \infty$ are considered in Theorem 1. In case (i) we consider $0 < \beta \le 1$ for $1 \le p \le 2$ and $0 < \beta \le 2/p$ for $p > 2$. In case (ii) we consider $2/p < \beta \le 1$ for $p > 2$ and in the case (iii) we consider all $\beta \ge 1$.

Note that (18) is based on Theorems 3.3.3 and 3.3.4 [F] and well known arguments which are given in [I].

REFERENCES

F Fernique, X. Fonctions aléatoires a valeurs vectorielles, Proceedings of the Sixth International Conference on Probability in Banach Space, 1987, Lecture Notes in Math., to appear.

I Iscoe, I., Marcus, M.B., McDonald, D., Talagrand, M., Zinn, J. Continuity of ℓ^2 valued Ornstein–Uhlenbeck Processes, Annals of Probability, to appear.

MS Marcus, M.B., Shepp, L.A. Continuity of Gaussian Processes, Trans. Amer. Math. Soc., 151, 1970, 377–391.

T Talagrand, M. Regularity of Gaussian processes, Acta Math., 159, 1987, 99–149.

Department of Mathematics, The City College of CUNY, New York, N.Y. 10031

On the Rate of Convergence for the Weighted Empirical Process

VYGANTAS PAULAUSKAS

Let $D = D[0,1]$ be the space of right continuous functions on $[0,1)$ having left limits on $(0,1]$, equipped with the well-known Skorohod topology, under which it becomes a Polish space. Let $X, X_1, \ldots, X_n$ be i.i.d. random elements with values in D, $EX(t) = 0$ and $EX^2(t) < \infty$ for all $t \in [0,1]$. We say that X satisfies the central limit theorem in D ($X \in CLT(D)$) if there exists a D-valued Gaussian random element Y, such that the distribution of $n^{-1/2}\sum_{i=1}^n X_i$ converges weakly to the distribution of Y. In [6] the following CLT in D was proved.

Theorem 1 [6]. *Let X be a D-valued random element such that as a process it is stochastically continuous and for all $t \in [0,1]$, $EX(t) = 0$, $EX^2(t) < \infty$. If there exist nondecreasing continuous functions F_1 and F_2 and numbers $\alpha_1 > 1/2$, $\alpha_2 > 1$ such that for all $0 \le s < t < u \le 1$*

$$E(X(t) - X(s))^2 \le (F_1(t) - F_1(s))^{\alpha_1}, \tag{1}$$

$$E(X(t) - X(s))^2(X(u) - X(t))^2 \le (F_2(u) - F_2(s))^{\alpha_2}, \tag{2}$$

then $X \in CLT(D)$ and the limit law satisfies $\mathcal{L}(Y)(C) = 1$, where $\mathcal{L}(Y)$ denotes the distribution of Y and $C \subset D$ is the usual space of continuous functions on $[0,1]$.

In our joint paper with D. Jukneviciene [10] we have proved the following theorem, which presents the first estimate of the rate of convergence in $CLT(D)$ and which can be regarded as the rate of convergence in the above formulated theorem. In order to formulate it we need the following notation

$$\Delta_n(\lambda) = |P\{\sup_{0\le t\le 1} |S_n(t)| < \lambda\} - P\{\sup_{0\le t\le 1} |Y(t)| < \lambda\}|,$$

$$\Delta_n = \sup_\lambda \Delta_n(\lambda),$$

where $S_n(t) = n^{-1/2}\sum_{i=1}^{n} X_i(t)$ and $Y(t)$ is the limiting Gaussian process, having the same covariance function as the process $X(t)$. For a continuous function F the modulus of continuity is defined as usual by

$$\omega_F(\delta) = \sup_{|s-t|<\delta} |F(s) - F(t)|.$$

Theorem 2 [10]. *Suppose that X is as in Theorem 1 and in addition satisfies the condition: there exists a nondecreasing function F_3 and a number $\alpha_3 > 1$ such that for all $0 \le s < t \le 1$*

$$E(X(t) - X(s))^4 \le (F_3(t) - F_3(s))^{\alpha_3}. \tag{3}$$

Suppose that $E\sup_{0\le t\le 1} |X(t)|^3 < \infty$ and there exist numbers $\beta_i > 0$ such that

$$\omega_{F_i}(\delta) \le C\delta^{\beta_i} \quad i = 1, 2, 3. \tag{4}$$

Here and in the sequel the letter C stands for a constant, which is not the same in different places.

Let the Gaussian random element Y satisfy the following condition: there exists a constant $C(T)$, depending on the covariance function T of the process Y, such that for all $s \ge 0$ and $\varepsilon > 0$

$$P\{s \le \sup_{0\le t\le 1} |Y(t)| < s + \varepsilon\} \le C(T)(1 + s)^{-3}\varepsilon. \tag{5}$$

Then

$$\Delta_n(\lambda) = 0(n^{-1/6}\lambda^{-\kappa}(\ln n + \ln\lambda)^{1+\delta}) \tag{6}$$

where $\kappa = 3$, if $\alpha_3 > 1$ and $\kappa = 12/5$, if $\alpha_3 = 1$; here $\delta > 0$ is arbitrary, but the constant appearing in the right-hand side of (6) and depending on the distributions of X and Y, tends to infinity if $\delta \to 0$.

Now let F_n be the empirical distribution function, based on the sample $x_1, x_2, \ldots, x_n$, taken from the uniform distribution on $[0, 1]$. Let $D_n(t) = n^{1/2}(F_n(t) - t)$. D_n can be regarded as a D-valued random element, and it is well-known that D_n converges weakly to the so-called Brownian bridge W_0 a Gaussian process with mean zero and covariance function $R(s,t) = EW_0(s)\, W_0(t) = s(1-t)$, $0 \le s \le t \le 1$. On the other side this convergence can be regarded as the central limit theorem in D, since D_n can be represented as the sum of i.i.d. D-valued random elements of a very simple structure

$$X(t) = U(t) - EU(t) \tag{7}$$
$$U(t) = 0 \text{ if } 0 \le t < \xi \quad \text{and} \quad = 1 \text{ if } \xi \le t \le 1,$$

where ξ is random variable, uniformly distributed on $[0, 1]$. Easy calculations show that the functions F_i and the numbers α_i from (1)–(3) can be taken as follows: $F_1(t) = F_2(t) = t$, $F_3(t) = 3t$, $\alpha_1 = \alpha_3 = 1$, $\alpha_2 = 3$. Since the density of $\sup_{0 \le t \le 1} |W_0(t)|$ is well-known and has the explicit expression, condition (5) holds too, and as a corollary from (6) we get a rate of convergence in the Kolmogorov–Smirnov theorem.

$$\left|P\{\sup_t |D_n(t)| < \lambda\} - P\{\sup_t |W_0(t)| < \lambda\}\right|$$
$$= 0(n^{-1/6}(1+\lambda)^{-12/5}(\ln n + \ln(1+\lambda))^{1+\delta}), \delta > 0$$

Of course, it is necessary to mention, that both exponents 1/6 and 12/5 are not optimal. Also it is worth mentioning that if in (3) we have $\alpha_3 > 1$ then by Theorem 12.4 from [2] it follows that X is concentrated on the subspace C. Therefore for the process X from (7) the exponent α_3 can not be larger than 1.

The aim of this note is to show that Theorem 2 can be applied to the so-called weighted empirical processes, too. Let Q be a class of functions $q: [0, 1] \to R$ with the properties: $q(t) \ge 0$, q is continuous and increasing on $[0, \frac{1}{2}]$, $q(t) = q(1-t)$ for $0 \le t \le 1/2$. For the simplicity of writing we denote $g(t) = (q(t))^{-1}$. By means of such functions we get the weighted empirical process $D_{n,g}(t) = D_n(t)g(t)$, and functionals of this process lead us to such statistics as the Andersen–Darling statistic $(q(t) = (t(1-t))^{1/2})$.

In [3] (see also [9]) it was proved that if $q \in Q$ then

$$\sup |D_{n,g}(t) - W_g(t)| \xrightarrow{P} 0$$

iff

$$\int_0^{1/2} \exp\{-\varepsilon q^2(t) t^{-2}\} t^{-1} dt < \infty \quad \text{for every} \quad \varepsilon > 0. \tag{8}$$

Here and in the sequel $W_g(t) = W_0(t)g(t)$.

During the last two decades, convergence of empirical processes and weighted empirical processes among them was investigated intensely in a very general setting and the reader may consult a comprehensive book on this topic [12] or the recent survey papers [1, 4, 5].

We consider the quantity

$$\Delta_{n,g}(\lambda) = \big|P\{\sup_{0\le t\le 1} |D_{n,g}(t)| < \lambda\} - P\{\sup_{0\le t\le 1} |W_g(t)| < \lambda\}\big|.$$

Theorem 3. *Suppose that $q \in Q$ and satisfies the following conditions*

(9) $$u^{1/4}(q(u))^{-1} \quad \textit{is increasing,}$$

(10) $$\sup_{0\le u\le 1/2} u^{(1-\gamma)/4}(q(u))^{-1} < C \quad \textit{for some } \gamma > 0.$$

Then for any $\delta > 0$ there exists a constant C_0, depending on q and δ, such that

$$\Delta_{n,g}(\lambda) \le C_0 n^{-1/6}\lambda^{-12/5}(\ln n + \ln(1+\lambda))^{1+\delta}.$$

Remark. Having in mind the symmetry of g around the point 1/2, we can estimate

$$\sup_{0\le t\le 1} E|U(t)-t|^3 g^3(t) \le \sup_{0\le t\le 1/2} t\, g^3(t),$$

therefore the condition

(11) $$\sup_{0\le t\le 1/2} t\, g^3(t) < \infty$$

or the condition

(12) $$\int_0^{1/2} g^3(t)dt < \infty,$$

which is a little bit stronger, should be natural to obtain a rate of convergence, at least of the order $n^{-1/6}$. It is an interesting, more difficult question if it is possible to get a rate of convergence for $\Delta_{n,g}$ of order $n^{-1/2}$ under one of the conditions (11) or (12).

Proof of Theorem 3. We shall show that the process

$$X_g(t) = (U(t)-t)g(t)$$

satisfies the conditions of Theorem 2. For $0 \le s < t \le 1/2$ simple calculations give

$$I(s,t) \equiv E(X_g(t)-X_g(s))^2 = g^2(t)(t-s) + (g(s)-g(t))^2 s - (tg(t)-sg(s))^2.$$

From (9) it follows that $u^{1/2}g(u)$ is increasing, therefore

$$\begin{aligned}\sqrt{s}(g(s)-g(t)) &= \sqrt{s}g(s)\left(1-\frac{g(t)}{g(s)}\right) \le \sqrt{t}g(t)\left(1-\left(\frac{s}{t}\right)^{1/2}\right) \\ &= g(t)(\sqrt{t}-\sqrt{s}).\end{aligned} \tag{13}$$

Thus we have

$$s(g(s)-g(t))^2 \le g^2(t)(\sqrt{t}-\sqrt{s})^2 \le g^2(t)(t-s) \le \int_s^t g^2(u)du.$$

Since from (10) it follows that

$$\int_0^{1/2} g^2(u)du \le \sup_{0\le u\le 1/2} u^{3/4}g(u)\int_0^{1/2} g(u)u^{-3/4}du < \infty,$$

therefore

$$I(s,t) \le 2\int_s^t g^2(u)du$$

and (1) is satisfied with $F_1(t) = 2\int_0^t g^2(u)du$ and $\alpha_1 = 1$. Now we estimate the quantity

$$J(s,t) = E(X_g(t) - X_g(s))^4.$$

By means of condition (9), similarly as in (13) we get

$$s^{1/4}(g(s)-g(t)) \le g(t)(t^{1/4}-s^{1/4}). \tag{14}$$

Therefore, using elementary estimates

$$E(U(t)-t)^4 \le Ct,\ E((U(t)-t)-(U(s)-s))^4 \le C(t-s)$$

and by (14) we have

$$\begin{aligned}J(s,t) &\le C[g^4(t)(t-s) + s(g(s)-g(t))^4] \\ &\le C\left(\int_s^t g^4(u)du + g^4(t)(t^{1/4}-s^{1/4})^4\right) \\ &\le C\int_s^t g^4(u)du.\end{aligned}$$

Again, since

$$\int_0^{1/2} g^4(t)dt \le \sup_{0\le u\le 1/2}\left(u^{\frac{1-\gamma}{4}}g(u)\right)^4 \int_0^{1/2} u^{\gamma-1}du < \infty,$$

we have that (3) is satisfied with $F_3(t) = c\int_0^t g^4(t)dt$ and $\alpha_3 = 1$. The most complicated estimate is for the quantity

$$L(s,t,u) \equiv E(X_g(u) - X_g(t))^2(X_g(t) - X_g(s))^2, \quad 0 \le s < t < u \le 1/2.$$

Considering the four events $\{0 \le \xi \le s\}$, $\{s < \xi \le t\}$, $\{t < \xi \le u\}$, $\{u < \xi \le 1\}$ and for shorter writing denoting $G(u) = (1-u)g(u)$, $H(u) = -u\,g(u)$, after some calculations we get

$$L(s,t,u) = L_1(s,t,u) + L_2(s,t,u), \tag{15}$$

where

$$L_1(s,t,u) = (G(u) - G(t))^2[(G(t) - G(s))^2 + (G(t) - G(s) + g(s))^2(t-s)];$$

$$L_2(s,t,u) = (H(t) - H(s))^2[(G(u) - H(t))^2(u-t) + (H(u) - H(t))^2(1-u)].$$

Now, using the monotonicity of g we have the following estimates:

$$(H(t) - H(s))^2 = (tg(t) - sg(s))^2 \le (g(t)(t-s))^2 \le \left(\int_s^t g(v)dv\right)^2, \tag{16}$$

$$\begin{aligned} s(g(u) - g(t))^2(g(s) - g(t))^2 &\le s(g(s) - g(u))^4 \\ &= (s^{1/4}(g(s) - g(u)))^4 \le (g(u)(u^{1/4} - s^{1/4}))^4 \\ &\le C\left(\int_s^u g(u)v^{-3/4}dv\right)^4 \\ &\le C\left(\int_s^u g(v)v^{-3/4}dv\right)^4, \end{aligned} \tag{17}$$

$$\begin{aligned} (g(t) - g(u))^2g^2(t)(t-s) &= (g^4(t)(t-s))^{1/2}(g(t) - g(u))^2(t-s)^{1/2} \\ &\le \left(\int_s^t g^4(v)dv\right)^{1/2}(t^{1/4}(g(t) - g(u)))^2 \\ &\le \left(\int_s^t g^4(v)dv\right)^{1/2}\left(\int_t^u g(v)v^{-3/4}dv\right)^2, \end{aligned} \tag{18}$$

$$(g(u) - g(t))^2t(tg(t) - sg(s)) \le tg^2(t)\int_s^t g(v)dv. \tag{19}$$

Since

$$L_1(s,t,u) \le 2[(g(u)-g(t))^2 + (ug(u)-tg(t))^2][t(tg(t)-sg(s))^2$$
$$+ \ (g(s)-g(t))^2 s + g^2(t)(t-s)],$$

from the estimates (16)–(19) it is not difficult to get the estimate

$$L_1(s,t,u) \le C\left(\int_s^u g(v)v^{-3/4}dv\right)^2. \tag{20}$$

Analogously,

$$L_2(s,t,u) \le 2(H(t)-H(s))^2[g^2(u)(u-t) + (ug(u)-tg(t))^2(u-t)$$
$$+ \ (H(u)-H(t))^2(1-u)].$$

Applying the same estimates (16)–(19), we have

$$L_2(s,t,u) \le C\left(\int_s^u g(v)v^{-3/4}dv\right)^2. \tag{21}$$

Thus (15), (20) and (21) imply (2) with $\alpha_2 = 2$ and

$$F_2(t) = \int_0^t g(v)v^{-3/4}dv.$$

Now we shall show that for these functions F_i, $i = 1,2,3$, condition (4) is satisfied. Namely, for $0 \le s < t \le 1/2$

$$F_1(t) - F_1(s) = \int_s^t g^2(v)dv \le \sup_{0\le v\le 1/2} (g(v)v^{1/4})^2 \int_s^t v^{-1/2}dv$$
$$\le C(t^{1/2} - s^{1/2}) \le C(t-s)^{1/2},$$

$$F_2(t) - F_2(s) = \int_s^t g(v)v^{-3/4}dv \le \sup_{0\le v\le 1/2} g(v)v^{\frac{1-\gamma}{4}} \int_s^t v^{\gamma-1}dv$$
$$\le C\gamma^{-1}(t^\gamma - s^\gamma) \le C\gamma^{-1}(t-s)^\gamma,$$

$$F_3(t) - F_3(s) = \int_s^t g^4(v)dv \le \sup_{0\le v\le 1/2} (g(v)v^{\frac{1-\gamma}{4}})^4 \int_s^t v^{\gamma-1}dv$$
$$\le C\gamma^{-1}(t-s)^\gamma.$$

What remains to be done is to verify condition (5) for the Gaussian process $W_g(t) = g(t)W_0(t)$. The essential step is to show that the density of the real random variable $\sup_{0\leq t\leq 1}|W_g(t)|$ is bounded. For this purpose we use a general result of M.A. Lifshits [8]. Let $\{Z(t), t \in T\}$ be a centered Gaussian process on some parameter set T, $\tau(s,t) = (E(Z(s) - Z(t))^2)^{1/2}$, $N(\varepsilon) = N(T, \tau, \varepsilon)$ be the smallest number of points in an ε-net for the pseudometric space (T, τ), $H(\varepsilon) = \log N(\varepsilon)$, $\Psi(\varepsilon) = \int_0^\varepsilon H^{1/2} u(du)$. Let $\nu(\varepsilon) \equiv \nu(T, \varepsilon)$ be the greatest length of a chain $t_1, t_2, \ldots, t_{\nu(\varepsilon)}$ for which

$$E(Z^2(t_k) \mid Z(t_1), \ldots, Z(t_{k-1})) - (E(Z(t_k) \mid Z(t_1), \ldots, Z(t_{k-1})))^2 \geq \varepsilon^2$$

for all $k \leq \nu(\varepsilon)$.

Theorem 4 [8]. *Let $E\,Z^2(t) \leq 1$, for all $t \in T$ and $\Psi(1) < \infty$. Suppose that for some $\beta > 0$ and any $\alpha > 0$*

$$\Psi(\varepsilon) = 0(\varepsilon^\beta) \quad \textit{as} \quad \varepsilon \to 0, \tag{22}$$

$$\log \nu(\varepsilon) = 0(\nu(\varepsilon^\alpha)) \quad \textit{as} \quad \varepsilon \to 0 \tag{23}$$

or

$$\nu(\varepsilon) \to \infty \quad \textit{and for some} \quad \kappa > 0 \; N(\varepsilon) = 0(\varepsilon^{-\kappa}). \tag{24}$$

Then the distribution density of $\sup_{t\in T}|Z(t)|$ is bounded.

By means of this theorem we are able to show that under a little stronger condition than (8) the density of $\sup_{0\leq t\leq 1}|W_g(t)|$ is bounded.

Proposition 5. *Let $W_g(t) = g(t)W_0(t)$, $g(t) = (q(t))^{-1}$, $q \in Q$ and satisfies the following additional conditions:*

$$q(t)\, t^{-1/2} \quad \textit{is decreasing,}$$

$$q(t) = t^\alpha |\ln t|^{(1+\delta)/2} \quad \textit{in the neighborhood of zero,} \tag{25}$$

where $0 < \alpha \leq 1/2$ and δ is arbitrary if $\alpha < 1/2$ and positive if $\alpha = 1/2$. Then the density function of $\sup_{0\leq t\leq 1}|W_g(t)|$ is bounded.

For the proof of this result (cf. [11]) we show that W_g satisfies the conditions of Theorem 4. Although the function g is not defined at points 0 and 1, but since

$$\lim_{t\to 0} W_g(t) = \lim_{t\to 1} W_g(t) = 0 \;\text{ a.s.}$$

we set $W_g(0) = W_g(1) = 0$ a.s. Let $\tau_g^2(s,t) = E(W_g(s) - W_g(t))^2$, $0 \le s < t \le 1/2$. Using the elementary properties of the Brownian bridge W_0 it is easy to check that

$$\tau_g^2(s,t) = (t-s)g^2(t)(1-(t-s)) + s(1-s)(g(s)-g(t))^2.$$

Since the function $g(t)\,t^{1/2}$ is increasing, we have

$$\begin{aligned} s^{1/2}(g(s)-g(t)) &= s^{1/2}g(s)\left(1-\frac{g(t)}{g(s)}\right) \\ &\le t^{1/2}g(t)\left(1-\left(\frac{s}{t}\right)^{1/2}\right) = g(t)(\sqrt{t}-\sqrt{s}), \end{aligned}$$

$$s(g(s)-g(t))^2 \le g^2(t)(t^{1/2}-s^{1/2})^2 \le g^2(t)(t-s),$$

therefore

$$\tau_g^2(s,t) \le 2g^2(t)(t-s). \tag{26}$$

Now we can get bounds for $N([0,1],\tau_g,\varepsilon)$. Namely, if $\alpha < 1/2$ and $\delta < -1$, we can find α', $\alpha < \alpha' < 1/2$, such that $g(t) \le C(\delta,\alpha)t^{-\alpha'}$ for all s, all t and then

$$N([0,1],\tau_g,\varepsilon) \le C(\delta,\alpha)\varepsilon^{-p}, p = \frac{2}{1-2\alpha'} > 0. \tag{27}$$

Here and in the sequel the letter $C(\ldots)$ stands for a constant, depending on the parameters in parenthesis. If $\alpha - 1/2$ we have

$$N([0,1],\tau_g,\varepsilon) \le C(\delta)\exp\{-2\varepsilon^{-2/1+\delta}\}.$$

What remains to be done is to estimate the function $\nu([0,1],\varepsilon)$. For the simplicity of writing denote $\eta_k = W_g(t_k)$. We choose $t_1 = 1/2$ (without loss of generality we can suppose that $E\eta_1^2 = g^2(t_1)t_1(1-t_1) = 1/4g^2(t_1) \ge \varepsilon^2$, since only small values of ε are of interest for us) and from the sequence of points $t_1 < t_2 < \ldots < t_{\nu(\varepsilon)}$ with the property

$$E(\eta_k^2 \mid \eta_1,\ldots,\eta_{k-1}) - (E(\eta_k \mid \eta_1,\ldots,\eta_{k-1}))^2 \ge \varepsilon^2 \text{ for all } k \le \nu(\varepsilon).$$

By the symmetry of the Brownian bridge and the function g with respect to the point $1/2$ we have a symmetric sequence of points $s_k = 1 - t_k$. For the construction of the t_k's we use the following fact (see exercise 17 on p. 38 of [12] or [7]): let x, y be real, $0 \le c \le t \le d \le 1$, then under the condition that $W_0(c) = x$, $W_0(d) = y$ the distribution of $W_0(t)$ is the same as that of the random variable

$$(d-c)^{1/2} W_0\left(\frac{t-c}{d-c}\right) + \frac{d-t}{d-c}x + \frac{t-c}{d-c}y.$$

Therefore, taking $c = 1/2$, $d = 1$, $y = 0$ and remembering that $W_0(t)$ is distributed as a normal random variable with mean zero and variance $t(1-t)$, it is easy to verify that

$$E(\eta_2^2 \mid \eta_1) - (E(\eta_2 \mid \eta_1))^2 = (t_2 - t_1) g^2(t_2) \frac{1-t_2}{1-t_1}.$$

Similarly, using the Markow property of the Brownian bridge (see [7]) we have

$$E(\eta_k^2 \mid \eta_1, \dots, \eta_{k-1}) - (E(\eta_k \mid \eta_1, \dots, \eta_{k-1}))^2 = E(\eta_k^2 \mid \eta_{k-1})$$

$$- (E(\eta_k \mid \eta_{k-1}))^2 = (t_k - t_{k-1}) g^2(t_k) \frac{1-t_k}{1-t_{k-1}}.$$

Since $t_k > t_{k-1}$, after comparing the last expression with formula (26) we can conclude that $\nu([0,1],\varepsilon)$ is of the same order as $N([0,1],\tau_g,\varepsilon)$. The fact that in the estimation of $\tau_g(t_k, t_{k-1})$ there is $g^2(t_{k-1})$ in (26) will not affect the order of $\nu(\varepsilon)$, since if we denote $\tilde{\tau}_g(s,t) = g^2(s)(t-s)$, $s < t$, then again using the property that $t^{1/2}g(t)$ increases we have

$$\tilde{\tau}^2(s,t) \le g^2(t)(t-s)\frac{t}{s}$$

and for the sequence t_k we always have $t_k t_{k-1}^{-1} \le 2$.

Now we can complete the proof of Proposition 5. If $\alpha < 1/2$, then by means of (27) we can use condition (24) and if $\alpha = 1/2$, it is easily verified that (22) and (23) are satisfied, and Proposition 5 is proved.

To verify the condition (5) for the process W_g we need one more result about the exponential decrease of the density of the distribution of the norm of a Gaussian random element.

Theorem 6 [13]. *Let η be a mean zero Gaussian random element in a separable Banach space. Let $F(x) = P\{\|\eta\| < x\}$ and $p(x) = F'(x)$. Then there exist constants C_1, C_2 and C_3 depending on the covariance operator T of the random element η such that for $x > C_3$*

$$p(x) \le C_1 \exp\{-C_2 x^2\}. \tag{28}$$

Now let $p_g(t)$ denote the distribution density of $||W_g|| = \sup_{0\le t\le 1}$ $|W_g(t)|$ and C_i, $i = 1, 2, 3$ be the constants from Theorem 6 of this particular Gaussian process W_g considered as an element of the space $C[0,1]$. Let C_4 be a constant for which $\sup_t p_g(t) \le C_4$. Then if $s < C_3$

$$P\{s \le ||W_g|| \le s+\varepsilon\} = \int_s^{s+\varepsilon} p_g(t)dt \le C_4\varepsilon \le (1+C_3)^3 C_4 (1+s)^{-3}\varepsilon$$

and if $s \ge C_3$

$$\int_s^{s+\varepsilon} p_g(t)dt \le C_1 \exp\{-C_2 s^2\}\varepsilon \le C_5(1+s)^{-3}\varepsilon.$$

Thus we have shown that all conditions of Theorem 2 are satisfied and the proof of Theorem 3 is completed.

Concluding Remarks. 1. For the estimation of the rate of convergence in the limit theorem for weighted empirical processes Proposition 5 is sufficient, since the conditions of Theorem 3 on the growth of g at the endpoints of the interval $[0,1]$ are much more restrictive. But the question whether the proposition holds true for all functions $q \in Q$ satisfying (8) remains open and seems rather interesting. One would like to believe that in the case of the monotone function g the density of $\sup|W_g(t)|$ would be bounded if W_g is continuous a.s. at the endpoints of the interval $[0,1]$.

2. Other questions related to Proposition 5 were considered in [11].

3. There is little hope to obtain the rate of convergence of order $n^{-1/2}$ in the limit theorem for weighted empirical processes as a consequence of a general result in the space $D[0,1]$. Therefore it should be interesting to try to get this order by more direct methods under one of the conditions (11) or (12).

REFERENCES

[1] Alexander, K.S., *The central limit theorem for empirical processes of Vapnik–Cervonenkis classes,* Ann. Probab. **15**, No. 1, (1987), 178–203.

[2] Billingsley, P., *Convergence of Probability Measures,* Wiley, New York, 1968.

[3] Cibisov, D.M., *Some theorems on the limiting behaviour of empirical distribution functions,* Selected Transl. Math. Statist. Probab. **6** (1964), 147–156.

[4] Dudley, R.M., *An extended Wichura theorem, definitions of Donsker class and weighted empirical distributions,* Lect. Notes in Math. **1153** (1985), 141–178.

[5] Gine, E. and Zinn, J., *Some limit theorems for empirical processes,* Ann. Probab. **12** (1984), 929–989.

[6] Hahn, M., *Central limit theoren in* $D([0,1])$, Wahrscheinlichkeitstheorie verw. Geb. **44** (1978), 89–101.

[7] Karlin, S. and Taylor, H.M., *A second course in stochastic processes,* Academic Press, New York, 1981.

[8] Lifshits, M.A., *The distribution of the maximum of a Gaussian process,* Teor. verojat. i primen., **31** No. 1, (1986), 134–142 (in Russian).

[9] O'Reilly, N., *On the convergence of empirical processes in supnorm metrics,* Ann. Probab. **2** (1974), 642–651.

[10] Paulauskas, V. and Jukneviciene, D., *On the rate of convergence in the central limit theorem in the space* $D[0,1]$, Liet. matem. rink. **28**, No. 3, (1988), 507–519 (in Russian).

[11] Paulauskas, V., *A note on the distribution of the supremum of some Gaussian processes,* University of Goteborg (1988), preprint no. 13.

[12] Shorack, G.R. and Wellner, J.A., *Empirical Processes with Applications to Statistics,* John Wiley & Sons, New York, 1986.

[13] Tsirel'son, B.S., *The density of the distribution of the maximum of a Gaussian process,* Theor. Probab. Appl. **20** 4, (1975), 847–855.

Vilnius University
Department of Mathematics
USSR, Vilnius 232006
Partizanu 24

Probability in Banach Spaces 7
Birkhauser Boston 1989

INFINITE–DIMENSIONAL DISTRIBUTIONS IN THE THERMODYNAMIC LIMIT OF GRAPH–VALUED MARKOV PROCESSES AND THE PHENOMENON OF POSTGELATION STICKING*

by

B. PITTEL** and **W.A. WOYCZYNSKI**

Department of Mathematics
Ohio State University
Columbus, Ohio 43210

Department of Mathematics and Statistics
Case Western Reserve University
Cleveland, Ohio 44106

1. INTRODUCTION

In the first four papers of this series (cf. Pittel, Woyczynski and Mann) (1987), (1989a), (1989b), and Pittel and Woyczynski (1989c)) we began a detailed asymptotic analysis of the Flory–Stockmayer–Whittle polymerization process (cf. e.g. Stockmayer (1944), Whittle (1986)). In the case considered in the present paper this process is modeled as a continuous time Markov process $\{M(t):\ t \geq 0\}$ whose finite state space $\mathscr{M}$ is a set of all forests M on a set V_n of n vertices labeled 1,2,...,n, and interpreted as basic structural units (monomers). The state of the process changes in time since bonds between the vertices may form, or break. Specifically, if

---

*Research supported in part by SRO ONR and NSF Grants.

**Research partially conducted during the author's visit at CWRU

Key words and phrases: Random trees, polymerization, Gaussian and 3/2–stable limit behavior, nearcritical, supercritical, phase transition, Markov process, stationary distribution.

AMS (1980) Classification: 60J25, 60K35, 82A51.

at a time t the process is in a state M, then the rate of bond formation between two vertices a and b, which belong to the same tree component of M, equals $\lambda A_{j+1} A_{k+1}/A_j A_k$, where j and k are, respectively, the degree of a and b in M, (in short, j = deg(a,M), k = deg(b,M)). Furthermore, the rate of bond breaking for two vertices a and b connected by a bond in M equals $\mu D_{j-1} D_{k-1}/D_j D_k$ if deg(a,M) = j, deg(b,M) = k. In these formulas $\lambda > \mu > 0$ and $\{A_j\colon j \geq 0\}$, $\{D_j\colon j \geq 0\}$ are such that $A_o, D_o, \geq 0$ and $A_j, D_j > 0$ for $1 \leq j \leq j_1$, $j_1 \geq 3$. (In case $j_1 \leq 2$, the trees are reduced to chains.) Concrete examples of natural choices of sequences (A_j) and (D_i) can be found in our papers mentioned above.

It was announced in Pittel, Woyczynski and Mann (1987) and proved (1989a) that the Markov process M(t) has a stationary distribution

$$P(M) = Q^{-1} q(M), \quad M \in \mathcal{M}, \tag{1.1}$$

where

$$q(M) = (\mu/\lambda)^{C(M)} \prod_{a \in V_n} H_{\deg(a,M)},$$

$$H_j =_{def} A_j D_j, \quad j \geq 0,$$

C(M) is the total number of trees in M, and Q is a normalizing factor. We sought information on asymptotic behavior of the stationary distribution for $n \to \infty$ under assumption that $\{H_j\colon j \geq 0\}$ is fixed, but μ, λ change with n in such a way that $\mu/\lambda = n/\sigma_n$, where σ_n is bounded away from both 0 and ∞. This corresponds to the thermodynamic passage to the limit in statistical mechanics.

We restrict our attention to a case when

(1) a series $H(y) = \Sigma_{j>0} H_j y^j/j!$ has a positive radius r of convergence, and

(2) the positive root $\bar{y}$ of

$$yH^{(2)}(y) - H^{(1)}(y) = 0, \quad y \in (0,r), \tag{1.2}$$

is its only root in the disk $|y| \le \bar{y}$.

Under these technical conditions, the equation

$$(1.3) \qquad y = xH^{(1)}(y)$$

determines a function $y = R_1(x)$ which is analytic for $|x| < \bar{x}$ and continuous for $|x| \le \bar{x}$, where $\bar{x}$ satisfies

$$(1.4) \qquad \bar{y} = \bar{x}\, H^{(1)}(\bar{y}), \quad 1 = \bar{x}\, H^{(2)}(\bar{y}),$$

see (1.2). It can be seen (Pittel, Woyczynski, Mann (1989a)) that

$$(1.5) \qquad R_s(x) =_{\text{def}} xH^{(s)}(R_1(x)), \quad s \ge 0,$$

is the exponential generating function (e.g.f.) of a nonnegative sequence $\{R_{sj}: j \ge 1\}$ given by the formula

$$R_{sj} = \sum_{T'} h_s(T').$$

Here the sum extends over all rooted trees on V_j and, if the root of T' is denoted by a,

$$h_s(T') = [\prod_{b \in V_j \setminus \{a\}} H_{\deg(b,T')}]\, H_{\deg(a,T') + s}.$$

We shall also need a sequence $\{f_j: j \ge 1\}$ determined by

$$(1.6) \qquad f_j = \sum_T h(T), \quad h(T) =_{\text{def}} \prod_{b \in T} H_{\deg(b,T)},$$

where T runs through $\mathcal{T}_j$ the set of all (free) trees on V_j. Again, by results in Pittel, Woyczynski, Mann (1989a), $F(x)$, the e.g.f. of this sequence, satisfies, for $|x| \le \bar{x}$,

$$R_o(x) = x\,F^{(1)}(x),$$

(1.7)

$$F(x) = R_o(x) - R_1^2(x)/2.$$

The present paper concentrates on infinite dimensional distributions arising in the thermodynamic limit for $M = M_n(t)$, and on the rigorous explanation of the phenomenon of post–gelation sticking discovered on a heuristic level by Stockmayer (1944).

2. PRELIMINARIES

In Pittel, Woyczynski, Mann (1989a–b) we studied the asymptotic behavior of the distribution $\{P(M)\}$. The study rigorously established the existence of three phases (resp. subcritical, nearcritical and supercritical) of polymerization corresponding to different domains of the ratio $\sigma_n = \sigma$ (resp. $< \overline{\sigma}$, $\sim \overline{\sigma}$, and $> \overline{\sigma}$, where $\overline{\sigma} = R_o(\overline{x}) = \overline{x}H(\overline{y})$) of association and dissociation rates of monomers. The following characteristics of the random forest M (or, more appropriately, M_n) were of primary interest there, and continue to be of importance in the present paper:

(a) the sequence $C_n = \{c_{nj}\colon\ j \geq 1\}$ where c_{nj} is the total number of tree components of size j, ($c_{nj} = 0$ for $j > n$, obviously), and

(b) $L_n^{(k)}$ and S_n, which are respectively the size of the k–th largest tree and the size of a tree which contains a randomly picked vertex.

To make comparisons for different domains easier, let us first recall four results from Pittel, Woyczynski, Mann (1989a–b). The first two theorems describe the situation in the subcritical case.

<u>*THEOREM A*</u>. *Let* $\sigma_n \equiv \sigma < \overline{\sigma}$, *and* ρ *be the positive root of* $R_o(x) = \sigma$, $x \in (0,\overline{x}\,)$. (*It exists since* $\overline{\sigma} = R_o(\overline{x})$.) *Introduce the numbers*

$$m_j = m_j(\sigma) = \sigma^{-1} f_j \rho^j / j!,\quad j \geq 1,$$

(f_j *is defined in* (1.6)), *and set*

$$C_n^* = \{c_{nj}^*: j \geq 1\} =_{def} \{(c_{nj} - nm_j)n^{-1/2}: j \geq 1\}.$$

The random sequence C_n^* *weakly converges in a Banach space* $\ell_1^{(\delta)}$, $\delta \in (0,1/2)$, *to a Gaussian sequence* $G = \{g_j: j \geq 1\}$ (*in short,* $C_n^* \overset{\ell_1^{(\delta)}}{\Longrightarrow} G$) *such that*

$$E(g_j) = 0, \ E(g_j g_k) = m_j \delta_{jk} - r_j r_k, \ j, \ k \geq 1,$$

where

$$r_j = a\, R_{oj} \rho^j / j!, \ a = [\sigma \rho R_o^{(1)}(\rho)]^{-1/2}.$$

Here $\ell_1^{(\delta)}$ *is the space of all sequences of reals* $s = \{s_j: j \geq 1\}$ *such that*

$$\|s\| =_{def} \sum_{j \geq 1} m_j^{-\delta} |s_j| < \infty.$$

In addition, for every $j \geq 1$, *all the moments of* c_{nj}^* *converge to those of* g_j.

<u>*THEOREM B.*</u> *Under the assumptions of Theorem* A,

(*a*) S_n *is bounded in probability, and, more precisely,* $S_n \Rightarrow S$, *such that*

$$P(S = j) = \sigma^{-1} R_{oj} \rho^j / j!, \quad j \geq 1,$$

$$\lim E(S_n) = E(S) = \sigma^{-1} \rho R_o^{(1)}(\rho) < \infty;$$

(*b*) *in probability,*

$$L_n^{(1)} = \eta^{-1}[\log n - (5/2) \log\log n] + 0(1), \quad \eta = \log(\overline{x}/\rho).$$

<u>*Remark 1*</u>. (i) It follows from (3.9) of Pittel, Woyczynski and Mann (1989a) that

$$(2.1)\qquad f_j/j! = (1 + o(1))\beta_0(\bar{x})^{-j}j^{-5/2}, \quad j \to \infty,$$

where

$$(2.2)\qquad \beta_0 =_{\text{def}} [\bar{x}\,\bar{y}/2\pi H^{(3)}(\bar{y})]^{1/2} H^{(1)}(\bar{y}).$$

Hence

$$(2.3)\qquad m_j = \sigma^{-1}\beta_0(\rho/\bar{x})^j j^{-5/2}(1 + o(1)), \quad j \to \infty,$$

where, as we recall, $\rho/\bar{x} < 1$. Thus, by Theorem A, the limiting process G lives in a rather thin space of sequences with tails decreasing exponentially fast. However, the rate of decrease diminishes while σ increases, since $\rho = \rho(\sigma)$ increases with σ. In fact, $\rho(\bar{\sigma})/\bar{x} = 1$, so in a nearcritical domain we should anticipate a much broader tail of a limiting process.

(ii) The limiting distribution of S_n depends on σ both explicitly (through σ^{-1}) and implicitly (through $\rho = \rho(\sigma)$).

(iii) Since $\eta = \eta(\sigma) \to 0$ as $\sigma \to \bar{\sigma}-$, the formula for $L_n^{(1)}$ signals that in the case when $\lim \sigma_n = \bar{\sigma}$, (not to mention the case when $\lim \sigma_n > \bar{\sigma}$) we must expect a faster growth of $L_n^{(1)}$.

The third theorem deals with the supercritical case.

<u>*THEOREM C*</u>. *Suppose that* $\sigma_n \equiv \sigma > \bar{\sigma}$. *Then, with probability approaching* 1 *as* $n \to \infty$ *(in short, almost surely (a.s.)), the forest* M_n *contains a single tree of size relatively close to* $n(1 - \bar{\sigma}/\sigma)$, *and all other trees are of order at most* $n^{2/3}$. *More precisely,*

(a) *The distribution of* $L_n^{(1)}$ *satisfies a local limit–type relation*

$$(2.4)\qquad \begin{gathered} P(L_n^{(1)} = j) = (1 + o(1))p(x_j)\Delta x_j, \\ x_j =_{\text{def}} [n(1 - \bar{\sigma}/\sigma) - j](\bar{\sigma}n/\sigma)^{-2/3}, \quad \Delta x_j =_{\text{def}} (\bar{\sigma}n/\sigma)^{-2/3}, \end{gathered}$$

uniformly over j *such that* x_j *belongs to a bounded interval. Here* $p(\cdot)$ *is the density of the* (3/2)*–stable (Holtsmark) distribution with canonical Levy* *measure concentrated on* $[0,\infty)$ *and assigning to an interval* $[0,x]$ *the measure equal to* $(\beta_0/\overline{\sigma})x^{1/2}$ *(see (2.2) for* β_0*); subsequently*

$$[n(1-\overline{\sigma}/\sigma) - L_n^{(1)}](\overline{\sigma}n/\sigma)^{-2/3} \Rightarrow X,$$

where

$$E[\exp(iuX)] = \exp[i\Psi(u)],$$

(2.5)

$$\Psi(u) = (4/3\overline{\sigma})\pi^{1/2}\beta_0 e^{-i3\pi/4}u^{3/2}, \quad u \geq 0, \quad \Psi(u) = \overline{\Psi}(-u) \text{ if } u < 0.$$

(*b*) *For every fixed* $x > 0$ *and* $k \geq 2$,

$$\lim P(L_n^{(k)} \leq x\, n^{2/3}) = e^{-\lambda(x)} \sum_{0 \leq j < k-2} \lambda^j(x)/j!,$$

(2.6)

$$\lambda(x) = (2\beta_0/3\sigma)x^{-3/2}.$$

Finally, the fourth result describes the behavior of M in the nearcritical case.

<u>THEOREM D</u>. Suppose that $\overline{\sigma}/\sigma_n = 1 - an^{-1/3}$, *where* $a \in (-\infty,\infty)$ *is fixed. Then, for every* $x > 0$ *and* $k \geq 1$,

$$\lim P(L_n^{(k)} \leq xn^{2/3}) = e^{-I(x)} \sum_{0 \leq j \leq k-1} I^j(x)/j!,$$

where

$$I(x) = \beta_0(\overline{\sigma}\, p(a))^{-1} \int_x^{\infty} y^{-5/2} p(a-y)dy.$$

In particular,

$$\lim P(L_n^{(1)} \leq xn^{2/3}) = e^{-I(x)}.$$

<u>*Remark 2.*</u> Thus (see Theorem B), the largest component of M_n is of size about $\log n$, or $n^{2/3}$, or n dependent upon whether σ_n is below, and bounded away from $\overline{\sigma}$, or close enough to $\overline{\sigma}$, or above $\overline{\sigma}$ and bounded away from it.

3. MAIN RESULTS

Our main result gives the infinite dimensional, joint, molecular weights distribution in the thermodynamic limit (as $n \to \infty$) in the nearcritical and supercritical domains.

THEOREM 1. Let $\sigma_n \equiv \sigma \geq \bar{\sigma}$. *Introduce a sequence*

$$C_n^* = \{c_{nj}^*: \ j \geq 1\} =_{\text{def}} \{(c_{nj} - nm_j)n^{-1/2}: \ j \geq 1\},$$

(2.7) $$m_j = m_j(\sigma) =_{\text{def}} \sigma^{-1} f_j \bar{x}^j / j!,$$

and let $\ell_1^{(\tau)}$, $\tau \in (0,1/4)$, *denote a Banach space of sequences* s *such that*

$$\|s\| =_{\text{def}} \sum_{j \geq 1} j^{\tau} |s_j| < \infty.$$

Then $C_n^* \xrightarrow{\ell_1^{(\tau)}} G$, *where* $G = \{g_j: \ j \geq 1\}$ *is a Gaussian sequence with* *independent components and*

(2.8) $$E(g_j) = 0, \ \ \text{var}(g_j) = m_j, \ \ j \geq 1.$$

Also, all the moments of c_{nj}^* *converge to those of* g_j, $j \geq 1$.

By this theorem, G is such that, with probability one, $\Sigma_{j > 0}\, j^{\tau} |g_j| < \infty$, for $\tau < 1/4$. This result can not be significantly improved since $\Sigma_{j > 0}\, j^{1/4} |g_j| = \infty$ with probability one. Indeed, $g_1, g_2,,,$ are independent, Gaussian and a simple estimate of two first moments of $\mathscr{S}_k = \Sigma_{1 \leq j \leq k}\, j^{1/4} |g_j|$ based on (2.8) and the asymptotics

(2.9) $$m_j = \sigma^{-1} \beta_0 j^{-5/2}(1 + o(1)), \ j \to \infty,$$

(cf. (2.7)) leads to the following limit behavior

$$(P)\lim_{k \to \infty} \mathscr{S}_k / \log k = \text{const} > 0.$$

<u>*COROLLARY 1.*</u> *Let, as before,* S_n *denote the size of a randomly picked tree in the forest* M_n. *Then*

$$\lim P(S_n = j) = \sigma^{-1} w_j, \quad w_j = R_{oj} \bar{x}^j / j!, \quad j \geq 1. \tag{2.9}$$

<u>*Remark 3*</u>. Observe that now

$$\sum_{j \geq 1} \sigma^{-1} w_j = \sigma^{-1} \sum_{j \geq 1} R_{oj} \bar{x}^j / j! = \sigma^{-1} R_o(\bar{x}) = \bar{\sigma}/\sigma.$$

Therefore, for $\sigma = \bar{\sigma}$ (the nearcritical domain), S_n is still bounded in probability, even though

$$\lim E(S_n) \geq \sigma^{-1} \bar{x}\, R_o^{(1)}(\bar{x}-) = \infty.$$

But if $\sigma > \bar{\sigma}$ (the supercritical domain), this is no longer the case. Moreover, from the proof of the Theorem C (cf. Pittel, Woyczynski, Mann (1989b)) it can be deduced that, for $\omega(n) \to \infty$, $\omega(n) = o(n)$,

$$P(S_n \leq \omega(n)) \to \bar{\sigma}/\sigma, \quad n \to \infty.$$

Thus we can interpret the deficit $1 - \bar{\sigma}/\sigma$ of $\{\sigma^{-1} w_j\}$ as the limit of a probability that a randomly selected vertex belongs to a very large component ("gel") whose size is of order n. Besides, for every $j \geq 1$,

$$\lim P(S_n = j \mid S_n \leq \lambda(n)) = w_j^* =_{\text{def}} w_j / \sum_{k \geq 1} w_k.$$

Hence, the limiting <u>conditional</u> distribution of S_n, given that it is relatively small compared to n, does not depend on σ! For the Flory

Stockmayer model of polymerization, when $H(y) = (1 + y)^f$, $f \geq 3$, this "*postgelation sticking*" phenomenon was already derived – with help of heuristic arguments – by Stockmayer (1944), and later computer simulations confirmed it as well. Still, our argument provides the first rigorous proof of the existence of this striking phenomenon.

A natural question is how many such giant components are there in the random forest M_n? In the case of the Flory–Stockmayer model, it has been long believed that – with high probability for large n – such a giant tree must be unique. In fact, Donoghue (1982) (see also Donoghue and Gibbs (1979)) undertook an asymptotic analysis of the Flory–Stockmayer distribution which did indicate uniqueness, but their arguments in this respect were not completely satisfying. Our Theorem C demonstrated that the uniqueness hypothesis is indeed true even for a general function $H(\cdot)$ and that, in the super–critical case, the size of the largest component has a Holtsmark distribution.

4. PROOFS.

4.1 An Outline. The structure of proof of Theorem 1 is somewhat complex so we include in this subsection a description of the composition of our arguments.

We begin (Subsection 4.2) with the formulation of two auxiliary Lemmas (1 & 2) which establish, under certain assumptions on σ_n, the asymptotic estimates for $Q_n E_n(z)/n!$ where Q_n is the normalizing factor in (1.1) and $E_n(z)$ is the multi–dimensional moment generating function of $C_n^* = \{c_{nj}^*;\ 1 \leq j \leq n\}$. (Notice that the special case of $z = 1$ was considered in Pittel, Woyczynski, Mann (1989b)). Then (Corollary 2) we establish the convergence of finite dimensional distributions of C_n^* to those of a Gaussian sequence. This will give the first part of Theorem 1. The proof of the second part (tightness condition) has to be postponed until later (Subsection 4.6).

The long tedious, and technical Subsections 4.3, and 4.4 contain proofs of auxiliary Lemmas 1 and 2.

As a particular case, Lemmas 1 and 2, contain information about asymptotic behavior of $Q_n/n!$ This (including the case $\liminf \delta_n > 0$) is summarized in Subsection 4.5 (Lemma 3) which also contains a discussion of connections of this type of result with local limit theorems for stable densities.

The paper concludes with the proof of the tightness conditions in Theorem 1 (Subsection 4.6).

4.2. *Proof of the finite–dimensional convergence in Theorem 1.* Fix μ, λ. Denote the normalizing factor in (1.1) by Q_n. If $z = \{z_j: j \geq 1\}$ is a sequence of reals such that $\sup\{|z_j|: j \geq 1\} < \infty$ then (Pittel, Woyczynski, Mann (1989a) (3.36)]

$$\sum_{n \geq 0} Q_n E_n(z) x^n/n! = \exp[(\mu/\lambda)\varphi(x,z)],$$

$$E_n(z) =_{\text{def}} E\Big(\prod_{j \geq 1} z_j^{c_{nj}}\Big), \tag{4.1}$$

$$\varphi(x,z) =_{\text{def}} \sum_{j \geq 1} z_j f_j x^j/j!$$

Here $E_n(z)$ is clearly the multidimensional generating function of the random sequence C_n, and (4.1) means that the function $\exp[(\mu/\lambda)\varphi(x,z)]$ happens to be the e.g.f. of $\{Q_n E_n(z): n \geq 0\}$, ($Q_o = E_o = 1$ by convention).

Fix an integer $k \geq 1$ and let $z_j = z_j(n)$ be such that $z_j = 1 + o(n^{-1/3})$, $n \to \infty$, for $1 \leq j \leq k$, and $z_j \equiv 1$ for $j > k$.

LEMMA 1. *If* $\liminf \sigma_n > \overline{\sigma}$ *then*

$$Q_n E_n(z)/n! = (1 + o(1)) n^{-3/2} q_n \exp[\mathscr{N}_n(z)],$$

where

$$q_n = q(\sigma_n) = \beta_o[\sigma_n(1 - \overline{\sigma}/\sigma_n)^{5/2}]^{-1}$$

(4.2)

$$\mathscr{N}_n(z) = n[\sigma_n^{-1}\varphi(\overline{x},z) - \log \overline{x}\,] = n[\sigma_n^{-1}F(\overline{x}\,) - \log \overline{x}\,] + n \sum_{j=1}^{k} m_j(z_j - 1),$$

and $F(\cdot)$, β_o, m. *are defined, respectively, in* (1.7), (2.2) *and* (2.7) *with* σ *replaced by* σ_n.

LEMMA 2. *If* $a_n = n^{1/3}(1 + \bar{\sigma}/\sigma_n) = 0(1)$ *then*

$$Q_n E_n(z)/n! = (1 + o(1))n^{-2/3}p(a_n)\exp[\mathcal{N}_n(z)],$$

where $p(\cdot)$ *is the density introduced in Theorem* C.

The following corollary establishes the convergence of finite–dimensional distribution in Theorem 1.

COROLLARY 2. Let $\sigma_n \equiv \sigma \geq \bar{\sigma}$. *Then the process* $C_n^* = \{c_{nj}^*: j \geq 1\} = \{(c_{nj} - n\ m_j)n^{-1/2}: j \geq 1\}$ *converges in terms of finite dimensional distributions to the Gaussian sequence* $G = \{g_j: j \geq 1\}$ *defined in Theorem* 1. *Moreover, all the moments of* c_{nj}^* *converge to those of* g_j, $j \geq 1$.

Proof of Corollary 2. To begin with suppose, that $\sigma > \bar{\sigma}$. Setting $z_j = \exp(u_j/n^{1/2})$, $u_j \in (-\infty,\infty)$, $(1 \leq j \leq k)$, and using Lemma 1, we have that

$$E_n(z) = E[(\sum_{1 \leq j \leq k} u_j c_{nj}/n^{1/2})]$$

$$(1 + o(1))\exp[\mathcal{N}_n(z) - \mathcal{N}_n(1)] = (1+o(1))\exp[n\sum_{1 \leq j \leq k} m_j(z_j - 1)]$$

$$= (1 + o(1))\exp[\sum_{1\leq j\leq k}(n\ m_j u_j/n^{1/2} + m_j u_j^2/2) + 0(n^{-1/2})],$$

or

$$\lim E[\exp(\sum_{1 \leq j \leq k} u_j c_{nj}^*)] = \exp(\sum_{1 \leq j \leq k} m_j u_j^2/2).$$

The case $\sigma = \bar{\sigma}$ is treated similarly with the help of Lemma 2.

The proof of the Theorem 1 will be complete when we check later

(Subsection 4.7) the tightness condition for C_n^* in $\ell_1^{(t)}$.

4.3. *Proof of Lemma 1.* Fix a sequence $\{z_j\colon\ j \geq 1\}$. Using the Cauchy integral formula, we write

$$Q_n E_n(z)/n! = (2\pi i)^{-1} \int_{\mathscr{C}} \exp[(\mu/\lambda)\varphi(x,z) - n \log x] x^{-1} dx,$$

where $\mathscr{C}$ is a contour surrounding the origin $x = 0$. Having arrived at this representation for $Q_n E_n(z)/n!$, we may and shall assume that both μ/λ and $\{z_j\}$ depend on n, namely $\mu/\lambda = n/\sigma_n$ and $\{z_j\}$ is such that $z_j = 1 + o(n^{-1/3})$, $n \to \infty$, $1 \leq j \leq k$, and $z_j \equiv 1$ for $j > k$. Choose $\mathscr{C} = \{x\colon x = \bar{x}e^{i\phi},\ -\pi \leq \phi < \pi\}$, where $\bar{x}$ is defined in (1.4). Then the identity above becomes

$$Q_n E_n(z)/n! = (2\pi)^{-1} \int_{-\pi}^{\pi} \exp[N_n(\phi,z)] d\phi, \tag{4.3}$$

where

$$\begin{aligned} N_n(\phi,z) = \mathscr{N}_n(z) + n\sigma_n^{-1}[F(\bar{x}e^{i\phi}) - F(\bar{x}) + \\ + \sum_{1 \leq j \leq k} (z_j - 1) f_j \bar{x}^j (e^{ij\phi} - 1)/j!] - in\phi, \end{aligned} \tag{4.4}$$

and $\mathscr{N}_n(z)$ is defined in (2.2). We need to find a sharp asymptotic estimate of the integral in (4.3). It turns out (not too surprisingly) that the dominant part of the integral corresponds to the small values of ϕ. To see it, we have to find an expansion for $N_n(\phi,z)$ in powers of ϕ. A key relation is: for $m \geq 4$, and $\phi \geq 0$,

$$\begin{aligned} F(\bar{x}e^{i\phi}) - F(\bar{x}) = i\,\bar{\sigma}\,\phi + \alpha_3\phi^{3/2} + \sum_{4 \leq \ell \leq m} \alpha_\ell \phi^{\ell/2} + 0(n\,\phi^{(m+1)/2}), \\ \alpha_3 = -(4/3)\pi^{1/2}\beta_0 e^{i\pi/4}. \end{aligned} \tag{4.5}$$

Let us outline the derivation of (4.5). According to (1.5), (1.7),

$$F(x) = R_o(x) - R_1^2(x)/2 = H(R_1)f(R_1) - R_1^2/2, \; R_1 = R_1(x),$$

where $f(y) = y/H^{(1)}(y)$. By (1.2), $f^{(1)}(\bar{y}) = 0$, so expanding F in powers of $R_1 - R_1(\bar{x}) = R_1 - \bar{y}$, we have

$$\text{(4.6)} \quad F(x) = F(\bar{x}) + 2^{-1}(R_1-\bar{y})^2 Hf^{(2)} + 6^{-1}(R_1-\bar{y})^3(2H^{(1)}f^{(2)} + Hf^{(3)}) + \ldots,$$

where H, f and their derivatives are evaluated at $y = \bar{y}$. In particular,

$$\text{(4.7)} \qquad f^{(2)}(\bar{y}) = -\bar{y}H^{(3)}(\bar{y})/[H^{(1)}(\bar{y})]^2.$$

According to (1.5), $R_1(x)$ satisfies $f(R_1) = x$. Since $f^{(1)}(\bar{y}) = 0$ and $f^{(2)}(\bar{y}) \neq 0$ (see (2.7)), by the implicit function theorem we have, for x close enough to $\bar{x}$ and $|x| \leq \bar{x}$, that

$$\text{(4.8)} \qquad R_1(x) - \bar{y} = \sum_{j \geq 1} r_j(x-\bar{x})^{j/2},$$

where we select the main branch of the square root function. Plugging (4.8) into $f(R_1) = x$, we get after some work:

$$\text{(4.9)} \qquad 2^{-1}f^{(2)}r_1^2 = 1, \; f^{(2)}r_1r_2 + 6^{-1}f^{(3)}r_1^3 = 0.$$

Subsequently (see (4.7)),

$$\text{(4.10)} \qquad r_1 = \pm\, i[2(H^{(1)})^2/\bar{y}\, H^{(3)}]^{1/2},$$

and we must choose $+$, since $R_1(x) < \bar{y}$ for $x < \bar{x}$. A combination of (4.6) – (4.10) leads to

$$F(x) = F(\bar{x}) + \sum_{j\geq 2} d_j (x-\bar{x})^{j/2},$$

where

$$d_2 = (2^{-1} f^{(2)} r_1^2) H = H,$$

$$d_3 = (f^{(2)} r_1 r_2 + 6^{-1} f^{(3)} r_1^3) H + 3^{-1} f^{(2)} r_1^3 \, H^{(1)}$$

$$3^{-1} f^{(2)} r_1^3 \, H^{(1)} = i \, 3^{-1} 2^{3/2} [(H^{(1)})^2 / \bar{y} \, H^{(3)}]^{1/2} H^{(1)}.$$

These formulas yield (4.5) (see (1.9) for β_o), since $\bar{x}e^{i\phi} - \bar{x} = \bar{x}i\phi + 0(\phi^2)$.

Using (4.4) and (4.5), we obtain: for $m \geq 2$ and $\phi \geq 0$,

(4.11) $$N_n(\phi,z) = \mathscr{N}_n(z) + n \sum_{2\leq j\leq m} \gamma_j(z)\phi^{j/2} + 0(n\phi^{(m+1/2)}),$$

where

$$\gamma_2(z) = i[\bar{\sigma}/\sigma_n - 1 + \sigma_n^{-1} \sum_{1\leq j\leq k} (z_j - 1) j f_j \bar{x}^j / j!]$$

(4.12)

$$= i[\bar{\sigma}/\sigma_n - 1 + o(1)], \quad n \to \infty,$$

$$\gamma_3(z) = \gamma_3 = (4/3\sigma_n) \, \pi^{1/2} \beta_o \exp(-3\pi i/4).$$

With the expansion (4.11) at hand, we can begin estimating the integral in (4.3). Choose s from $(0,2/3)$, set $\phi_o = n^{-s}$, and rewrite (4.3) as

$$Q_n E_n(z)/n! = (2\pi)^{-1} \int_1 + (2\pi)^{-1} \int_2,$$

(4.13)

$$\int_1 = 2\mathrm{Re}\{\int_0^{\phi_o} \exp[N_n(\phi,z)]d\phi\}, \quad \int_2 = 2\mathrm{Re}\{\int_{\phi_o}^{\pi} \exp[N_n(\phi,z)]d\phi\}.$$

(Re(u) is the real part of a complex number u.)

First, let us estimate $\int_2$. By (4.4),

$$\mathrm{Re}[N_n(\phi,z)] = \mathcal{N}_n(z) + n\sigma_n^{-1} \sum_{j \geq 1} z_j f_j(\bar{x}^j/j!)(\cos j\phi - 1)$$

$$\leq \mathcal{N}_n(z) - x\, n\, \phi^2, \quad \phi \in (-\pi,\pi].$$

(Here and below we use the letter c – with, or without, a sub(super) script to denote various positive numbers <u>independent</u> of n.) In addition (see (4.11), (4.12)), for <u>small</u> $\phi \geq 0$,

$$\mathrm{Re}[N_n(\phi,z)] \leq \mathcal{N}_n(z) - c_1 n\, \phi^{3/2}.$$

Since $\phi_o = n^{-s}$, $s \in (0,2/3)$, from these estimates if follows that

$$\int_2 = 0(\exp[\mathcal{N}_n(z) - c_1 n^{\nu}]), \quad \nu = 1 - 3s/2 > 0. \tag{4.14}$$

Second, let us asymptotically evaluate $\int_1$. By (4.12), for a fixed $m \geq 2$ and $\phi \in [0,\phi_o]$,

$$\mathrm{Re}[n \sum_{2 \leq j \leq m} \gamma_j(z)\phi^{j/2}] = -c\, n\, \phi^{3/2}[1 + 0(n^{-s/2})] \leq 0. \tag{4.15}$$

Select m so large that $m > (2/s) - 1$; then

$$n\, \phi_o^{(m+1)/2} = n^{-s(m+1)/2 + 1} = o(1). \tag{4.16}$$

Introducing

$$\tilde{N}_n(\phi,z) = \mathcal{N}_n(z) + A_n(\phi,z), \quad A_n(\phi,z) = n \sum_{2 \leq j \leq m} \gamma_j(z)\phi^{j/2}, \tag{4.17}$$

and using (4.15), (4.16), we obtain

$$|\exp[N_n(\phi,z)] - \exp[\tilde{N}_n(\phi,z)]|$$

(4.18)
$$= \exp(\mathrm{Re}[\tilde{N}_n(\phi,z)]|\exp[0(n\,\phi_o^{(m+1)/2})] - 1|$$

$$= 0(\exp[\mathcal{N}_n(z)]n^{-s(m+1)/2\,+\,1}).$$

Therefore

(4.19)
$$\int_0^{\phi_o} \exp[N_n(\phi,z)]d\phi = \int_0^{\phi_o} \exp[\tilde{N}_n(\phi,z)]d\phi + o(n^{-3/2}\exp[\mathcal{N}_n(z)]),$$

provided that

(4.20)
$$s + s(m+1)/2 - 1 > 3/2, \ \text{ or } \ m > (5/s) - 3,$$

which we may, and shall, assume. Thus, it remains to show that

$$\mathrm{Re}(J) = (1 + o(1)) \text{ const } n^{-3/2}, \quad J =_{\text{def}} \int_0^{\phi_o} \exp[A_n(\phi,z)]d\phi.$$

For this purpose, set first $u = \phi^{1/2}$ and write

$$J = 2\int_0^{u_o} u \exp[\mathcal{A}_n(u,z)]du, \quad \mathcal{A}_n(u,z) =_{\text{def}} A_n(u^2,z), \; u_o = \phi_o^{1/2}.$$

Second, introduce in the complex plane u a closed (clockwise oriented) contour $\mathcal{D} = \mathcal{D}_1 \cup \mathcal{D}_2 \cup \mathcal{D}_3$. Here $\mathcal{D}_1$ is the interval $[0,u_o]$ of the real line, $\mathcal{D}_2$ is the arc $u = u_o e^{i\psi}$, $0 \geq \psi \geq -\pi/4$, and $\mathcal{D}_3$ is the line segment connecting the points $u_o e^{-i\pi/4}$ and the origin 0. Since the integrand $2u \exp[\mathcal{A}_n(u,z)]$ is an analytic function of u, denoting it by $w(u)$ we have:

$$J = -\int_{\mathscr{D}_2} w(u)du - \int_{\mathscr{D}_3} w(u)du.$$

Consider the integral along $\mathscr{D}_2$. On $\mathscr{D}_2$, according to (4.12),

$$n\,\mathrm{Re}[\gamma_2(z)u^2] = -n\,\phi_o[(1-\bar{\sigma}/\sigma_n) + o(1)]\sin 2|\psi| \le 0,$$

$$n\,\mathrm{Re}(\gamma_3 u^3) \le -c\,n\,\phi_o^{3/2}\cos[(\pi/4) + 3\,\psi].$$

Hence,

$$\mathrm{Re}[\mathscr{A}_n(u,z)] = n\,\mathrm{Re}[\gamma_2(z)u^2] + n\,\mathrm{Re}(\gamma_3 u^3) + 0(n\phi_o^2)$$

$$\le \begin{cases} -c_1 n\phi_o^{3/2}, & \text{if } -\pi/8 \le \psi \le 0, \\ -c_2 n\phi_o, & \text{if } -\pi/4 \le \psi \le -\pi/8. \end{cases}$$

Subsequently,

$$\int_{\mathscr{D}_2} w(u)du = 0[\phi_o^{1/2}\exp(-c_1 n\,\phi_o^{3/2})] =$$

(4.21)

$$0[\exp(-c_1 n^{1-(3/2)s})] = o(n^{-3/2}).$$

Turn to the integral along $\mathscr{D}_3$. On $\mathscr{D}_3$, $u = ve^{-\pi/4}$ where $v \ge 0$ runs from u_o to 0. Hence (see again (4.12)),

$$\int_{\mathscr{D}_3} w(u)du = -2i\int_{u_o}^{0} v\,\exp[\mathscr{A}_n(ve^{-i\pi/4},z)]dv$$

$$= \int_0^{u_o} 2iv\,\exp[-a_2 nv^2 + ia_3 nv^3 + 0(nv^4)]dv = \int_0^{u_o},$$

where

$$a_2 = 1 - \overline{\sigma}/\sigma_n + o(1), \quad a_3 = (4/3\sigma_n)\pi^{1/2}\beta_o. \tag{4.22}$$

Break $[0,u_o]$ into $[0,u_1]$ and $[u,u_o]$ where $u_1 = n^{-t}$, $t \in (s/2, 1/2)$. Observe that

$$|\int_{u_1}^{u_o}| \le u_o^2 \exp(-cn^{1-2t}) = o(n^{-3/2}). \tag{4.23}$$

It remains to evaluate asymptotically the integral $\int_0^{u_1}$, or rather its real part. Let us confine t to a subinterval of $(s/2, 1/2)$; namely, let $t \in (t_1, 1/2)$, where $t_1 =_{def} \max(s/2, 1/3) = 1/3$. (Recall that $s \in (0,2/3)$.) For this t, $nu_1^3 = n^{1-3t} = o(1)$; therefore

$$\begin{aligned} \int_0^{u_1} &= \int_0^{u_1} 2iv \exp(-a_2nv^2)[1 + i\, a_3nv^3 + 0(nv^4 + n^2v^6)]dv \\ &= 2i\int_0^{u_1} v \exp(-a_2nv^2)dv - 2a_3n \int_0^{u_1} v^4\exp(-a_2nv^2)dv + \mathcal{R}_n. \end{aligned} \tag{4.24}$$

Here

$$\begin{aligned} \mathcal{R}_n &= 0[n \int_0^{u_1} v^5\exp(-a_2nv^2)dv + n^2 \int_0^{u_1} v^7\exp(-a_2nv^2)dv] \\ &= 0[n/(n^{1/2})^6 + n^2/(n^{1/2})^8] = 0(n^{-2}) = o(n^{-3/2}). \end{aligned} \tag{4.25}$$

Putting together (4.21) – (4.25), we arrive at

$$\mathrm{Re}(j) = o(n^{-3/2}) - \mathrm{Re}(\int_{\mathscr{D}_3}) = o(n^{-3/2}) - \mathrm{Re}(\int_0^{u_1})$$

$$= o(n^{-3/2}) + 2a_3 n \int_0^{u_1} v^4 \exp(-a_2 n v^2) dv$$

(4.26)

$$= o(n^{-3/2}) + (1 + o(1)) 2a_3 a_2^{-5/2} n^{-3/2} \int_0^{\infty} v^4 \exp(-v^2) dv$$

$$= (1 + o(1)) \pi \beta_o [\sigma_n (1 - \bar{\sigma}/\sigma_n)^{5/2}]^{-1} n^{-3/2}.$$

A combination of (4.13), (4.14), (4.19) and (4.26) leads to

$$Q_n E_n(z)/n! = (1 + o(1)) \exp[\mathscr{N}_n(z)] \beta_o [\sigma_n (1 - \bar{\sigma}/\sigma_n)^{5/2}]^{-1} n^{-3/2}.$$

4.4. *Proof of Lemma 2.* Consider again the relation (4.13). No changes are needed to obtain the estimate (4.14) and an analogue of (4.19), that is

(4.27) $$\int_0^{\phi_o} \exp[N_n(\phi,z)] d\phi = \int_0^{\phi_o} \exp[\tilde{N}_n(\phi,z)] d\phi + o(n^{-2/3} \exp[\mathscr{N}_n(z)]),$$

provided that

$$s + s(m+1)/2 - 1 > 2/3 \quad \text{or} \quad m > (10/3s) - 3.$$

Restricting s to $(5/9, 2/3)$, we can use therefore (4.27) with $m = 3$. It remains to evaluate (see (4.17))

$$\int_0^{\phi_o} \exp[\tilde{N}_n(\phi,z)]d\phi = \exp[\mathscr{N}_n(z)]\int_0^{\phi_o} \exp[n\gamma_2(z)\phi + n\gamma_3(z)\phi^{3/2}]d\phi$$

(4.28)

$$= \exp[\mathscr{N}_n(z)]\ I_n.$$

Substitute $u = (n\bar{\sigma}/\sigma_n)^{2/3}\ \phi$ and write (see (4.12))

$$I_n = (n\bar{\sigma}/\sigma_n)^{-2/3}\int_0^{u_o} \exp[-ia_n(z)u + \gamma u^{3/2}]du, \tag{4.29}$$

where $u_o = n(\bar{\sigma}/\sigma_n)^{-2/3}\phi_o$, and, $(z_j - 1 = o(n^{-1/3}), 1 \le j \le k, \sigma_n/\bar{\sigma} = 1 + 0(n^{-1/3}))$,

$$a_n(z) = -n^{1/3}[(\bar{\sigma}/\sigma_n - 1) + \sigma_n^{-1}\sum_{1\le j\le k} (z_j-1)jf_j\bar{x}^j/j!](\sigma_n/\bar{\sigma})^{2/3}$$
$$= n^{1/3}(1-\bar{\sigma}/\sigma_n) + o(1) = a_n + o(1),$$

and

$$\gamma = (4/3\bar{\sigma})\pi^{1/2}\exp(-3\pi i/4).$$

Introduce a (3/2)–stably distributed random variable X such that $E[\exp(iuX)] = \exp(\gamma u^{3/2})$, $u \ge 0$, and denote its density by $p(\cdot)$. According to the inversion formula, we have

$$p(a_n(z)) = (2\pi)^{-1}\int_{-\infty}^{\infty}\exp[ia_n(z)u]E[\exp(iuX)]du$$

(4.30)

$$= (\pi)^{-1}\mathrm{Re}\{\int_0^{\infty}\exp[-ia_n(z)u + \gamma u^{3/2}]du\}$$

$$= (\pi)^{-1}\mathrm{Re}\{\int_0^{u_o} \exp[-ia_n(z)u + \gamma u^{3/2}]du\} + 0[\exp(-\delta u_o^{3/2})],$$

$\delta = -\mathrm{Re}(\gamma) > 0$. In view of (4.27) – (4.30), we conclude

$$Q_n E_n(z)/n! = (1 + o(1))n^{-2/3} p(a_n)\exp[\mathscr{N}_n(z)].$$

4.5. Asymptotics of $Q_n/n!$ *and local limit theorems.* The Lemmas 1 and 2 contain, in particular, the asymptotic formulas for $Q_n/n!$ in the nearcritical and supercritical cases. We shall also need below an estimate for $Q_n/n!$ in case $\liminf \sigma_n > 0$. Its proof is close to , but much simpler than, the proof of the Lemma 2. For easiness of references, let us collect these results together in

LEMMA 3. (i) *If* $a_n = n^{1/3}(1-\bar{\sigma}/\sigma_n) = 0(1)$ *then*

$$Q_n/n! = (1 + o(1))n^{-2/3}[\exp(\sigma_n^{-1}F(\bar{x}))/\bar{x}]^n p(a_n).$$

(ii) *If* $\liminf \sigma_n > 0$ *then*

$$Q_n/n! = 0\{n^{-2/3}[\exp(\sigma_n^{-1}F(\bar{x}))/\bar{x}]^n\}.$$

(iii) *If* $\liminf \sigma_n > \bar{\sigma}$ *then*

$$Q_n/n! = (1 + o(1))n^{-3/2}[\exp(\sigma_n^{-1}F(\bar{x}))/\bar{x}]^n \cdot \beta_o[\sigma_n(1 - \bar{\sigma}/\sigma_n)^{5/2}]^{-1}.$$

Remark 6. The assertion (i) of the last lemma strongly suggests that there ought to be a local limit theorem – type connection lurking beneath the surface. This is actually true. To see it, confine ourselves to $\sigma_n \equiv \sigma$. Observe that then, according to (2.1),

$$Q_n/n! = \mathrm{coeff}_{x^n} \exp[n\sigma^{-1}F(x)].$$

Introduce a random variable $Y = \Sigma_{j>0}\, jy_j$, where $\{y_j\colon j \geq 1\}$ are independent, Poisson, with parameters $\{\sigma^{-1} f_j \bar{x}^j / j!: \ j \geq 1\}$. Then $E(z^Y) = \exp[\sigma^{-1}(F(\bar{x}z) - F(\bar{x}))]$, hence the formula for $Q_n/n!$ becomes

$$\text{(a)} \qquad Q_n/n! = [\exp(\sigma^{-1} F(\bar{x}))/\bar{x}]^n P(Z_n = n), \ Z_n = \sum_{1 \leq m \leq n} Y_m.$$

Now, it is easy to check that, by (2.5), $Z_n^* = (Z_n - n\bar{\sigma}/\sigma)n^{-2/3} \Rightarrow Z^*$, such that

$$E[\exp(iuZ^*)] = \exp[-(4/3\sigma)\pi^{1/2}\beta_0 \, e^{i\pi/4} u^{3/2}], \ u \geq 0,$$

so that Z^* is (3/2)–stably distributed. Since the distribution of Y turns out to be attracted to the one of the stable Z^*, by a local limit theorem for the stable distributions (Ibragimov and Linnik (1965), we have

$$\text{(b)} P(Z_n = n) = P(Z_n^* = n^{1/3}(1-\bar{\sigma}/\sigma)) = n^{-2/3}[p_\sigma(n^{1/3}(1-\bar{\sigma}/\sigma)) + o(1)],$$

where $p_\sigma(\cdot)$ is the density of Z^*. Now, if $\sigma = \bar{\sigma}$, then we get the statement (i) of the Lemma 3 for $\sigma_n \equiv \bar{\sigma}$, since $p(0) = p(0) > 0$. If $\sigma > \bar{\sigma}$ then $n^{1/3}(1-\bar{\sigma}/\sigma) \to \infty$, and using an asymptotic formula for the stable density (Ibragimov and Linnik (1965)), we get

$$\text{(c)} \qquad p_\sigma(n^{1/3}(1-\bar{\sigma}/\sigma)) = (1 + o(1))\beta_0[\sigma(1-\bar{\sigma}/\sigma)^{5/2}]^{-1} n^{-5/6}.$$

The relations (a) – (c) would imply the statement (iii) of the Lemma 3, if we could claim that o(1) in (c) is, in fact, $o(n^{-5/6})$. Such a formula actually can be obtained from a more general result for the stable distributions due to Tkachuk (1973).

4.6. Proof of the tightness condition in Theorem 1. Now we can finish the proof of the Theorem 1. What has remained is to prove that the sequence $\{C_n^*: \ n \geq 1\}$ satisfies the tightness condition in $\ell_1^{(\tau)}$. The proof is close, in essence, to the corresponding argument for the subcritical case (cf. Pittel, Woyczynski and Mann (1989a)), but is more delicate. We present an outline of the necessary steps with accent on the new elements.

(1) Suppose that $\sigma > \bar{\sigma}$. (1) Take $\epsilon_o \in (0,1-\bar{\sigma}/\sigma)$ and set $j_o = n[(1-\bar{\sigma}/\sigma) - \epsilon_o]$. Observe that

$$\sum_{j_o \le j \le n} j^\tau |c^*_{nj}| \le n^{1/2} \sum_{j_o \le j \le n} j^\tau m_j + n^{-1/2} \sum_{j_o \le j \le n} j^\tau c_{nj}$$

$$= n^{1/2} \sum_{j_o \le j \le n} j^\tau m_j + 0(n^{\tau-1/2}) \text{ a.s.,}$$

because a.s. $\sum_{j_o \le j \le n} c_{nj} = 1$. (Indeed, we know from Pittel Woyczynski and Mann (1989b) that, almost surely, the random forest M_n has a unique giant component of order n.) Furthermore, by (2.9),

$$n^{1/2} \sum_{j_o \le j \le n} j^\tau m_j = 0(n^{1/2} \sum_{j \ge j_o} j^{\tau-5/2}) = 0(n^{\tau-1}).$$

So, in probability,

$$\sum_{j_o \le j \le n} j^\tau |c^*_{nj}| \to 0, \quad n \to \infty. \tag{4.31}$$

(2) Furthermore, from the proof of Lemma 4 in the above quoted paper, we can see that

$$E(c_{nj}) = 0(nj^{-5/2}) = 0(nm_j), \quad 1 \le j \le j_o. \tag{4.32}$$

(For the sake of completeness we shall outline a proof of this inequality at the end of this section.) Therefore, for those j's,

$$E(|c^*_{nj}|) = 0(n^{-1/2} nm_j) = 0(n^{1/2} j^{-5/2}).$$

Introduce $j_1 = n^{2/5}$; by the last estimate,

$$E\left(\sum_{j_1 \le j < j_o} j^\tau |c^*_{nj}|\right) = 0\left(n^{1/2} \sum_{j \ge j_1} j^{s-5/2}\right)$$

$$= 0[n^{-(2/5)(1/4-\tau)}] \longrightarrow 0, \quad n \to \infty;$$

recall that $\tau < 1/4$. Consequently, in probability,

$$\sum_{j_1 \le j < j_o} j^\tau |c^*_{nj}| \longrightarrow 0, \ n \to \infty. \tag{4.33}$$

(3) In view of (4.31) and (4.33), it will be sufficient to check the tightness for $C_n^{**} = \{c_{nj}^{**}: j \ge 1\}$, where $c_{nj}^{**} = c_{nj}^*$ for $j \le j_1$, and $c_{nj}^{**} \equiv 0$ for $j > j_1$. For this, we need to look closely at c^*_{nj}, $1 \le j \le j_1$. We want to show that, for every $v_o > 0$, there exists a positive constant c such that

$$E[\exp(vc^*_{nj})] = 0[\exp(cm_jv^2)], \tag{4.34}$$

uniformly over $v \in [-v_o, v_o]$ and $j \in [1, j_1]$. (One may wish to call those c^*_{nj} uniformly sub Gaussian.) To prove (4.34), we begin with an identity (cf. (4.3), (4.4))

$$Q_n E\{\exp[u(c_{nj} - nm_j)]\}/n!$$

$$= \{\exp[\sigma^{-1}F(\bar{x})]/\bar{x}\}^n \exp[nm_j(e^u - 1 - u)] \cdot$$

$$(2\pi)^{-1}\int_{-\pi}^{\pi} \exp[\mathscr{H}_{nj}(\phi)]d\phi;$$

here $u = v/n^{1/2}$, $|v| \le v_o$, and

$$\mathscr{H}_{nj}(\phi) = n\sigma^{-1}[F(\bar{x}e^{i\phi}) - F(\bar{x})] + nm_j(e^u - 1)(e^{ij\phi} - 1) - in\phi$$

(4.35)

$$= n\sigma^{-1}\sum_{k\geq 1} e^{u_k}(f_k\bar{x}^k/k!)(e^{ik\phi} - 1) - in\phi,$$

($u_k \equiv 0$ for $k \neq j$, $u_j = u$). Introduce $\phi_o = n^{-s}$, $s \in (5/9, 2/3)$. Since $e^u - 1 = 0(u)$ and $j\phi_o = 0(n^{2/5 - 2/3}) = o(1)$ for $j \leq j_1$, we get from (4.35): uniformly over $\phi \in [0, \phi_o]$, $j \in [1, j_1]$,

$$\mathscr{H}_{nj}(\phi) = n\sum_{2\leq k\leq 6}\gamma_{kj}\phi^{k/2} + 0(n\phi^{7/2}), \tag{4.36}$$

where

$$\begin{aligned}\gamma_{2j} &= i[\bar{\sigma}/\sigma - 1 + o(1)],\\ \gamma_{3j} &= -(4/3\sigma)\pi^{1/2}\beta_o e^{-i\pi/4},\\ \gamma_{kj} &= 0(1),\ \ 2\leq k\leq 6.\end{aligned} \tag{4.37}$$

(To get (4.36), we used an inequality $|e^{ia} - \sum_{0\leq k\leq 3}(ia)^k/k!| = 0(a^4)$.)

Observe that $s \in (5/9, 2/3)$ together with $m = 6$ satisfy the inequality (4.20). So, acting in exactly the same way as in the proof of the Lemma 1 (the relation (4.11) and on), in view of (4.35)–(4.37), we obtain

$$\begin{aligned}&Q_n E\{\exp[u(c_{nj} - nm_j)]\}/n!\\ &= (1 + o(1))\{\exp[\sigma^{-1}F(\bar{x})]/\bar{x}\}^n \exp[n\, m_j(e^u - 1 - u)]\\ &\cdot\beta_o[\sigma(1-\bar{\sigma}/\sigma)^{5/2}]^{-1}n^{-3/2}.\end{aligned}$$

Therefore, by Lemma 3 (iii),

$$E\{\exp[u(c_{nj} - nm_j)]\} = (1 + o(1))\exp[nm_j(e^u - 1 - u)],$$

which implies (4.34), since $u = v/n^{1/2}$ and $e^u - 1 - u = 0(u^2)$.

We are almost finished. According to (4.34), uniformly for $\delta > 0$, $a > 0$ and $j \in [1, j_1]$, we have that

$$P(|cn_j^*| \geq nm_j^{\delta}) = 0(\exp(cm_j v^2 - v\ am_j^{\delta}))$$

for every $v > 0$. So, choosing the best v, i.e. $v = a/(2cm_j^{1-\delta})$, we obtain that

$$P(|c_{nj}^*| \geq am_j^{\delta}) = 0[\exp(-ba^2/m_j^{1-2\delta})],\ b = -(4c)^{-1}.$$

Since $c_{nj}^{**} = c_{nj}$ for $j \leq j_1$ and $c_{nj}^{**} = 0$ for $j > j_1$ we have then that

(4.38) $$P(|c_{nj}^{**}| \geq a\ m_j^{\delta}) = 0[\exp(-ba^2/m_j^{1-2\delta})],\ j \geq 1.$$

With the estimate (4.38) at hand, we can now find, for a given $\epsilon > 0$, a compact subset K of $\ell_1^{(\tau)}$ such that $P(C_n^{**} \in K_\epsilon) > 1 - \epsilon$ for all n. Let $\delta \in (0,1/2)$ be fixed. For a given a, introduce a set K(a) of all sequences $y = \{y_j: j \geq 1\}$ such that

$$K(a) = \{y: |y_j| \leq am_j^{\delta},\ j \geq 1\}.$$

The set K(a) belongs to $\ell_1^{(\tau)}$ if the series $\sum_{j>0} j^{\tau} m_j^{\delta}$ converges. Since m_j is of order $j^{-5/2}$ $(j \to \infty)$, the last condition holds true provided that $\delta > (2/5)(\tau+1)$. Now, $\delta < 1/2$ too; the two conditions can be satisfied simultaneously since $\tau < 1/4$. (This is the second time the condition $\tau < 1/4$ comes into play!) Compactness of K(a) is verified easily. Finally, see (4.38),

$$P(C_n^{**} \notin K(a)) \leq \sum_{j \geq 1} P(|c_{nj}^{**}| > am_j^{\delta})$$
$$= 0[\sum_{j\geq 1} \exp(-ba^2/m_j^{1-2\delta})] = 0[a^{-2\nu}\sum_{j\geq 1} m_j^{(1-2\delta)\nu}]\ (\forall\ \nu\ \lambda\ 1)$$
$$= 0[a^{-2\nu}\sum_{j \geq 1} j^{-(5/2)(1-2\delta)\nu}] = 0(a^{-2\nu}),$$

provided that ν is so large that $(5/2)(1-2\delta)\ \nu > 1$. Thus, if a is sufficiently large, K(a) is a compact we need. (In case $\sigma = \bar{\sigma}$, the proof of tightness is basically the same, and we omit it.)

(4) Finally, as promised, we provide an outline of proof of formula (4.32). Recall that c_{nj} denotes the total number of trees of size j. By (1.1) (see also (1.6))

$$E(c_{nj}) = [\binom{n}{j} Q_n^{-1} (\mu/\lambda) \sum_{M_{n-j}} (\mu/\lambda)^{C(M_{n-j})} \prod_{T \in M_{n-j}} h(T)] \cdot \sum_{T_j} h(T_j). \tag{4.39}$$

Here M_{n-j} is a forest on the vertices $j+1,\dots,n$, $C(M_{n-j})$ is the number of trees in it, the product $\prod$ is taken over all the trees of M_{n-j}, and the summation inside the square brackets is over all such forests, while the summation outside is over all trees on the vertices $1,2,\dots,j$. By the definition of f. and Q., we can simplify (4.39) to

$$E(c_{nj}) = (\mu/\lambda) f_j \binom{n}{j} (Q_{n'}/Q_n), \quad n' = n - j.$$

If $j \le j_2$ then $n' \ge n(\bar{\sigma}/\sigma + \epsilon)$, whence $n/\sigma = \mu/\lambda =_{\text{def}} n'/\sigma_n'$, where $\sigma_n = (\bar{n}/n)\sigma \ge \sigma + \epsilon\sigma > \sigma$. Thus, we can use the Lemma 3(iii) to evaluate both Q_n' and Q_n. For $j \to \infty$, there is also available the asymptotic formula (2.1) for $f_j/j!$. Subsequently,

$$E(c_{nj}) = (1 + o(1)) n \sigma^{-1} \beta_0(\bar{x})^{-j} j^{-5/2} j! [n!/(n')! j!]$$

$$\cdot (n')!(n')^{-3/2} \exp[(\mu/\lambda) F(\bar{x})] (\bar{x})^{-n'} q(\sigma_n')$$

$$\cdot \{n! n^{-3/2} \exp[(\mu/\lambda) F(\bar{x})] (\bar{x})^{-n} q(\sigma)\}^{-1}$$

$$= 0\,(nj^{-5/2}) = 0(nm_j).$$

REFERENCES

[1] PITTEL, B., WOYCZYNSKI, W.A. and MANN, J.A. (1987), From Gaussian subcritical to Holtsmark (3/2–Levy stable) supercritical asymptotic behavior in "rings forbidden" Flory–Stockmayer model of polymerization, *Graph Theory and Topology in Chemistry* (R. King and D. Rouvray, Eds.), Studies in Physical and Theoretical Chemistry (Elsevier) 51, 362–370.

[2] PITTEL, B., WOYCZYNSKI, W.A. and MANN, J.A. (1989a), Random tree–type partitions as a model for acyclic polymerization: Gaussian behaviour of the subcritical sol phase, *Random Graphs* '87 (J. Jaworski, M. Karonski and A. Rucinski, Eds.), Wiley 64 pp., to appear.

[3] PITTEL, B., WOYCZYNSKI, W.A. and MANN, J.A. (1989b), Random tree–type partitions as a model for acyclic polymerization: Holtsmark (3/2–stable) distribution of the supercritical gel, *Annals of Probability,* 42 pp, to appear.

[4] PITTEL, B. and WOYCZYNSKI, W.A. (1989c), A graph–valued Markov process as rings–allowed polymerization model: subcritical behavior, *SIAM J. Applied Math.*, 42 pp., to appear.

[5] WHITTLE, P. (1986), *Systems in Stochastic Equilibrium*, Wiley, New York, New York.

[6] STOCKMAYER, W.H. (1944), Theory of molecular size distribution and gel formation in branched polymers II. General crosslinking, *J. Chemical Physics* 12, 125–131.

[7] DONOGHUE, E. (1982), Analytic solutions of gelation theory for finite closed systems, *J. Chemical Physics* 77, 4326–4246.

[8] DONOGHUE, E. and GIBBS, J.H. (1979), Mean molecular size distributions and the sol–gel transition in finite, polycondensing systems, *J. Chemical Physics* 70, 2346–2356.

[9] IBRAGIMOV, I.A. and LINNIK, J.V. (1965), *Independent and stationary dependent random variables*, Nauka, Moscow.

[10] TKACHUK, S.G. (1973), Local limit theorems and large deviations for stable limit distributions, *Izvestia Ak.Nauk Uzb. S.S.R., Phys.–Math. Ser.* (2), 30–33.

AN APPLICATION OF SERIES REPRESENTATIONS TO ZERO - ONE LAWS FOR INFINITELY DIVISIBLE RANDOM VECTORS[1]

JAN ROSINSKI

Department of Mathematics
University of Tennessee
Knoxville, TN 37996-1300

0. Introduction.

In this paper we give an application of series representations of infinitely divisible random vectors to the zero-one laws for measurable subgroups. In Proposition 2 of Section 1 we provide a series representation of Poissonian-type random vectors suitable for this purpose. Then the zero-one dichotomy is a consequence of the observation that either all terms of the series representing an infinitely divisible measure μ are members of a subgroup G or infinitely many of them lie outside of G . In the first case the zero-one law follows from Hewitt-Savage zero-one law and in the second case the whole sum lies outside of G, so $\mu(G) = 0$. The latter assertion is justified by a generalization of a theorem of P. Lévy which we prove in Section 2. This provides a transparent probabilistic argument for the zero-one law due to A. Jannsen [**2**] (Section 3 of this paper). Jannsen's result, which gives a complete answer to the zero-one dichotomy problem for infinitely divisible measures, relies on special topological and algebraic techniques which are not needed here, due to series representations originated by Ferguson - Klass [**1**] and LePage [**4**], and developed by the author [**7**].

Our method can also be used for stochastic processes and, together with a Karhunen-Loève type expansion of the Gaussian part, it leads to a unified approach to zero-one laws for infinitely divisible processes.

Finally we notice that Proposition 2 as well as Theorem 1 admit extensions to more general topological groups so that Theorem 2 can be proven for generalized Poissonian

[1] Research supported in part by AFOSR Grant No. 87-0136

measures using the same method. We refer the reader to [**3**] for historical accounts of the development of zero-one laws for infinitely divisible measures.

1. Series Representations of Infinitely Divisible Random Vectors without Gaussian Components

Throughout this paper E will denote a separable Banach space equipped with a norm $\|\cdot\|$ and E will be the dual of E. All measures considered on E' will be defined on $\mathcal{B}_E$ (the Borel σ-field of E). We recall that a measure F on E is said to be a Lévy measure if, for every $x' \epsilon E'$, $\int_E (\langle x', x\rangle^2 \wedge 1)\, F(dx) < \infty$ and the function ϕ given by

$$\phi(x') = \exp\left\{ \int_E \left[e^{i\langle x', x\rangle} - 1 - i\,\langle x', x\rangle \;\; I(\|\dot{x}\| \leq 1)\right] \; F(dx) \right\}, \quad x' \epsilon E' ,$$

is characteristic function of a probability measure on E. The probability measure with characteristic function ϕ will be denoted by c-Pois(F) and called a centered Poissonian measure with Lévy measure F.

Let $(D, \mathcal{D})$ be a measurable space and $H : (0, \infty) \times D \to E$ be a Borel measurable map. Let $\{\xi_n\}$ be a sequence of i.i.d. random elements in $(D, \mathcal{D})$, which is defined on a probability space $(\Omega, \mathcal{F}, P)$. Set

$$F_H(B) = \int_0^\infty P\big(H(s, \xi_1) \epsilon\; B \backslash \{0\}\big)\; ds\;, \quad B \epsilon\; \mathcal{B}_E\;.$$

Then F_H is a measure on E such that $F_H(\{0\}) = 0$. Put

$$A_H(t) = \int_0^t E\big[H(s, \xi_1);\;\; \|H(s, \xi_1)\| \leq 1\big] ds\;, \quad t \geq 0\;.$$

Finally, let $\{e_n\}$ be a sequence, independent of $\{\xi_n\}$, of i.i.d. exponential with parameter 1 random variables and set $\tau_n = e_1 + \cdots + e_n$. The following result has been proven in [**7**] (Theorem 2.4 and Theorem 3.4).

PROPOSITION 1. *$\sum[H(\tau_n, \xi_n) - C_n]$ converges a.s. if and only if F_H is a Lévy measure on E, where $C_n = A_H(\tau_n) - A_H(\tau_{n-1})$. If F_H is a Lévy measure, then*

$$\mathcal{L}\left(\sum_{n=1}^\infty \big[H(\tau_n, \xi_n) - C_n\big]\right) = c\text{-Pois}\;(F_H)\;.$$

If additionally $\int_E(\|x\|^2 \wedge 1)F(dx) < \infty$, *then all the above holds with nonrandom centers* $C_n = A_H(n) - A_H(n-1)$.

Now let X be a random vector in E whose distribution is infinitely divisible without Gaussian component. Then $\mathcal{L}(X) = \delta_a * \text{c-Pois}(F)$, for some $a \epsilon E$ and a Lévy measure F. By choosing H and $\{\xi_n\}$ such that $F_H = F$ one obtains, applying Proposition 1, a series representation of X:

$$X \stackrel{d}{=} a + \sum_{n=1}^{\infty} \left[H(\tau_n, \xi_n) - C_n\right] , \tag{1}$$

where "$\stackrel{d}{=}$" means "equal in distribution". LePage [4] gave a series representation of this type using a polar decomposition of F with respect to the unit sphere. For our purposes a different form of (1) will be useful. Namely, let λ be a probability measure on E such that F is absolutely continuous with respect to λ. Put $f(x) = (dF/d\lambda)(x)$. Let $\{V_n\}, \{e_n\}$ and $\{\delta_n\}$ be mutually independent sequences of random variables such that

- V_n are i.i.d. random vectors in E with the common distribution λ;
- e_n are as above and $\tau_n = e_1 + \cdots + e_n$;
- δ_n are i.i.d. symmetric random variables such that $P(|\delta_n| = 1) = P(\delta_n = 0) = 1/2$.

PROPOSITION 2. *Under the above notation,*

$$X \stackrel{d}{=} a + \sum_{n=1}^{\infty} \left[I(f(V_n) > \tau_n)V_n - C_n\right] , \tag{2}$$

where the series converges a.s. and $C_n = \int_{\|x\| \le 1}(\tau_n \wedge f(x) - \tau_{n-1} \wedge f(x))x\lambda(dx)$. *Further, if X is symmetric, then*

$$X \stackrel{d}{=} \sum_{n=1}^{\infty} \delta_n I(2f(V_n) > \tau_n)V_n \tag{3}$$

PROOF: (2) and (3) follow from Proposition 1 applied to, $H(s, \xi_n) = I(f(V_n) > s)V_n$ and to $H(s, \xi_n) = \delta_n I(2f(V_n) > s)V_n$, respectively. ∎

2. A Generalization of a Theorem of P. Lévy.

P. Lévy ([**5**], Thòréme XIII) proved that if a series of independent random variables $\sum X_n$ converges and $\{X_n\}$ is not equivalent to a sequence of constants then the distribution of the series is continuous. The following generalization of this result will be needed for the proof of the zero-one law in Section 3 but it may also be of independent interest. Our theorem reduces to Lévy's theorem when $G = \{0\}$.

THEOREM 1. *Let $\{X_n\}$ be a sequence of independent random vectors in E such that the series $\sum X_n$ converges a.s.. Suppose that for some measurable subgroup G of E and $x_0 \epsilon E$*

$$P(\Sigma X_n \epsilon\, G + x_0) > 0 \, .$$

Then there exists a sequence $\{x_n\} \subset E$ such that $\sum x_n$ converges, $\sum x_n \epsilon\, G + x_0$ and

$$P(X_n \epsilon G + x_n \textit{ eventually}) = 1 \, .$$

PROOF: Put $S = \sum_1^\infty X_n$, $\alpha = P(S \epsilon G + x_0) > 0$, $S_n = \sum_1^n X_k$. First we shall show that for large n, $S - S_n$ takes values in a coset, say $G + u_n$ of G with the probability arbitrarily close to one. Let $\varepsilon \epsilon (0, 2^{-1}\alpha)$. There exists a closed subset $C \subset G$ such that

$$\alpha - \varepsilon < P(S \epsilon\, C + x_0) \leq \alpha$$

Set $C_\eta = \{x \epsilon E : \|x - y\| \leq \eta \textit{ for some } y \epsilon\, C\}$. Since C is closed there is an $\eta > 0$ such that

$$P(S \epsilon\, C_{2\eta} + x_0) < \alpha + \varepsilon \tag{4}$$

Now choose n_0 such that for $n > n_0$

$$P(\|S - S_n\| > \eta) < \varepsilon \, . \tag{5}$$

Since

$$P(S_n \epsilon\, C_\eta + x_0)\; P(\|S - S_n\| \leq \eta) =$$
$$P(S_n \epsilon\, C_\eta + x_0,\; \|S - S_n\| \leq \eta) \leq P(S \epsilon\, C_{2\eta} + x_0)$$

we get by (4) and (5)

$$P(S_n \epsilon C_\eta + x_0) < (\alpha + \varepsilon)\,(1 - \varepsilon)^{-1}\,, \tag{6}$$

for $n > n_0$. On the other hand, we have

$$\alpha - \varepsilon < P(S \epsilon\, C + x_0) < P(S \epsilon\, C + x_0,\ \|S - S_n\| \le \eta) + \varepsilon =$$
$$\int_{\|u\| \le \eta} P(S_n \epsilon\, C + x_0 - u)\, P(S - S_n \epsilon\, du) + \varepsilon\,.$$

Hence, for every $n > n_0$, there exists $u_n \epsilon\, E$, $\|u_n\| \le \eta$ such that

$$P(S_n \epsilon C + x_0 - u_n) > \alpha - 2\varepsilon > 0\,. \tag{7}$$

Set, for $n > n_0$,

$$B_n = (G + x_0 - u_n) \cap (C_\eta + x_0)\,.$$

Since $C \subset G$ and $C + x_0 - u_n \subset C_\eta + x_0$ we get by (6) and (7)

$$\alpha - 2\varepsilon < \alpha_n \stackrel{\text{def}}{=} P(S_n \epsilon\, B_n) < (\alpha + \varepsilon)\,(1 - \varepsilon)^{-1}\,. \tag{8}$$

Put

$$\beta_n = P(S - S_n \epsilon\, G + u_n)\,.$$

We have

$$P(S \epsilon\, C + x_0,\ S_n \notin B_n) \ge$$
$$P(S \epsilon\, C + x_0) - P(S \epsilon\, G + x_0,\ S_n \epsilon\, B_n) >$$
$$\alpha - \varepsilon - P(S_n \epsilon\, B_n)\, P(S - S_n \epsilon\, G + u_n) >$$
$$\alpha - \varepsilon - \beta_n (\alpha + \varepsilon)\,(1 - \varepsilon)^{-1}\,,$$

where the last inequality follows from (8). On the other hand, using (6), (8) and (5) we get

$$P(S \epsilon\, C + x_0,\ S_n \notin B_n) \le$$
$$P(S_n \notin B_n, S_n \epsilon\, C_\eta + x_0) + P(S \epsilon\, C + x_0,\ S_n \notin C_\eta + x_0) \le$$
$$P(S_n \epsilon\, C_\eta + x_0) - P(S_n \epsilon\, B_n) + P(\|S - S_n\| > \eta) <$$

$$3\,\varepsilon + \varepsilon(1+\alpha)\,(1-\varepsilon)^{-1}\ .$$

Combining the above bounds we obtain

$$\alpha - \varepsilon - \beta_n(\alpha+\varepsilon)(1-\varepsilon)^{-1} < P(S\epsilon\, C + x_0, S_n \notin B_n) < 3\varepsilon + \varepsilon(1+\alpha)(1-\varepsilon)^{-1}\ ,$$

which yields the lower bound for β_n:

$$(9) \qquad P(S - S_n \epsilon\, G + u_n) = \beta_n > 1 - \varepsilon(2 + 6\alpha^{-1})\ ,$$

for $n > n_0$. This proves the claim that $S - S_n$ takes values in a coset of G with the probability arbitrarily close to 1 if n is sufficiently large.

Let $\{X_n'\}$ be an independent copy of $\{X_n\}, \tilde{X}_n = X_n - X_n'$, $\tilde{S}_n = \sum_1^n \tilde{X}_k$ and $\tilde{S} = \sum_1^\infty \tilde{X}_n$. We shall prove now the conclusion of Theorem 1 for random vectors $\tilde{X}_n$ with $x_n = 0$. Using (9) we get

$$P(\tilde{S} - \tilde{S}_n \epsilon\, G) \geq P\left(\sum_{n+1}^{\infty} X_k \epsilon\, G + u_n,\ \sum_{n=1}^{\infty} X_k' \epsilon\, G + u_n\right) > [1 - \varepsilon(2 + 6\alpha^{-1})]^2\ ,$$

for $n > n_0$. Hence there exist positive integers $n_1 < n_2 < \ldots$ such that

$$P(\tilde{S} - \tilde{S}_n \epsilon\, G) > 1 - 2^{-j-1} \quad \text{for all} \quad n \geq n_j\ .$$

Therefore , by Bonferroni's inequality, we get

$$P(\tilde{S}_{n_{j+1}} - \tilde{S}_n \epsilon\, G) \geq P(\tilde{S} - \tilde{S}_n \epsilon\, G, \tilde{S} - \tilde{S}_{n_{j+1}} \epsilon\, G) > 1 - 2^{-j}\ ,$$

for $n_j \leq n \leq n_{j+1}$. Let

$$\tau_j = \inf\{n : n_j < n \leq n_{j+1},\ \tilde{S}_n - \tilde{S}_{n_j} \notin G\}, \qquad (\inf \emptyset = \infty)\ .$$

Then we have

$$\begin{aligned} 2^{-j} > & P(\tilde{S}_{n_{j+1}} - \tilde{S}_{n_j} \notin G) \geq \sum_{n=n_j}^{n_{j+1}} P(\tilde{S}_{n_{j+1}} - \tilde{S}_{n_j} \notin G,\ \tau_j = n) \geq \\ & \sum_{n=n_j}^{n_{j+1}} P(\tilde{S}_{n_{j+1}} - \tilde{S}_n \epsilon\, G)\ P(\tau_j = n) > \\ & (1 - 2^{-j})\ P(\tilde{S}_n - \tilde{S}_{n_j} \notin G \textit{ for some } n_j < n \leq n_{j+1})\ , \end{aligned}$$

which yields

$$P(\tilde{S}_n - \tilde{S}_{n_j} \notin G \text{ for some } n_j < n \leq n_{j+1}) < 2^{-j+1} .$$

Hence

$$P(\tilde{X}_n \notin G \text{ for some } n_j < n \leq n_{j+1}) \leq$$
$$P(\tilde{S}_n - \tilde{S}_{n_j} \notin G \text{ for some } n_j < n \leq n_{j+1}) < 2^{-j+1} ,$$

and, by the Borel - Cantelli Lemma,

$$P(\tilde{X}_n \in G \text{ eventually }) = 1$$

To complete this proof note that by Fubini's Theorem there exists ω_0 such that for $x_n = X_n'(\omega_0)$, $\sum x_n \in G + x_0$ and

$$P(X_n - x_n \in G \text{ eventually}) = 1 . \quad \blacksquare$$

REMARK 1: In the above proof we adapted the original ideas of Lévy's proof of Thèoréme XIII [**5**] given for real discrete random variables and $G = \{0\}$. Since in our case G is only a measurable subgroup (not closed neither open in many interesting cases of applications) and since the original technique of utilizing inequalities for concentration functions can not be applied here, this adaptation is not immediate.

REMARK 2: The centers x_n in Theorem 1 are unavoidable even when X_n are symmetric. Indeed, let E be an infinite dimensional Hilbert space with a C.O.N.S. $\{h_n\}$. Put

$$G = \left\{ \sum_1^\infty a_j \, 3^{-j} \, h_j : a_j \text{ are even integers with } \sup_j |a_j| < \infty \right\}$$

and define $X_{2k-1} = \varepsilon_{2k-1} \, 3^{-k} \, h_k$ and $X_{2k} = \varepsilon_{2k} \, 3^{-k} \, h_k$, $k = 1, 2, \ldots$, where $\{\varepsilon_n\}$ is a Rademacher sequence. Then, clearly, $\sum X_n \in G$ and $X_n \notin G$ for any n.

COROLLARY 1. *Let X_n be independent symmetric random vectors in E such that $\sum X_n$ converges. Suppose that, for some $x_0 \in E$ and a measurable subgroup G of E, $P(\sum X_n \in G + x_0) > 0$. Then $P(2X_n \in G$ eventually$) = 1$.*

The following corollary, which is an immediate consequence of the Borel-Cantelli Lemma, may be helpful in certain cases to establish $x_n = 0$ in Theorem 1.

COROLLARY 2. *Under the assumptions and notation of Theorem 1,* $\lim_{n\to\infty} P(X_n \in G + x_n) = 1$.

PROPOSITION 3. *Let* X_n *be independent random vectors in* E *such that* $\sum X_n$ *converges. Suppose that, for some measurable subgroup* G *of* E, $P(X_n \notin G$ *infinitely often*$)= 1$ *and* $P(X_n \in G) \geq p$ *for certain* $p > 0$. *Then*

$$P(\sum X_n \in G + x_0) = 0 \quad \textit{for every} \quad x_0 \in E \ .$$

PROOF: A contrario. If $P(\sum X_n \in G + x_0) > 0$, then by Theorem 1, $P(X_n - x_n \in G$ *eventually*$) = 1$ for some $x_n \in E$. Since $P(X_n \in G) \geq p$, using Corollary 2, we get that $x_n \in G$ eventually. Thus $P(X_n \in G$ eventually$) = 1$ which contradicts our assumption. ∎

3. The Zero-One Laws.

We recall a useful symmetrization lemma ([**2**], Lemma 4) which reduces the proof of zero-one laws to symmetric random vectors. Its short proof will be repeated here for completeness.

LEMMA 1. *Let* μ *be a probability measure on* E *and let* $|\mu|^2(A) = \int_E \mu(A + x)\ \mu(dx)$. *For any measurable subgroup* G *of* E *the following assertions are valid:*

a) $|\mu|^2(G) = 1$ *if and only if there exists a coset* $G + x$ *satisfying* $\mu(G + x) = 1$ *and* $\mu(G + y) = 0$ *for each coset* $G + y \neq G + x$.

b) $|\mu|^2(G) = 0$ *if and only if* $\mu(G + y) = 0$ *for all* $y \in E$.

PROOF: a) is obvious. b) follows from the inequality

$$|\mu|^2(G) \geq \int_{G+y} \mu(G + x)\ \mu(dx) \geq [\mu(G + y)]^2$$

which holds for arbitrary $y \in E$.

Now we are ready to give a new and simple proof of the zero-one law for infinitely divisible measures (due to A. Janssen [**2**]).

THEOREM 2. *Let μ be an infinitely divisible probability measure on E with Lévy measure F and without Gaussian component. Let G be a measurable subgroup of E. Then*

(i) *If $F(G^c) = 0$, then either $\mu(G+y) = 0$ for all $y \in E$ or $\mu(G+x) = 1$ for some $x \in E$. In the latter case $x = 0$ if μ is symmetric.*

(ii) *If $F(G^c) = \infty$, then $\mu(G+y) = 0$ for all $y \in E$.*

(iii) *If $0 < F(G^c) < \infty$, then $\mu(G+y) < 1$ for all $y \in E$. Moreover, the following alternative is true: either $\mu(G+y) = 0$ for all $y \in E$ or $\mu(G+z) \geq \exp(-F(G^c))$ for some $z \in E$. The last inequality holds with $z = 0$ provided μ is symmetric.*

PROOF: (i). Observe that the Lévy measure $\tilde{F}$ of $|\mu|^2$ satisfies $\tilde{F}(G^c) = F(G^c) + F(-G^c) = 0$. Therefore, using Lemma 1, it is enough to prove that $\mu(G) = 0$ or 1, if μ is symmetric and $F(G^c) = 0$. Now we use the fact that μ can be represented as distribution of the series specified in (3) of Proposition 2. Moreover we can choose λ such that F and λ are mutually absolutely continuous. Put

$$A = \left\{ \omega : \sum_{n=1}^{\infty} \delta_n \; I(2f(V_n) > \tau_n) V_n \in G \right\} ,$$

$\mu(G) = P(A)$. Since $P(V_n \in G) = \lambda(G) = 1$, without loss of generality we may assume that $V_n(\omega) \in G$ for each ω. Consequently, A is an exchangeable event on the i.i.d. sequence $((\delta_1, e_1, V_1), (\delta_2, e_2, V_2), \ldots, (\delta_n, e_n, V_n), \ldots)$, hence $P(A) = 0$ or 1 by Hewitt-Savage zero-one law (see, e.g. [**6**] p.374).

(ii). Similarly as in (i) we assume that μ is symmetric and use Proposition 3 to represent μ . Let A be as given above. Without loss of generality we may assume that Ω is a product space, $\Omega = \Omega_1 \times \Omega_2$, $P = P_1 \times P_2$, and $\{e_n\}$ depends on $\omega_1 \in \Omega_1$ while the rest of random sequences depends on $\omega_2 \in \Omega_2$. By Fubini's Theorem it is enough to show that $P_2(A^{\omega_1}) = 0$ for $P_1 - a.a.$ $\omega_1 \in \Omega_1$. By the Strong Law of Large Numbers

$n^{-1}\tau_n(\omega_1) \to 1$ for $P_1 - a.a.\ \omega_1 \epsilon\ \Omega_1$, therefore, fixing such an ω_1, we get

$$\begin{aligned}&\sum P_2(\delta_n\ I(2f(V_n) > \tau_n(\omega_1))V_n \notin G) = \\ &\sum P_2(\delta_n = 1,\ 2f(V_n) > \tau_n(\omega_1),\ V_n \notin G) = \\ &\frac{1}{2}\sum P_2(2f(V_1) > \tau_n(\omega_1),\ V_1 \notin G) = \infty\end{aligned}$$

because

$$\sum_0^\infty P_2(2f(V_1) > n,\ V_1 \notin G) \geq E[2f(V_1);\ V_1 \notin G] = 2F(G^c) = \infty\ .$$

Hence

$$P_2(\delta_n\ I(2f(V_n) > \tau_n(\omega_1))V_n \notin G\ \textit{infinitely often}) = 1\ . \tag{10}$$

Also,

$$P_2(\delta_n\ I(2f(V_n) > \tau_n(\omega_1))V_n \epsilon\ G) \geq \frac{1}{2}\ , \tag{11}$$

since $P_2(\delta_n = 0) = \frac{1}{2}$. In view of (10) and (11) Proposition 3 yields $P_2(A^{\omega_1}) = 0$ which ends the proof of (ii).

(iii). Let $\mu_1 = c\text{-Pois}(F_{|G})$ and $\mu_2 = c\text{-Pois}(F_{|G^c})$, $\mu = \delta_a * \mu_1 * \mu_2$. Applying (i) we get $\mu_1(G + y) = 0$ for all $y \epsilon\ E$ or $\mu_1(G + x) = 1$ for some $x \epsilon\ E$. In the first case $\mu(G + y) = 0$ for all $y \epsilon\ E$. In the second case $\mu(G + z) = \mu_2(G + z - x - a)$ and since μ_2 is a shifted compound Poisson measure generated by $F_{|G^c}$, (iii) follows easily. This completes the proof of Theorem 2.

COROLLARY 3. *Suppose that the Lévy measure F of μ can be written in the form*

$$F(B) = \int_K \int_{(0,\infty)} I_B(tx)\ \rho(dt, x)\sigma(dx),\quad B \epsilon\ \mathcal{B}_E\ ,$$

where σ is a measure on a Borel set $K, 0 \notin K \subset E$, and $\{\rho(\cdot, x)\}_{x \epsilon K}$ is a measurable family of measures on $(0, \infty)$ such that, for each $x \epsilon\ K$, $\rho(\cdot, x)$ is equivalent to the Lebesgue measure and $\rho((0, \infty), x) = \infty$. Then for every measurable subgroup G of E and $x \epsilon\ E$, $\mu(G + x) = 0$ or 1.

PROOF: Put $G_x = \{s\epsilon\, \boldsymbol{R} : sx\epsilon\, G\}$. G_x is a measurable subgroup of $\boldsymbol{R}$, hence the Lebesgue measure of G_x is zero or $G_x = \boldsymbol{R}$. In the first case $\rho(\boldsymbol{R}^+ \setminus G_x, x) = \infty$ and in the second case $\rho(\boldsymbol{R}^+ \setminus G_x, x) = 0$. Thus

$$F(G^c) = \int_K \rho(\boldsymbol{R}^+ \setminus G_x, x)\, \sigma(dx) = \infty \quad \text{or} \quad 0 ,$$

and the result follows from Theorem 2.

As far as we know, the above simple criterium for the zero-one law has not been noticed in the literature. It provides a whole class of examples of infinitely divisible measures which satisfy the zero-one law; in particular selfdecomposable measures belong to this class. Other examples of infinitely divisible measures whose Lévy measures satisfy the zero-infinity dichotomy for certain subgroups, and consequently, they obey the zero-one law for these subgroups, are given in the survey article [**3**]. Finally, we notice that the condition that $\rho(\cdot, x)$ is equivalent to the Lebesgue measure can be removed from the assumptions of Corollary 3 if G is a measurable subspace of E.

REFERENCES

[1] Ferguson, T.S. and Klass, M.J. (1972). A representation of independent increment processes without Gaussian components. Ann. Math. Stat. 43, 1634-1643.
[2] Janssen, A.(1982). Zero-one laws for infinitely divisible measures on groups. Z. Wahr. verw. Gebiete 60, 119-138.
[3] Janssen, A.(1984). A survey about zero-one laws for probability measures on linear spaces and locally compact groups. Lecture Notes in Math. 1064, Springer-Verlag, 551 563.
[4] LePage (1980). Multidimensional infinitely divisible variables and processes. Part II. Lecture Notes in Math.860, Springer-Verlag, 279-284.
[5] Lévy, P.(1931). Sur les séries dont les termes sont des variables éventuelles indépendantes. Studia Math. 3, 119-155.
[6] Loève, M. Probability Theory I. 4th Ed. Springer-Verlag, New York, 1977.
[7] Rosinski, J.(1989). On series representations of infinitely divisible random vectors, Ann. Probability (to appear).

STRONG APPROXIMATIONS FOR SET–INDEXED PARTIAL–SUM PROCESSES AND EMPIRICAL PROCESSES OF MIXING RANDOM FIELDS

Wolfram Strittmatter

Institut für Mathematische Stochastik, Universität Freiburg

Hebelstr. 27, D–7800 Freiburg, West Germany

0. Introduction

Let $(x_j, j \in \mathbb{N}^q)$ be a mixing family of vector–valued random variables, $\mathcal{A}$ a class of subsets of the q–dimensional unit cube. In this paper we will give theorems on the strong approximation, uniformly in $A \in \mathcal{A}$, of the partial–sum process $(\Sigma_{j \in nA}\, x_j,\ A \in \mathcal{A},\ n \in \mathbb{N})$ by a partial–sum process $(\Sigma_{j \in nA}\, y_j,\ A \in \mathcal{A},\ n \in \mathbb{N})$ of some family of i.i.d. Gaussian random vectors.

As usual in strong approximation theory for weakly dependent random variables, we will impose conditions on higher moments and on the mixing rate of the $(x_j, j \in \mathbb{N}^q)$. Here, in addition, we will have to control the size of $\mathcal{A}$ by a metric entropy type condition. Such metric entropy conditions play a central role in limit theory for partial sum processes indexed by large classes $\mathcal{A}$. Pyke (1983) and Bass and Pyke (1984) gave a functional law of the iterated logarithm and a uniform central limit theorem for these processes in the independent case, and Morrow and Philipp (1986) proved strong approximation results for this case. In the mixing case a strong approximation theorem has only been given for small classes $\mathcal{A}$ of sets by Berkes and Morrow (1981). For large classes $\mathcal{A}$ and mixing random fields functional central limit theorems have been proved by Goldie and Greenwood (1986), but no strong approximation theorems have been given in the literature. So our results at the same time put together and unify the limit theorems of [BM], [MPh] and [GG] and generalize each of them.

A key technique in proving limit theorems for partial–sum processes is the so–called chaining (or nesting) technique, which consists of a step–by–step replacement of the partial sums over nA with A in the class $\mathcal{A}$ by partial sums over smaller and smaller finite classes $\mathcal{A}_i$ of sets which approximate the sets $A \in \mathcal{A}$ in a certain sense. This replacement process ends up with partial sums over unions of certain rectangles. For this last step a "boundary smooth" condition is needed for the approximating sets. The successive replacement is done for the x_j as well as for the Gaussian y_j process. Over the rectangles the partial sums of both processes can be put together sufficiently close by using the method of [BM].

For obtaining a bound on the error made in the replacement, an exponential inequality is essential. We give such an inequality for strongly mixing random fields (Lemma 5.1). In the absolutely regular case we use another approach: We replace certain blocks of $(x_j, j \in \mathbb{N}^q)$ by independent blocks, and apply an exponential inequality for independent variables. For the error made in this replacement we use a lemma of Eberlein (1984) on the total variation distance of the distribution of the original and the replacement processes. The advantage of this second approach (which only works for absolutely regular random fields) has been pointed out to the author by Charles Goldie.

A second aim of this paper is to give generalizations of our results to mixing families $(X_j, j \in \mathbb{N}^q)$ of random elements in a non–separable Banach space. These random elements need not be measurable. In the independent case such elements have been introduced by Dudley and Philipp (1983), who noticed that the measurability problems in limit theory for empirical processes can be avoided by regarding empirical processes as non–measurable processes in non–separable Banach spaces and proving limit theorems for such processes. As there, we need (in addition to the finite–dimensional case) assumptions that the random elements X_j may be approximated by finite–dimensional images at a certain rate.

Finally, we apply our results on random elements in Banach spaces to obtain strong approximation theorems for empirical processes of absolutely regular random fields. Let us note that our results for Banach space valued random elements and for empirical processes are new even for $q = 1$, i.e. for the case of sequences. For example, as a corollary we obtain an improvement of a strong approximation theorem of Dhompongsa (1984) for the empirical process of vector–valued, absolutely regular sequences.

Let us now make a few remarks on the organization of this paper. In sections 1, 2 and 3 we state our results for $\mathbb{R}^N$–valued resp. Banach space valued random fields resp. empirical processes. In section 4 we give bounds for the second and higher moments of partial sums. Sections 5 and 6 consist of the proofs of the results for $\mathbb{R}^N$–valued random fields (Theorems 1 and 2). Section 7 contains central limit theorems for strongly mixing $\mathbb{R}^N$–valued random fields, stating the dependence of the convergence rate on N. In section 8, moment and maximal inequalities for strongly mixing, non–measurable random elements in non–separable Banach spaces are given. In section 9 we prove one of the theorems for Banach space valued random fields (Theorem 3), while for the proofs of Theorems 4 and 5 we will only give a few hints, but no details (which may be found in [Str]). The results for empirical processes (Theorems 6 and 7) are proved in section 10.

Some notation: $\mathbb{N}$, $\mathbb{N}_0 = \mathbb{Z}_+$, $\mathbb{R}_+$ denote the sets $\{1, 2, 3, ...\}$, $\{0, 1, 2, ...\}$, $[0,\infty)$. $\mathbb{1}$ denotes the vector in $\mathbb{R}^q$ all whose coordinates are one. For $n = (n_1,...,n_q) \in \mathbb{N}^q$ we write $\Pi(n) := \Pi_{i=1}^q n_i$. For $a = (a_1,...,a_q)$, $b = (b_1,...,b_q) \in \mathbb{R}^q$, $a \le b$ is equivalent with $a_i \le b_i$ for $i = 1,...,q$; put $(a,b] := \{x \in \mathbb{R}^q : a_i < x_i \le b_i$ for $i = 1,...,q\}$, similarly $[a,b]$ etc. For a topological space T, $\mathcal{B}(T)$ denotes its Borel σ–algebra. $c, c_1, c_2, ...$ are constants, not necessarily the same at different places of occurrence. For a signed measure μ we denote its total variation norm by $\|\mu\|$. For a random variable X, $\mathcal{L}(X)$ denotes its distribution. "... << ..." has the same meaning as the Landau symbol "... $=O(...)$", and $f \asymp g$ means that $f << g$ and $g << f$. For two sets A and B, $A \triangle B$ is their symmetric difference. P_* resp. P^* denote the inner resp. outer measure belonging to a probability measure P.

1. Results for Strongly Mixing and Absolutely Regular $\mathbb{R}^N$–Valued Random Fields

Let $N, q \in \mathbb{N}$ and $(x_j, j \in \mathbb{N}^q)$ be an $\mathbb{N}^q$–indexed family of $\mathbb{R}^N$–valued random variables (or, as we will also say, a random field). For $\phi \neq S \subset \mathbb{N}^q$ denote by $\sigma(S)$ the σ–algebra generated by the x_j, $j \in S$. Let $\mathbb{R}^q$ be equipped with the metric $d(u,v) := \max\{|u_i - v_i| : 1 \leq i \leq q\}$ $(u = (u_1,\dots,u_q),\ v = (v_1,\dots,v_q) \in \mathbb{R}^q)$, for $\phi \neq S_i \subset \mathbb{R}^q$ $(i = 1, 2)$ let $d(S_1,S_2) := \inf\{d(u,v) : u \in S_1, v \in S_2\}$. We say $(x_j, j \in \mathbb{N}^q)$ is strongly mixing, if there is a nonincreasing function $\alpha: \mathbb{R}_+ \to \mathbb{R}_+$ with $\alpha(t) \downarrow 0$ for $t \to \infty$ such that

$$\text{(1.1)} \qquad \sup\{|P(A \cap B) - P(A)P(B)| : A \in \sigma(S_1), B \in \sigma(S_2)\} \leq \alpha(d(S_1,S_2))$$

for all disjoint, finite $S_1, S_2 \subset \mathbb{N}^q$. $(x_j, j \in \mathbb{N}^q)$ is called absolutely regular, if for some non–increasing function $\beta: \mathbb{R}_+ \to \mathbb{R}_+$ with $\beta(t) \downarrow 0$ for $t \to \infty$ we have

$$\text{(1.2)} \qquad \tfrac{1}{2}\|\mathcal{L}(x_j, j \in S_1 \cup S_2) - \mathcal{L}(x_j, j \in S_1) \otimes \mathcal{L}(x_j, j \in S_2)\| \leq \beta(d(S_1,S_2))$$

for all disjoint, finite $S_1, S_2 \subset \mathbb{N}^q$. Obviously, absolute regularity implies strong mixing and we can choose $\alpha \leq \beta$.

Now assume $Ex_j = 0$ and $E\|x_j\|^2 < \infty$ for all $j \in \mathbb{N}^q$. $(x_j, j \in \mathbb{N}^q)$ is called weakly stationary, if for $j, k \in \mathbb{N}^q$ the covariance matrix $\mathrm{Cov}(x_j,x_k)$ depends only on $j - k$, i.e. if there is a function $r: \mathbb{Z}^q \to \mathbb{R}^{N\times N}$ into the space of real $N\times N$ matrices such that

$$\text{(1.3)} \qquad \mathrm{Cov}(x_j,x_k) = (Ex_j^{(i)}x_k^{(m)})_{i,m=1}^N = r(j-k)$$

for all $j, k \in \mathbb{N}^q$. Here $x_j = (x_j^{(1)},\dots,x_j^{(N)})$.

For $j \in \mathbb{N}^q$ let $C_j := (j-1, j]$. Let μ be a Borel measure on $\mathbb{R}_+^q$ with

(1.4) $$\mu(C_j) = 1 \quad \text{for all } j \in \mathbb{N}^q.$$

Let $\mathcal{A}$ be a subclass of the Borel σ–algebra of $[0,1]^q$. For $A \subset \mathbb{R}^q_+$ and $\epsilon > 0$ let $A^\epsilon := \{y \in \mathbb{R}^q_+ : d(x,y) \le \epsilon \text{ for some } x \in A\}$ and $\tilde{A} := \mathbb{R}^q_+ \setminus A$. We will work with the following conditions for the class $\mathcal{A}$.

(1.5) For every $\epsilon > 0$ there is a subclass $\mathcal{A}(\epsilon) \subset \mathcal{B}([0,1]^q)$ with $N(\epsilon) = N(\mu,\epsilon) = N(\mu,\mathcal{A},\epsilon) := \text{card } \mathcal{A}(\epsilon) < \infty$, such that for each $A \in \mathcal{A}$ there is an $A(\epsilon) = A(\epsilon,A) \in \mathcal{A}(\epsilon)$ with $\sup\{n^{-q}\mu(nA \,\triangle\, nA(\epsilon)) : n \ge \epsilon^{-1/q}\} \le \epsilon$.

(If μ is q–dimensional Lebesgue measure, the last condition reduces to $\mu(A \,\triangle\, A(\epsilon)) \le \epsilon$, and so $\log N(\mu,\epsilon)$ is an upper bound for the metric entropy of $\mathcal{A}$ w.r.t. the quasi–metric $d_\mu(A,B) = \mu(A \,\triangle\, B)$.)

(1.6) $b(\epsilon) = b(\mu,\epsilon) = b(\mu,\mathcal{A},\epsilon) :=$
$\sup\{n^{-q}\, \mu((nA)^{n\epsilon} \cap (n\tilde{A})^{n\epsilon} : A \in \cup_{\eta>0} \mathcal{A}(\eta),\ n \ge 1/\epsilon\} \to 0$
as $\epsilon \to 0$.

Theorem 1. Let $(x_j, j \in \mathbb{N}^q)$ be a weakly stationary strongly mixing family of $\mathbb{R}^N$–valued random vectors with $Ex_j = 0$. Assume that for some $\delta > 0$ and $C_1 < \infty$

(1.7) $$E\|x_j\|^{2+\delta} < C_1 \quad \text{for all } j \in \mathbb{N}^q$$

and that (1.1) holds in such a way that

(1.8) $\alpha(t) \le C_2 t^{-s}$ for a constant C_2 and some $s > 1 + 2\cdot(17/2)^{q-1}/\min(1,\delta)$.

Let μ and $\mathcal{A}$ be as above and assume that (1.5) and (1.6) hold with

(1.9) $$N(\epsilon) \le c\epsilon^{-u} \quad \text{for some } u < \frac{\left(\frac{1}{2} - \frac{1}{2+\delta}\right)s - 1}{q\left(\frac{1}{2} + \frac{1}{2+\delta}\right)} - 2 = \frac{\delta s - 4 - 2\delta}{q(4+\delta)} - 2$$

and

(1.10) $$b(\epsilon) \le c\epsilon^h \quad \text{for some } 0 < h \le 1.$$

Then the series of matrices

$$T = \sum_{j \in \mathbb{Z}^q} r(j) \tag{1.11}$$

is absolutely convergent, T is positive semidefinite, and without changing its joint distribution the family $(x_j, j \in \mathbb{N}^q)$ can be defined on a new probability space on which there exists a family $(y_j, j \in \mathbb{N}^q)$ of i.i.d. centered Gaussian random vectors with covariance matrix T, such that for some constant γ (which depends only on u, s, q, h, δ, and N) and for suitable measurable functions u_n $(n \in \mathbb{N})$ we have

$$\sup \{\| \sum_{j \in \mathbb{N}^q} \mu(C_j \cap nA)(x_j - y_j)\| : A \in \mathcal{A}\} \leq u_n << n^{q/2 - \gamma} \tag{1.12}$$

almost surely as $n \to \infty$.

Theorem 2. Assume that all hypotheses of Theorem 1 (including (1.1) and (1.8)) are satisfied, except (1.9). Let in addition $(x_j, j \in \mathbb{N}^q)$ be absolutely regular satisfying (1.2) with

$$\beta(t) \leq C_3 t^{-s'} \text{ for some } C_3 < \infty \text{ and } s' > 2(1+q)(1+2/\delta) - q. \tag{1.13}$$

Instead of (1.9) assume that (1.5) holds with

$$\log N(\epsilon) \leq c\epsilon^{-u} \text{ for some } u < \frac{\delta - 2(2+\delta)(q+1)/(q+s')}{4+\delta}. \tag{1.14}$$

Then the assertion of Theorem 1 holds.

Example for Theorem 1. Let $E \in \mathbb{N}$ be fixed and $\mathcal{A}$ the set of all polytopes contained in $[0,1]^q$ which have no more than E vertices. Let μ be the Lebesgue measure on $\mathbb{R}^q_+$. Then elementary considerations show that $N(\mu,\mathcal{A},\epsilon) \leq c\epsilon^{-qE}$. Furthermore we know from the theory of convex bodies that (1.10) is satisfied with $h = 1$. Hence we can apply Theorem 1 if $u = qE$ and s and δ satisfy (1.8) and (1.9). For further classes of sets which satisfy (1.9) for some u see Gaenssler (1984), proof of Theorem 2.3 and the examples following it.

Example for Theorem 2. The hypotheses of Theorem 2 are satisfied if $q = 2$, $\mathcal{A}$ is the set of all closed convex sets contained in $[0,1]^2$, μ is the Lebesgue measure on $\mathbb{R}_+^2$ or the counting measure $\#$ on $\mathbb{N}^2$ and s, s' and δ are sufficiently large such that (1.8), (1.13) and (1.14) hold with $u = 1/2$ (resp. some $u > 2/3$ in the case $\mu = \#$), see the proof of Corollary 1 in [MPh].

Remark 1. Although for functional CLT's a boundary smoothness condition is unnecessary (see e.g. [GG]), it seems that we cannot do without it for strong approximation theorems.

Remark 2. In this paper we only considered polynomial mixing rates. However, for random fields with exponential rate of strong mixing, our methods presumably would yield a strong approximation result under a polynomial bound for the metric entropy (instead of the logarithmic bound (1.9)).

Remark 3. For invariance principles in distribution or in probability or in L^1, the quantitative hypotheses of our theorems could be weakened somewhat. The reason is that our truncation (5.3) could then be replaced by another one, see [GM], proof of Theorem 3.2.

2. Results for Banach Space Valued Random Fields

Let (Ω,Σ,P) be a probability space and $f\colon \Omega \to [-\infty,\infty]$ a function not necessarily measurable. Let us first recall the definition of the P–upper envelope (measurable cover function) f^* of f, as it is given e.g. in [DuPh]: Let $\mathcal{L}^0(\Omega,\Sigma,P)$ be the set of all Σ–measurable functions $j\colon \Omega \to [-\infty,\infty]$, and $f\colon \Omega \to [-\infty,\infty]$ arbitrary. Then put

(2.1) $$f^* := P\text{–ess inf}\,\{j \in \mathcal{L}^0(\Omega,\Sigma,P) : j \geq f\}.$$

By Lemma 2.1 of [DuPh] we can choose $f^* \geq f$ everywhere, and f^* is P–a.s. uniquely determined. Note that f^* depends on P.

For a measurable space $(E,\mathcal{F})$ and a random variable $\varphi: (\Omega,\Sigma) \to (E,\mathcal{F})$ with distribution $\mu = P_\varphi$ we say, following Andersen(1985), that φ is P–perfect, if

(2.2) $$f^* \circ \varphi = (f \circ \varphi)^* \quad P\text{–a.s. for all } f: E \to [-\infty,\infty].$$

(Of course here f^* is the μ–upper envelope of f and $(f \circ \varphi)^*$ is the P–upper envelope of $f \circ \varphi$.)

Let us further recall that the probability measure P is called perfect, if for any real–valued (Borel measurable) random variable $\xi: (\Omega,\Sigma,P) \to (\mathbb{R},\mathcal{B}(\mathbb{R}))$ with distribution $\mu = P_\xi$ on $\mathcal{B}(\mathbb{R})$ we have $\mu_*(\xi(\Omega)) = 1$. We also mention that any Borel probability measure on a Polish space is perfect, see e.g. [Pt], and if $(E,\mathcal{F})$ is a measurable space with countably generated σ–algebra $\mathcal{F}$ which contains all singletons of E, then there exists an injective measurable function $g: (E,\mathcal{F}) \to (\mathbb{R},\mathcal{B})$. (E.g. the Marczewski function $\Sigma_{n=1}^{\infty} 2\cdot 3^{-n}\, 1_{E_n}$ where $\{E_1, E_2,...\}$ is a countable generator of $\mathcal{F}$.) Now we quote the following result from [An] (Proposition 2.5, p. II.16):

Lemma 2.1. All random variables $\varphi: (\Omega,\Sigma,P) \to (E,\mathcal{F})$ are P–perfect, if P is perfect and there exists an injective measurable function from $(E,\mathcal{F})$ into $(\mathbb{R},\mathcal{B}(\mathbb{R}))$.

Now we are going to modify the definition of an independent resp. strongly mixing family of Banach space valued random elements as it has been given in [DuPh], [MPh] and [Ph84]. (Ω,Σ,P) always denotes a perfect probability space and $(E,\mathcal{F})$ a measurable space with countably generated σ–algebra $\mathcal{F}$ which contains all singletons of E. From now on we consider the index set $\mathbb{Z}^q$ (instead of $\mathbb{N}^q$ in section 1) for easier notation of the covariance structure of the approximating Gaussian random vectors, see (2.14). Let $(x_j, j \in \mathbb{Z}^q)$ be a strictly stationary family of $(E,\mathcal{F})$–valued r.v.'s on (Ω,Σ,P). Let $(S,\|.\|)$ be an arbitrary (not necessarily separable) Banach space and $h: E \to S$ a mapping. Write $X_j := h(x_j)$ for $j \in \mathbb{Z}^q$, then we call $(X_j, j \in \mathbb{Z}^q)$ a <u>strictly stationary random field of S–valued random elements</u>. It is called <u>strongly mixing at rate α</u> if (1.1) holds for $(x_j, j \in \mathbb{Z}^q)$; and <u>absolutely regular at rate β</u> if (1.2) holds for $(x_j, j \in \mathbb{Z}^q)$.

As a motivation of the perfectness assumption let us note that by Lemma 2.1, $(x_j)_{j\in J}$ is a P–perfect E^J–valued r.v. for every $J \subset \mathbb{Z}^q$, and so e.g. $\|\Sigma_{j\in J} X_j\|^*$ is measurable w.r.t. the σ–algebra generated by $\{x_j, j \in J\}$. This and similar observations (which we will not always state explicitly in the proofs) will allow to exploit the conditions of strong mixing resp. absolute regularity.

Finally, we will always assume in the following that on (Ω,Σ,P) there exists a uniformly [0,1]–distributed r.v. which is independent of $(x_j, j \in \mathbb{Z}^q)$. (If necessary this can be achieved by passing over from (Ω,Σ,P) to the product of this space with [0,1] and Lebesgue measure. This does not destroy perfectness, see e.g. [Pl], p. 335.) This hypothesis ensures that the underlying probability space is rich enough so that the approximating Gaussian random vectors to be constructed later on will live on the given space.

Theorem 3. Assume the hypotheses described above to hold, let $(X_j, j \in \mathbb{Z}^q)$ be a strictly stationary random field of S–valued random elements. Assume that there exist $C_1 < \infty$ and $\delta > 0$ such that

$$E\|X_j\|^{*2+\delta} \le C_1 \text{ for all } j \in \mathbb{Z}^q. \tag{2.3}$$

Let (1.1) and (1.2) hold for $(x_j, j \in \mathbb{Z}^q)$ with (1.13) and

(2.4) $$\alpha(t) \le C_2 t^{-s} \text{ for some } C_2 < \infty \text{ and some}$$
$$s \ge (2+\epsilon)\cdot(1 + 2\cdot(17/2)^{q-1}/\min(1,\delta)) \quad \text{with } 0 < \epsilon \le 1.$$

Assume that for every $m \in \mathbb{N}$ there is a linear operator $\Lambda_m: S \to S$ such that

(2.5) $$\sup_m \|\Lambda_m\| < \infty,$$

(2.6) $$\dim \Lambda_m S \le C_3 m$$

for some $C_3 < \infty$, and that for some $\vartheta \in (0, 1/2]$ there is for each $m \in \mathbb{N}$ some

(2.7) $$N_o(m) \le C_4 \exp(m^{1-\vartheta}),$$

such that for all $d_j \in (0,1]$ $(j \in \mathbb{N}^q)$ and all $J \subset \mathbb{N}^q$ with $\text{card } J \ge N_o(m)$

(2.8) $$P^*((\text{card } J)^{-\frac{1}{2}} \| \sum_{j \in J} d_j(X_j - \Lambda_m X_j)\| \ge m^{-\frac{1}{2}}) \le m^{-q(1+\vartheta)}.$$

Here it suffices to require (2.8) only for such J which are a rectangle or which are disjoint unions of Q_o cubes with length of edges p_o, where Q_o and p_o satisfy the relation

(2.9) $$Q_o \beta(p_o) \le p_o^{-b} \text{ for some fixed } b \in (0,1].$$

Furthermore assume that $\Lambda_m \circ h: E \to \Lambda_m S$ is a measurable mapping into the finite-dimensional Banach space $\Lambda_m S$ and that

(2.10) $$E\Lambda_m X_j = 0$$

for all $m \in \mathbb{N}$, $j \in \mathbb{Z}^q$.

Moreover let μ be a Borel measure on $\mathbb{R}_+^q$ with (1.4) and let $\mathcal{A} \subset \mathcal{B}([0,1]^q)$ such that (1.5) and (1.6) are satisfied with

(2.11) $$b(\epsilon) \le c\epsilon^h \text{ for some } h \in (0,1]$$

and

(2.12) $$\log N(\epsilon) \le c\epsilon^{-u} \text{ for some } u > 0 \text{ with}$$

(2.12a) $$u < \frac{\delta - 2(2+\delta)(q+1)/(q+s')}{4+\delta}$$

and

(2.12b) $$u < \left(1 - \frac{q+1}{q+s'}\right) \frac{2(2+\delta)}{q(4+\delta)} h.$$

Let T be the closure of the linear span of $\cup_{m\geq 1} \Lambda_m S$, so T is a separable Banach space. Then there exist i.i.d. T–valued Gaussian centered random vectors Y_j for $j \in \mathbb{N}^q$ such that for some $\alpha > 0$ and for measurable u_n $(n \in \mathbb{N})$ we have

(2.13) $$\sup_{A\in\mathcal{A}} \| \Sigma_{j\in\mathbb{N}^q} \mu(C_j \cap nA)(X_j - Y_j)\| \leq u_n << n^{q/2}(\log n)^{-\alpha}$$

P–a.s. as $n \to \infty$. The covariance structure of $Y_{\mathbb{1}}$ is given by

(2.14) $$E(t(Y_{\mathbb{1}})t'(Y_{\mathbb{1}})) = \lim_{m\to\infty} \Sigma_{j\in\mathbb{Z}^q} E(t(\Lambda_m X_0)t'(\Lambda_m X_j))$$

for $t, t' \in T'$, the topological dual of T.

Theorem 4. Modify the hypotheses in Theorem 3 as follows: Instead of (2.6), (2.7) and (2.8) we assume

(2.15) $$\dim \Lambda_m S \leq C \exp(m^\eta)$$

and

(2.16) $$N_o(m) \leq C\, m^\kappa$$

for constants $C, \eta, \kappa \in (0,\infty)$ and

(2.17) $$P^*((\text{card } J)^{-\frac{1}{2}} \| \Sigma_{j\in J} d_j(X_j - \Lambda_m X_j)\| \geq \tfrac{1}{m}) \leq \frac{1}{m}$$

for all $d_j \in [0,1]$ and all $J \subset \mathbb{N}^q$ with card $J \geq N_o(m)$. This is required only for sets J of the shape described in Theorem 3, however with $b = \frac{1}{4}(s' + 1 - 2q(1+2/\delta))$ in (2.9). Moreover we require in addition to (1.13) that

(2.18) $$s' > 2q(1 + 2/\delta) - 1.$$

Then the assertion of Theorem 3 remains valid.

Theorem 5. Now we modify the assumptions in Theorem 3 as follows: Instead of (2.6), (2.7) and (2.8) we assume that

(2.19) $$\dim \Lambda_m S \leq C\, m^{\eta}$$

for constants η, $C \in (0,\infty)$ and that (2.16) and (2.17) hold (again for all $J \subset \mathbb{N}^q$ with card $J \geq N_0(m)$ which are of the shape described in Theorem 3). Then the assertion of Theorem 3 can be improved as follows: There exist i.i.d. T–valued centered Gaussian random vectors Y_j for $j \in \mathbb{N}^q$ such that for some $\gamma > 0$ and measurable u_n $(n \in \mathbb{N})$

(2.20) $$\sup_{A\in\mathcal{A}} \| \sum_{j\in\mathbb{N}^q} \mu(C_j \cap nA)\,(X_j - Y_j)\| \leq u_n \ll n^{q/2-\gamma}$$

P–a.s. as $n \to \infty$. The covariance structure of $Y_{\mathbb{1}}$ is again given by (2.14).

Remark. If S is a separable Hilbert space and $h: E \to S$ measurable, then we do not need the condition (2.12b) in Theorems 3, 4, and 5; and (2.4) can be weakened to (1.8) in the case $q \geq 2$. (For $q = 1$ we need $s > 1 + 2/\min(\frac{2}{3},\delta)$ because Theorem 1 of Dehling (1983) is used.) The reason for this is that instead of Lemmas 8.5 and 9.2 we can use the better inequality of Lemma 4.1 in the proof of Lemmas 9.3 through 9.6. (In the proof of Lemma 9.4 note that the number β there can be chosen arbitrarily close to 1, so one has again $u < 2\alpha$.)

3. Results for Empirical Processes

By an approach analogous to that of Dudley and Philipp (1983) (which has also been used in [Ph84] and [MPh]) we can deduce strong approximation theorems for empirical processes from Theorems 4 and 5. (This is also possible for Theorem 3, however, it would yield worse results than Theorems 4 and 5.)

Again let (Ω,Σ,P) be a perfect probability space and $(E,\mathcal{F})$ a measurable space with countably generated σ–algebra $\mathcal{F}$ which contains all singletons of E. Let $(x_j, j \in \mathbb{Z}^q)$ be a strictly stationary family of $(E,\mathcal{F})$–valued random variables on (Ω,Σ,P). Again we always assume that on (Ω,Σ,P) there exists a uniformly [0,1]–distributed random variable which is independent of $(x_j, j \in \mathbb{Z}^q)$.

Let $\mathcal{A}$ be a family of Borel sets in $[0,1]^q$ and $\mathcal{G}$ be a family of $\mathcal{F}$–measurable functions $f\colon E \to [0,1]$. Denote the common distribution of the x_j by $P' = \mathcal{L}(x_j)$. For P'–integrable f write $P'(f) := \int f\, dP'$. Let μ again be a Borel measure on $\mathbb{R}^q_+$ with (1.4). We define the <u>n–th empirical measure</u>

(3.1) $$P'(n,f,A) := n^{-q} \sum_{j \in \mathbb{N}^q} \mu(nA \cap C_j)\, f(x_j), \quad f \in \mathcal{G},\ A \in \mathcal{A},\ n \in \mathbb{N},$$

and the <u>normalised empirical measure</u>

(3.2) $$\nu(n,f,A) := n^{-q/2} \sum_{j \in \mathbb{N}^q} \mu(nA \cap C_j)\, (f(x_j) - P'(f)), \quad f \in \mathcal{G},\ A \in \mathcal{A},\ n \in \mathbb{N}.$$

For each $y > 0$ let $N_I(y) := N_I(y,\mathcal{G},P')$ be the smallest number $d \in \mathbb{N}$ of $\mathcal{F}$–measurable functions $f_1,\dots,f_d\colon E \to [0,1]$ with the following property: For every $f \in \mathcal{G}$ there exist f_a, f_b, $1 \le a,b \le d$, with $f_a \le f \le f_b$ and $P'(f_b - f_a) < y$. Recall (e.g. from [Du78], [Du84]) that $\log N_I(y)$ is called a <u>metric entropy with bracketing</u> (resp. <u>with inclusion</u> if $\mathcal{G}$ consists only of indicators of $\mathcal{F}$–measurable sets; in this case w.l.o.g. the $f_1,\dots,f_d$ can also be chosen as indicators, see [Du84]).

Now we have the following results.

Theorem 6. Let $(x_j, j \in \mathbb{Z}^q)$, μ, $\mathcal{A}$, and $\mathcal{G}$ satisfy the assumptions above. Let (1.1) hold with

(3.3) $$\alpha(t) \leq C_2 t^{-s}$$

for some constant C_2 and some

(3.4) $$s > 2\cdot(1 + 2\cdot(17/2)^{q-1}).$$

Let (1.2) hold with

(3.5) $$\beta(t) \leq C_2 t^{-s'} \text{ with}$$

(3.6) $$s' > \max(q+2, 2q-1).$$

Assume that $\mathcal{A}$ satisfies (1.5) and (1.6) with

(3.7) $$b(\mu,\mathcal{A},\epsilon) \leq c\epsilon^h \text{ for some } h \in (0,1]$$

and

(3.8) $$\log N(\mu,\mathcal{A},\epsilon) \leq c\epsilon^{-u} \text{ for some } u > 0 \text{ with}$$

(3.9) $$u < 1 - 2(q+1)/(q+s')$$

and

(3.10) $$u < (1 - \frac{q+1}{q+s'})\frac{2h}{q}.$$

Assume that the class $\mathcal{G}$ satisfies

(3.11) $$\log N_I(y,\mathcal{G},P') \leq cy^{-\tau} \text{ for some } \tau > 0 \text{ with}$$

(3.12) $$\tau < 1 - \frac{2q}{s'+1}.$$

Then on (Ω,Σ,P) there exists a collection of i.i.d. Gaussian processes Y_j $(j \in \mathbb{N}^q)$ with index set $\mathcal{G}$ such that

(3.13) $$EY_{\mathbb{1}}(f) = 0$$

and

(3.14) $$E(Y_{\mathbb{1}}(f)\, Y_{\mathbb{1}}(g)) = \sum_{j\in\mathbb{Z}^q} (E(f(x_0)g(x_j)) - P'(f)P'(g))$$

$(f, g \in \mathcal{G})$ and for some $\alpha > 0$ and measurable functions u_n $(n \in \mathbb{N})$

$$(3.15)\quad \sup_{\substack{A\in\mathcal{A}\\ f\in\mathcal{G}}} \left| \sum_{j\in\mathbb{N}^q} \mu(nA \cap C_j)\,(f(x_j) - P'(f) - Y_j(f))\right| \le u_n << n^{q/2} (\log n)^{-\alpha}$$

a.s. for $n \to \infty$.

Each Y_j can be chosen with a.s. uniformly continuous sample paths w.r.t. the L^2 norm on $\mathcal{G}$ $(\subset L^2(E,\mathcal{F},P'))$.

Theorem 6 is a generalisation of Theorem 7 of [MPh] to the absolutely regular case, except for the fact that (3.10) may be a stronger restriction than $u < 1$ even for independent $(x_j, j \in \mathbb{Z}^q)$ where s' can be chosen arbitrarily large. On the other hand, Theorem 6 is a variant of Corollary 6.1 of [Ph84] in that is requires the stronger hypothesis of absolute regularity (instead of strong mixing) but at the same time by (3.11) and (3.12) it allows much larger classes $\mathcal{G}$ than [Ph84, (6.1)] and it deals with the more general situation of a multidimensional index set $\mathbb{N}^q$ and larger class $\mathcal{A}$. Furthermore the following Theorem 7 shows that in the absolutely regular case one can get a sharper convergence rate than in Corollary 6.1 of [Ph84] if the metric entropy with inclusion of $\mathcal{G}$ satisfies a logarithmic bound (as in [Ph84, (6.1)]).

Theorem 7. In Theorem 6 replace (3.11) by the stronger hypothesis

$$(3.16)\qquad N_I(y,\mathcal{G},P') \le cy^{-H} \text{ for some } H \in (0,\infty).$$

Then in (3.15) the bound on the right–hand side can be sharpened to $O(n^{q/2-\gamma})$ a.s. as $n \to \infty$, for some constant $\gamma > 0$.

We give a corollary to each of Theorems 6 and 7. The first one generalises Corollary 2 of [MPh], and it can be deduced from Theorem 6 and the proof of [MPh, Corollary 1]. Let $\mathcal{C}$ denote the class of all measurable convex sets in $[0,1]^2$.

Corollary 1. Let $(x_j, j \in \mathbb{Z}^2)$ be a strictly stationary random field defined on (Ω,Σ,P), with values in $[0,1]^2$ and with common distribution $\mathcal{L}(x_0) = P'$ having a bounded density w.r.t. Lebesgue measure. Assume that $(x_j, j \in \mathbb{Z}^2)$ satisfies the conditions of absolute regularity (1.2) and (3.5) with $s' > 36$. Then there exists a family $(Y_j, j \in \mathbb{N}^2)$ of i.i.d. Gaussian processes on (Ω,Σ,P) with index set $\mathcal{C}$ such that $EY_{\mathbb{1}}(C) = 0$,

$$E(Y_{\mathbb{1}}(C)Y_{\mathbb{1}}(D)) = \sum_{j\in\mathbb{Z}^2} (P(x_0 \in C, x_j \in D) - P(x_0 \in C)P(x_0 \in D))$$

for $C, D \in \mathcal{C}$ and for some $\alpha > 0$ and measurable u_n $(n \in \mathbb{N})$

$$\sup_{C,D\in\mathcal{C}} \Big| \sum_{j\in nC} (1_D(x_j) - P'(D) - Y_j(D)) \Big| \le u_n << n(\log n)^{-\alpha} \quad \text{a.s.} \tag{3.17}$$

Corollary 2. Let $(x_j, j \in \mathbb{Z})$ be a strictly stationary sequence of $\mathbb{R}^d$–valued r.v.'s, defined on (Ω,Σ,P) and with distribution $P' = \mathcal{L}(x_0)$. Assume that $(x_j, j \in \mathbb{Z})$ is absolutely regular with rate $\beta(n) << n^{-s'}$ for some $s' > 6$. Then there is a sequence $(Y_j, j \in \mathbb{N})$ of i.i.d. Gaussian processes on (Ω,Σ,P) with index set $\mathbb{R}^d$ such that for some $\gamma > 0$ and measurable u_n $(n \in \mathbb{N})$

$$\sup_{0\le k\le n} \sup_{x\in\mathbb{R}^d} \Big| \sum_{j=1}^{k} (1_{\{x_j\le x\}} - P'((-\infty,x]) - Y_j(x)) \Big| \le u_n << n^{\frac{1}{2}-\gamma} \quad \text{a.s.}$$

This is a variant of Theorem 1 of Dhompongsa (1984). There $s' > d + 2$ is needed. Here we can drop the continuity assumption on the common distribution function F of the x_j of [Dh, p. 116]. However, we do not assert that the Y_j have continuous paths w.r.t. the natural metric on $\mathbb{R}^d$. We only get continuity w.r.t. the quasi-metric $\rho(u,v) = P'((-\infty,u] \triangle (-\infty,v])$ $(u, v \in \mathbb{R}^d)$. Of course, if F is continuous, this implies the continuity of the Y_j w.r.t. the natural topology on $\mathbb{R}^d$.

For the proof of Corollary 2 use Theorem 7 and note that for $\mathcal{G} = \{(-\infty,x]: x \in \mathbb{R}^d\}$ we have $N_I(y,\mathcal{G},P') \le cy^{-d}$ (see [Ph84, p. 872]).

4. Upper Bounds for Moments of Partial Sums

For the special case of real valued random fields the following lemma has been given by Guyon and Richardson [GR, Lemme 5.0]. It is immediately seen that it remains valid in the Hilbert space case if one uses Lemma 2.2 of Dehling and Philipp (1982) for the proof.

Lemma 4.1. Let $(x_j, j \in \mathbb{N}^q)$ be a random field with values in a separable Hilbert space H with norm $\|.\|$, assume that for some $C < \infty$ and $\delta > 0$ we have for all $j \in \mathbb{N}^q$, $Ex_j = 0$ and $E\|x_j\|^{2+\delta} \le C$ and that $(x_j, j \in \mathbb{N}^q)$ satisfies a strong mixing condition (1.1) with

(4.1) $$C_2 := \sum_{r \in \mathbb{N}} r^{q-1} \alpha(r)^{\delta/(2+\delta)} < \infty.$$

Let $d_j \in [0,1]$ for $j \in \mathbb{N}^q$, $d_j \ne 0$ for only finitely many $j \in \mathbb{N}^q$. Then

(4.2) $$E\|\sum_{j \in \mathbb{N}^q} d_j x_j\|^2 \le \sum_{j \in \mathbb{N}^q} d_j \cdot A$$

with $A = (1 + 15q \cdot 3^q C_2) C^{2/(2+\delta)}$.

Lemma 4.2. Let (Ω,Σ,P) be a probability space, $(E,\mathcal{F})$ a measurable space, for $j \in \mathbb{N}^q$ let x_j be an $(E,\mathcal{F})$-valued r.v. on (Ω,Σ,P), $h_j: E \to S$ a mapping into the Banach space S, and $X_j := h_j \circ x_j$. Assume

(a) $(x_j)_{j\in J}$ is P–perfect for all $J \subset \mathbb{N}^q$, or

(b) S is separable and all h_j are $\mathcal{F}$–$\mathcal{B}(S)$–measurable.

Moreover assume that for some $C < \infty$ and $\delta \in (0,1]$, $E\|X_j\|^{*2+\delta} \leq C$ for all $j \in \mathbb{N}^q$, $(x_j, j \in \mathbb{N}^q)$ satisfies a strong mixing condition (1.1) with

$$\alpha(r) \leq C_1 r^{-(1+\epsilon)(1 + 2(17/2)^{q-1}/\delta)}$$

for some $C_1 < \infty$ and $\epsilon \in (0,1]$, and for some $A < \infty$

$$E\|\sum_{a+1\leq j\leq a+n} X_j\|^{*2} \leq \Pi(n)\cdot A$$

for all $a \in \mathbb{N}_0^q$, $n \in \mathbb{N}^q$.

Then there is a constant C_3, which depends only on ϵ, δ, q, C_1, A, C, such that for all $a \in \mathbb{N}_0^q$, $n \in \mathbb{N}^q$ and $0 \leq \alpha \leq (2/17)^q\epsilon\delta$

$$E\|\sum_{a+1\leq j\leq a+n} X_j\|^{*2+\alpha} \leq C_3\Pi(n)^{1+\alpha/2}.$$

For $q = 1$ and real–valued measurable h_j this lemma has been proved by Sotres and Ghosh (1977). Dehling and Philipp (1982) remarked that it remains valid under the condition (b) (for $q = 1$). A careful analysis of the proof of [SG] and of Serfling's (1968) paper on which their proof rests, shows that for $q = 1$ the lemma remains valid in the case (a). (Note that in this case $\|\sum_{j\in J} X_j\|^*$ is measurable w.r.t. the σ–algebra generated by $\{x_j, j \in J\}$ for all $J \subset \mathbb{N}^q$, and use Lemma 2.2 of [DuPh].) Finally, for $q > 1$ the proof can be given by induction on q as in [BM], Lemma 4 (with the correction in Berkes (1984), p. 158 and p. 160, Lemma 3).

5. Proof of Theorem 1

The following lemma gives an exponential inequality for strongly mixing random fields. It is a generalisation of Theorem 4 of Philipp (1984). As there the proof consists of the following steps: building blocks, approximating the sequence of the partial sums over the blocks (with appropriate gaps between them) by a martingale, and applying an exponential inequality for (super–)martingales of Stout (1974, p. 299). Since the proof follows closely the lines of Philipp (1984), we do not give the details here. They may be found in [Str].

Lemma 5.1. Let $(x_j,\ j \in \mathbb{N}^q)$ be a family of real–valued r.v.'s with $Ex_j = 0$ and $|x_j| \le M$ for all j and for some fixed $M < \infty$. Assume $(x_j,\ j \in \mathbb{N}^q)$ satisfies a strong mixing condition (1.1) with rate α, as well as (4.2) for some $A < \infty$.

Let $D_1,\dots,D_Q$ be mutually disjoint cubes in $\mathbb{R}^q_+$ of the form $D_k = (n_k - 2p\mathbb{1}, n_k]$ with $n_k \in 2p\mathbb{N}^q$ for some fixed $p \in \mathbb{N}$ for all $k = 1,\dots,Q$. Let $0 \le d_j \le 1$ for $j \in \mathbb{N}^q$. Put $D := \cup_{k=1}^{Q} D_k$, $F_k := \sum_{j \in D_k \cap \mathbb{N}^q} d_j$, $F := \sum_{j \in D} d_j$. Then for all $K > 0$

$$P(|\Sigma_{j\in D} d_j x_j| > 2^{q+1}K) \le c\Big(K^{-2}M^2F^2\alpha(p) + p^{2q}A^{-2}M^4\alpha(p) + \exp(-\frac{K^2}{8AF}) + \exp(-2^{-q-2}\frac{K}{Mp^q})\Big).$$

Here the constant c depends only on q.

Now assume the hypotheses of Theorem 1. (1.9) implies that there are real numbers $\tau, \zeta > 0$ such that

$$\zeta < \frac{1}{2} - \frac{1+\tau}{2+\delta} \tag{5.1}$$

and

(5.2) $$u < \frac{\zeta s - 1}{q\left(\frac{1}{2} + \frac{1+\tau}{2+\delta}\right)} - 2.$$

Define for $j \in \mathbb{N}^q$

(5.3) $$x_j' := x_j \cdot 1\{\|x_j\| \le \Pi(j)^{(1+\tau)/(2+\delta)}\},$$

(5.4) $$\bar{x}_j := x_j' - Ex_j',$$

(5.5) $$x_j'' := x_j - x_j'.$$

Then $Ex_j' = -\,Ex_j''$, hence by the Hölder and Markov inequalities (cf. [MPh, proof of Lemma 2.3])

(5.6) $$\|Ex_j'\| = \|Ex_j''\| \le E\|x_j''\| \le c\, P(x_j'' \ne 0)^{(1+\delta)/(2+\delta)}$$
$$\le c\, \Pi(j)^{-(1+\tau)(1+\delta)/(2+\delta)}.$$

<u>Remark 5.2.</u> a) $\|\bar{x}_j\| \le \|x_j'\| + \|Ex_j'\| \le \Pi(j)^{(1+\tau)/(2+\delta)} + c \le c\Pi(j)^{(1+\tau)/(2+\delta)}$ for all $j \in \mathbb{N}^q$, so up to a constant factor $\|\bar{x}_j\|$ has the same bound as $\|x_j'\|$.

b) $\|\bar{x}_j\|^{2+\delta} \le c(\|x_j'\|^{2+\delta} + \|Ex_j'\|^{2+\delta}) \le c\|x_j\|^{2+\delta} + c\Pi(j)^{-(1+\tau)(1+\delta)}$ by (5.6). Therefore $E(\|\bar{x}_j\|^{2+\delta}) \le cE(\|x_j\|^{2+\delta})$.

Remark 5.2 ensures that the Lemmas 4.1, 4.2, and 5.1 can be applied to $(\bar{x}_j,\, j \in \mathbb{N}^q)$.

<u>Remark 5.3.</u> $\|x_j - \bar{x}_j\| = \|x_j - x_j' + Ex_j'\| \le \|x_j - x_j'\| + \|Ex_j''\|$, hence by [MPh, Lemma 2.3] $\Sigma_{\mathbb{1} \le j \le n\mathbb{1}} \|x_j - \bar{x}_j\| = O(n^{q/(2+\delta)})$ a.s. for $n \to \infty$.

For a constant β to be specified later define

(5.7) $$\psi(m) := \sum_{k=1}^{m} k^{\beta} \quad \text{for } m \in \mathbb{N}.$$

Assume that m and n are linked by

(5.8) $$\psi(m) < n \leq \psi(m+1).$$

We adopt some notations from [MPh]: For $r = (r_1,\dots,r_q) \in \mathbb{N}^q$ write

(5.9) $$R_r := \{(v_1,\dots,v_q) \in \mathbb{R}_+^q \colon \psi(r_i) < v_i \leq \psi(r_i+1) \ \forall i \leq q\}.$$

For $A \subset \mathbb{R}_+^q$ let

(5.10) $$A_* := \cup_{R_r \subset A} R_r ,$$

and for $A \in \mathcal{B}([0,1]^q)$

(5.11) $$S_n(A) := \sum_{j \in \mathbb{N}^q} \mu(nA \cap C_j)\, x_j ,$$

(5.12) $$\overline{S}_n(A) := \sum_{j \in \mathbb{N}^q} \mu(nA \cap C_j)\, \overline{x}_j ,$$

(5.13) $$\overline{V}_n(A) := \sum_{j \in \mathbb{N}^q} \mu(((nA) \setminus (nA)_*) \cap C_j)\, \overline{x}_j .$$

Lemma 5.4. There are $\kappa > q\left(\frac{1}{2} + \frac{1+\tau}{2+\delta}\right)$ (with $\kappa \leq q$), $\gamma', \gamma'' > 0$ and $c < \infty$ such that for all $n \in \mathbb{N}$

$$P(\sup\{\|\overline{V}_n(A(n^{-\kappa}))\| \colon A(n^{-\kappa}) \in \mathcal{A}(n^{-\kappa})\} > cn^{q/2-\gamma'}) \leq cn^{-1-\gamma''}.$$

Proof. Define $p := [n^{\zeta}]$ and for fixed $A \in \mathcal{B}([0,1]^q)$ cover $(nA) \setminus (nA)_*$ with cubes in $\mathbb{R}_+^q$ with length of edges equal to $2p$ and of the form explained in Lemma 5.1; as there let D be the union of these cubes. For $j \in \mathbb{N}^q$ let $d_j := \mu(((nA) \setminus (nA)_*) \cap C_j)$. By (1.6) and (1.10) we then have (cf. [MPh, Lemma 2.1])

(5.14) $$F := \sum_{j \in D} d_j = \mu((nA) \setminus (nA)_*) \leq n^q\, b((\psi(m+1) - \psi(m))/n)$$
$$\leq cn^{q-h}\, (\psi(m+1) - \psi(m))^h \leq cn^{q-h}\, n^{h\beta/(\beta+1)} = cn^{q-\gamma}$$

where $\gamma = h/(\beta+1)$. There is a $\gamma' > 0$ such that

(5.15) $$\gamma' < \gamma/2 ,$$

$$\gamma' < q\left(\frac{1}{2}-\frac{1+\tau}{2+\delta}-\zeta\right). \tag{5.16}$$

Then by Lemma 4.1, (5.14), Remark 5.2 and Lemma 5.1 (which is also valid for $\mathbb{R}^N$–valued random fields if the factor 2^{q+1} is replaced by $2^{q+1}N$)

$$\begin{aligned} P(\|\overline{V}_n(A)\| > c_1 n^{q/2-\gamma'}) \le c\Big(& n^{2q\zeta + 4q(1+\tau)/(2+\delta) - \zeta s} \\ &+ n^{-q+2\gamma'+2q(1+\tau)/(2+\delta)+2q-2\gamma-\zeta s} \\ &+ \exp(-c_2 n^{q-2\gamma'-q+\gamma}) \\ &+ \exp(-c_3 n^{q/2-\gamma'-q(1+\tau)/(2+\delta)-\zeta q})\Big). \end{aligned}$$

For the fourth summand we use (5.16), for the second and third ones (5.15), for the first one (5.1), and obtain

$$P(\|\overline{V}_n(A)\| > c_1 n^{q/2-\gamma'}) \le c n^{-\zeta s + q(1+2(1+\tau)/(2+\delta))}. \tag{5.17}$$

There exist ϑ and κ with $0 < \vartheta < \zeta s - 1 - q(1+2(1+\tau)/(2+\delta))$, $q \ge \kappa > q\left(\frac{1}{2}+\frac{1+\tau}{2+\delta}\right)$ and $\vartheta/\kappa = u$ (use (5.2)). We multiply both sides of (5.17) by $N(\mu, n^{-\kappa}) \le cn^{\kappa u} = cn^{\vartheta}$ and obtain

$$P(\sup_{A(n^{-\kappa})} \|\overline{V}_n(A(n^{-\kappa}))\| > c_1 n^{q/2-\gamma'}) \le cn^{-1-\gamma''}$$

for some $\gamma'' > 0$. □

Lemma 5.5. We have for all $A \in \mathcal{A}$

$$\|\overline{S}_n(A) - \overline{S}_n(A(n^{-\kappa}))\| \le cn^{q/2-\gamma_3}$$

for some $\gamma_3 > 0$ where $A(n^{-\kappa}) = A(n^{-\kappa}, A)$ according to (1.5), and κ is chosen as in Lemma 5.4.

Proof. As in the proof of [MPh, Lemma 2.4] we obtain for the left–hand side the bound $cn^{q(1+\tau)/(2+\delta)} n^q n^{-\kappa} = cn^{q/2-\gamma_3}$, where $\gamma_3 = \kappa - q(1+\tau)/(2+\delta) - q/2 > 0$. □

Set

(5.18) $$G := \bigcap_{i=1}^{q} \{j=(j_1,\dots,j_q) \in \mathbb{N}^q\colon\ j_i^{8q} \geq \Pi(j)\}$$

$$(= G_{1/(8q-1)} \text{ in the notation of [BM]}),$$

(5.19) $$L := \{r \in \mathbb{N}^q\colon (\psi(r_1),\dots,\psi(r_q)) \in G\}.$$

Lemma 5.6. If in (5.7) $\beta \geq 6q$ then

$$P\Big(\sum_{\substack{r \notin L \\ \psi(r_i) \leq n\ \forall i}} \| \sum_{j \in R_r} x_j \| > n^{q/2 - 1/16} \Big) \leq cm^{-2}$$

(where n and m are linked by (5.8)), hence by the Borel–Cantelli–Lemma

$$\sum_{\substack{r \notin L \\ \psi(r_i) \leq n\ \forall i}} \| \sum_{j \in R_r} x_j \| << n^{q/2 - 1/16} \quad \text{a.s. for } n \to \infty.$$

Proof. For $r \in \mathbb{N}^q \setminus L$ with $\psi(r_i) \leq n$ for $i = 1,\dots,q$ we have by definition of L for some $i \leq q$ $\psi(r_i)^{8q} < \Pi_{i=1}^{q} \psi(r_i) \leq n^q$, hence $\psi(r_i) \leq n^{1/8}$, and therefore

(5.20) $$\mathrm{card}(R_r \cap \mathbb{N}^q) \leq n^{1/8} \cdot n^{q-1} = n^{q-7/8}.$$

Furthermore we have $\mathrm{card}\{t \in \mathbb{N}\colon \psi(t) \leq n\} \leq \mathrm{card}\{t \in \mathbb{N}\colon ct^{\beta+1} \leq n\}$ $\leq \mathrm{card}\{t \in \mathbb{N}\colon t \leq cn^{1/(\beta+1)}\} \leq cn^{1/(\beta+1)}$ and hence

(5.21) $$\mathrm{card}\{r \in \mathbb{N}^q\colon \psi(r_i) \leq n \ \ \forall i \leq q\} \leq cn^{q/(\beta+1)}.$$

By Lemma 4.1 and Chebyshev's inequality this implies

$$P\Big(\sum_{\substack{r \notin L \\ \psi(r_i) \leq n\ \forall i}} \| \sum_{j \in R_r} x_j \| > n^{q/2 - 1/16} \Big) \leq$$

$$\leq cn^{\frac{q}{\beta+1}} \max_{r \notin L,\ \psi(r_i) \leq n\ \forall i} P(\| \sum_{j \in R_r} x_j \| \geq cn^{\frac{q}{2} - \frac{1}{16} - \frac{q}{\beta+1}})$$

$$\le cn^{q/(\beta+1)} n^{q-7/8} n^{-q + 1/8 + 2q/(\beta+1)}$$

$$= cn^{-3/4 + 3q/(\beta+1)}$$

$$\le cm^{-3(\beta+1)/4 + 3q}$$

$$\le cm^{-2}. \square$$

For the construction of the approximating Gaussian random vectors y_j, $j \in \mathbb{N}^q$, we can now follow [BM]: First of all, [BM, (2.1) and Lemma 3] imply that the series (1.11) is absolutely convergent and T is positive semidefinite. Now choose the constant $\beta \in \mathbb{N}$ in (5.7) as in [BM, (4.1), p. 25] (with $\rho = 1/(8q-1)$ there; then in particular $\beta \ge 6q$ as required for Lemma 5.6 here). By the proof of Theorem 1 of [BM, p. 30f] (and Lemma 8 there) there is (after changing to a rich enough probability space) an i.i.d. family $(y_j, j \in \mathbb{N}^q)$ of Gaussian random vectors with mean 0 and covariance matrix T such that for some constant $\gamma > 0$

(5.22) $$\sum_{\substack{r \in L \\ \psi(r_i) \le n \, \forall i}} \|\sum_{j \in R_r} (x_j - y_j)\| << n^{q/2 - \gamma} \quad \text{a.s. as } n \to \infty.$$

Now as in [MPh] define for $j \in \mathbb{N}^q$

(5.23) $$y_j' := y_j \cdot 1\{\|y_j\| \le \Pi(j)^{(1+\tau)/(2+\delta)}\},$$

and for $A \in \mathcal{A}$, $n \in \mathbb{N}$

(5.24) $$T_n(A) := \sum_{j \in \mathbb{N}^q} \mu(nA \cap C_j)\, y_j ,$$

(5.25) $$T_n'(A) := \sum_{j \in \mathbb{N}^q} \mu(nA \cap C_j)\, y_j' ,$$

(5.26) $$U_n'(A) := \sum_{j \in \mathbb{N}^q} \mu(((nA)\setminus(nA)_*) \cap C_j)\, y_j' .$$

Then $Ey_j' = 0$ and Remarks and Lemmas 5.2 to 5.6 are also valid with y_j', T_n', U_n' instead of $\bar{x}_j$, $\bar{S}_n$, $\bar{V}_n$. For each $A \in \mathcal{A}$ we then have

$$\|S_n(A)-T_n(A)\| \le \|S_n(A)-\bar{S}_n(A)\| + \|T_n(A)-T'_n(A)\| \tag{5.27}$$
$$+ \|\bar{S}_n(A)-\bar{S}_n(A(n^{-\kappa}))\| + \|T'_n(A)-T'_n(A(n^{-\kappa}))\|$$
$$+ \|\bar{V}_n(A(n^{-\kappa}))\| + \|U'_n(A(n^{-\kappa}))\|$$
$$+ \Sigma_{1\le j\le n1} (\|x_j-\bar{x}_j\| + \|y_j-y'_j\|)$$
$$+ \|\Sigma_{j\in\mathbb{N}^q} (x_j-y_j)\, \mu((nA(n^{-\kappa}))_* \cap C_j)\|.$$

The following holds uniformly in $A \in \mathcal{A}$. The expressions in the first and fourth lines are $O(n^{q/(2+\delta)})$ a.s. as $n \to \infty$ by Remark 5.3, the expression in the second line is $\le cn^{q/2-\gamma_3}$ for constants $c < \infty$, $\gamma_3 > 0$ by Lemma 5.5, the term in the third line is $O(n^{q/2-\gamma'})$ a.s. by Lemma 5.4 and the Borel–Cantelli–Lemma, and the last expression is bounded by

$$D_n := \sum_{\substack{r\notin L \\ \psi(r_i)\le n\ \forall i}} \left(\|\Sigma_{j\in R_r} x_j\| + \|\Sigma_{j\in R_r} y_j\|\right) \tag{5.28}$$
$$+ \Sigma_{r\in L,\ \psi(r_i)\le n} \|\Sigma_{j\in R_r} (x_j-y_j)\|.$$

Here we used (1.4). The expression on the right–hand side of the first line of (5.28) is $O(n^{q/2-1/16})$ a.s. as $n \to \infty$ by Lemma 5.6, the expression in the second line is $O(n^{q/2-\gamma})$ a.s. as $n \to \infty$ for some $\gamma > 0$ by (5.22). Now the assertion of Theorem 1 follows by taking the supremum over $A \in \mathcal{A}$ on both sides of (5.27):

$$\sup_{A\in\mathcal{A}} \|S_n(A)-T_n(A)\| \le 2 \sum_{1\le j\le n1} (\|x_j-\bar{x}_j\| + \|y_j-y'_j\|)$$
$$+ cn^{q/2-\gamma_3} + \sup_{A(n^{-\kappa})} (\|\bar{V}_n(A(n^{-\kappa}))\| + \|U'_n(A(n^{-\kappa}))\|) + D_n$$
$$=: u_n << n^{q/2 \quad \gamma_4} \quad \text{a.s. as } n \to \infty \text{ for some } \gamma_4 > 0. \ \square$$

6. Proof of Theorem 2

Here we won't apply the exponential inequality from section 5 which was essential in the proof of Theorem 1, but we will approximate certain blocks of $(x_j, j \in \mathbb{N}^q)$ by independent blocks and then apply an exponential inequality for independent r.v.'s. For this we need the following lemmas.

Lemma 6.1. ([Eb], Lemma 2]) Let $x_1,\dots,x_n$ be r.v.'s. in arbitrary measurable spaces. Assume for all $k = 1,\dots,n-1$

$$\|\mathcal{L}(x_1,\dots,x_k) \otimes \mathcal{L}(x_{k+1},\dots,x_n) - \mathcal{L}(x_1,\dots,x_n)\| \leq \epsilon.$$

Then

$$\|\mathcal{L}(x_1,\dots,x_n) - \bigotimes_{k=1}^{n} \mathcal{L}(x_k)\| \leq (n-1)\epsilon.$$

In view of the proof of Theorems 3, 4, and 5 it is convenient to use the following exponential inequality which also holds for Banach space valued random elements of the type described in section 2.

Lemma 6.2. ([DuPh], Lemma 2.6) Let $x_1,\dots,x_n$ be independent $\mathbb{R}^N$–valued random vectors. Let $S_n := \Sigma_{j=1}^n x_j$ and $v \geq \Sigma_{j=1}^n E\|x_j\|^2$. Assume $\|x_j\| \leq M$ for $j = 1,\dots,n$. Then for all $K > 0$

$$P(\|S_n\| > K) \leq \max\left\{\exp\left(-\frac{(K-E\|S_n\|)^2}{12v}\right), \exp\left(-\frac{K-E\|S_n\|}{4M}\right)\right\}.$$

Now let the hypotheses of Theorem 2 be satisfied. $s' > 2(1+q)(1+2/\delta) - q$ is equivalent to $\frac{s'+q}{1+q} > 2 \cdot \frac{2+\delta}{\delta}$, hence to

$$\frac{1+q}{s'+q} < \frac{1}{2} - \frac{1}{2+\delta} = \frac{\delta}{2(2+\delta)}\,. \tag{6.1}$$

Therefore and by (1.14) there are real numbers τ, r, $z > 0$ with

$$\frac{1+q}{s'+q} < r < \frac{1}{2} - \frac{1+\tau}{2+\delta} \tag{6.2}$$

and

$$q\left(\frac{1}{2} + \frac{1+\tau}{2+\delta}\right) < z < \min\left\{q, \frac{q}{u}\left(\frac{1}{2} - \frac{1+\tau}{2+\delta} - r\right)\right\}. \tag{6.3}$$

Having once found such τ, r, z we can always replace τ by some smaller, arbitrarily small $\tau' > 0$ without changing r and z and without violating the validity of (6.2) and (6.3). Therefore we can require repeatedly: "Let τ be sufficiently small." With such τ define x_j', $\bar{x}_j$ and $\bar{S}_n(A)$ by (5.3), (5.4) and (5.12). Let β be chosen as in section 5, i.e. as in [BM, (4.1), p. 25], h according to (1.10). For $n \in \mathbb{N}$ given let b, d be the smallest integers such that

$$2^d \geq n^{h/(2(\beta+1))}, \tag{6.4}$$

$$2^b \geq n^z. \tag{6.5}$$

For fixed $A \in \mathcal{A}$ let

$$w_n := w_n(i) := \bar{S}_n(A(2^{-i})) - \bar{S}_n(A(2^{-i+1})), \quad i = 1,\dots,b. \tag{6.6}$$

Lemma 6.3. For τ sufficiently small we have for some $\gamma > 0$, as $n \to \infty$,

$$P(\sup\{\|w_n(i)\| \cdot 2^{i\tau}: A(2^{-i}), A(2^{-i+1}), d \leq i \leq b\} \leq cn^{q/2}) << n^{-1-\gamma}.$$

<u>Proof.</u> Let $p := [n^r]$. Cover $(0,1]^q$ by disjoint half-open cubes $D_1,\dots,D_Q$ of length 2p and of the form $D_k = (n_k - 2p\mathbb{1}, n_k]$ $(k = 1,\dots,Q)$. We partition each of the cubes D_k into 2^q disjoint cubes of length p, $D_k = H_k^0 \cup \dots \cup H_k^{2^q-1}$, in the following way. Let $b_1 \dots b_q$ be the binary representation of i $(i = 0,\dots,2^q-1)$, then set

$H_k^i := (n_k - 2p\mathbb{1} + p(b_1,\dots,b_q),\ n_k - p\mathbb{1} + p(b_1,\dots,b_q)]$. Then e.g. H_k^0 always lies in the "left lower corner" of D_k. By construction, for $k \neq k'$ we have $d(H_k^i, H_{k'}^i) \geq p$ for all i. Obviously it suffices to prove Lemma 6.3 for each of the r.v.'s

$$(6.7) \qquad w_n^{(t)}(i) := \bar{S}_n(A(2^{-i}) \cap n^{-1} \bigcup_{k=1}^{Q} H_k^t) - \bar{S}_n(A(2^{-i+1}) \cap n^{-1} \bigcup_{k=1}^{Q} H_k^t)$$

$(t = 0,1,\dots,2^q-1)$ instead of the r.v.'s $w_n(i)$. We will do this for $t = 0$. Let $H_k := H_k^0$, $H := \cup_{k=1}^Q H_k$. By Lemma 6.1, (1.13) and (6.2) we have for some $\gamma > 0$

$$(6.8) \qquad \|\mathcal{L}(x_j,\ j \in H \cap \mathbb{N}^q) - \bigotimes_{k=1}^{Q} \mathcal{L}(x_j,\ j \in H_k \cap \mathbb{N}^q)\| \leq c\, Q\, \beta(p)$$

$$\leq c\, n^q p^{-q} p^{-s'} \leq c\, n^{q-r(q+s')} << n^{-1-\gamma}.$$

For easier notation let z_j $(j \in H \cap \mathbb{N}^q)$ be $\mathbb{R}^N$–valued random vectors on some probability space (Ω',Σ',P') such that $\mathcal{L}(x_j,\ j \in H_k \cap \mathbb{N}^q) = \mathcal{L}(z_j,\ j \in H_k \cap \mathbb{N}^q)$ for $k = 1,\dots,Q$ and $(z_j,\ j \in H_{k+1} \cap \mathbb{N}^q)$ is independent of $(z_j,\ j \in \cup_{k'=1}^k H_{k'} \cap \mathbb{N}^q)$ for $k = 1,\dots,Q-1$. Define $\bar{z}_j$, $\tilde{w}_n(i)$ and $\tilde{w}_n^{(t)}(i)$ analogously to $\bar{x}_j$, $w_n(i)$ and $w_n^{(t)}(i)$, with x_j replaced by z_j. By (6.8) and the arguments above, it suffices to show that

$$(6.9) \qquad P'(\sup\{\|\tilde{w}_n^{(t)}(i)\| \cdot 2^{i\tau}\colon A(2^{-i}), A(2^{-i+1}), d \leq i \leq b\} \geq n^{q/2}) << n^{-1-\gamma}.$$

For this we will apply Lemma 6.2. Since

$$(6.10) \qquad |\mu(A \cap C) - \mu(B \cap C)| \leq \mu((A \triangle B) \cap C)$$

and by Remark 5.2, Lemma 4.1 and (1.5) we have

$$(6.11) \qquad E(\|\tilde{w}_n^{(0)}(i)\|^2) \leq c\mu(nA(2^{-i}) \triangle nA(2^{-i+1})) \leq cn^q \cdot 2^{-i},$$

and so by Jensen's inequality

$$(6.12) \qquad E\|\tilde{w}_n^{(0)}(i)\| \leq cn^{q/2} \cdot 2^{-i/2}.$$

$\tilde{w}_n^{(0)}(i)$ is the sum of the independent random vectors

$$\Sigma_{j\in H_k} (\mu(nA(2^{-i}) \cap C_j) - \mu(nA(2^{-i+1}) \cap C_j))\, \bar{z}_j\,,$$

$k = 1,\ldots,Q$. By Remark 5.2 they are bounded by some $M \le c\, p^q\, n^{q(1+\tau)/(2+\delta)}$. Put $K := n^{q/2} \cdot 2^{-i\tau}$. W.l.o.g. we may assume $\tau < 1/2$, so by (6.12) $E\|\tilde{w}_n^{(0)}(i)\| = o(K)$ as $n \to \infty$, so for n sufficiently large $K - E\|\tilde{w}_n^{(0)}(i)\| \ge K/2$. Therefore by Lemma 6.2 we have for fixed i, $A(2^{-i})$, $A(2^{-i+1})$

$$P'(\|\tilde{w}_n^{(0)}(i)\| > K) << \exp\left(-c\,\frac{n^q\, 2^{-2i\tau}}{n^q\, 2^{-i}}\right) + \exp\left(-c\,\frac{n^{q/2}\, 2^{-i\tau}}{p^q\, n^{q(1+\tau)/(2+\delta)}}\right)$$

$$<< \exp(-c2^{i(1-2\tau)}) + \exp\left(-c2^{-i\tau}\, n^{q(\frac{1}{2} - \frac{1+\tau}{2+\delta} - r)}\right).$$

Multiplication with $N(2^{-i})\, N(2^{-i+1})\, b << \exp(c_2 2^{iu}) \cdot \log(n^z) << \exp(c_2 2^{iu})$ shows that the probability in (6.9) is bounded by

$$c_1 \exp(-c2^{i(1-2\tau)} + c_2 2^{iu}) + c_1 \exp\left(-c2^{-i\tau}\, n^{q(\frac{1}{2} - \frac{1+\tau}{2+\delta} - r)} + c_2 2^{iu}\right).$$

Since $u < 1$, we have $u < 1 - 2\tau$ for τ sufficiently small, and therefore the first expression is $<< \exp(-c2^{i(1-2\tau)}) << \exp(-c2^{d(1-2\tau)}) << \exp(-cn^{\gamma})$ for some $\gamma > 0$. By (6.3) and (6.5) the second term is

$$<< \exp\left(c_2 2^{bu} - c2^{-b\tau}\, n^{q(\frac{1}{2} - \frac{1+\tau}{2+\delta} - r)}\right) << \exp(-cn^{\gamma})$$

for some $\gamma > 0$ if τ is sufficiently small. This implies (6.9) and so the assertion of Lemma 6.3. □

Now we use the notations defined in (5.7) and (5.9) to (5.13). Let m and n be linked by (5.8).

Lemma 6.4. If τ is sufficiently small, then we have for some $\gamma_1, \gamma_2 > 0$

$$P(\sup\{\|\overline{V}_n(A(2^{-d}))\| : A(2^{-d})\} \geq n^{\frac{q}{2}-\gamma_1}) << n^{-1-\gamma_2}.$$

Proof. We adopt the construction of the cubes D_k and H_k^i (with the same p) from the proof of Lemma 6.3, as well as the definition of the z_j $(j \in H \cap \mathbb{N}^q)$. Set

$$\nabla_n^{(i)}(A) := \sum_{j \in \mathbb{N}^q} \mu(((nA) \setminus (nA)_*) \cap C_j \cap \bigcup_{k=1}^{Q} H_k^i)\, \bar{z}_j$$

for $n \in \mathbb{N}$, $A \in \mathcal{A}$, $i = 0,1,\ldots,2^q-1$. By (6.8) it suffices to prove the asserted inequality in Lemma 6.4 for $\nabla_n^{(i)}(A(2^{-d}))$ $(i = 0,1,\ldots,2^q-1)$ instead of $\overline{V}_n(A(2^{-d}))$. For instance for $i = 0$ we have by (1.10) and Lemma 4.1

$$\begin{aligned} E\|\nabla_n^{(0)}(A(2^{-d}))\|^2 &\leq cn^q\, b(\tfrac{1}{n}(\psi(m+1) - \psi(m))) \\ &\leq cn^{q-h}(\psi(m+1) - \psi(m))^h \\ &\leq cn^{q-h}\, n^{h\beta/(\beta+1)} \\ &= cn^{q - h/(\beta+1)}. \end{aligned} \tag{6.13}$$

As in the proof of Lemma 6.3 we conclude using Lemma 6.2 that for fixed $A(2^{-d})$

$$\begin{aligned} P'(\|\nabla_n^{(0)}(A(2^{-d}))\| \geq n^{q/2-\gamma'}) &<< \exp(-cn^{q-2\gamma'-q+h/(\beta+1)}) \\ &\quad + \exp(-cn^{q/2-\gamma'}\, p^{-q}\, n^{-q(1+\tau)/(2+\delta)}) \\ &<< \exp(-cn^{h/(\beta+1)-2\gamma'}) + \exp\left(-cn^{\gamma' + q(\frac{1}{2} - r - \frac{1+\tau}{2+\delta})}\right). \end{aligned}$$

By (6.4) and since $u \leq 1$ we have $N(2^{-d}) \leq c\, \exp(cn^{h/(2(\beta+1))})$; $2^d \leq 2^b$ implies $N(2^{-d}) \leq c\, \exp(c2^{bu}) \leq c\, \exp(cn^{zu})$, so we obtain the asserted inequality (with some $\gamma_2 > 0$) by multiplying the above bound by $N(2^{-d})$, if τ and γ' are small enough. □

Beginning with Lemma 5.5 we now can adopt the proof of Theorem 1 literally, except that n^{κ} there has to be replaced by 2^b and (5.27) has to be replaced by

$$\begin{aligned}
\|S_n(A) - T_n(A)\| &\leq \|S_n(A) - \bar{S}_n(A)\| + \|T_n(A) - T_n'(A)\| \\
&+ \|\bar{S}_n(A) - \bar{S}_n(A(2^{-b}))\| + \|T_n'(A) - T_n'(A(2^{-b}))\| \\
&+ \sum_{i=d+1}^{b} (\|\bar{S}_n(A(2^{-i})) - \bar{S}_n(A(2^{-i+1}))\| + \|T_n'(A(2^{-i})) - T_n'(A(2^{-i+1}))\| \\
&+ \|\bar{V}_n(A(2^{-d}))\| + \|U_n'(A(2^{-d}))\| \\
&+ \sum_{\mathbb{1}\leq j\leq n\mathbb{1}} (\|x_j - \bar{x}_j\| + \|y_j - y_j'\|) \\
&+ \|\sum_{j\in\mathbb{N}^q} (x_j - y_j)\, \mu((nA(2^{-d}))_* \cap C_j)\|.
\end{aligned}$$

For the second line here Lemma 5.5 holds analogously, the term in the third line is by Lemma 6.3 a.s. bounded by

$$cn^{q/2} \sum_{i>d} 2^{-i\tau} \leq cn^{q/2}\, 2^{-d\tau} \leq cn^{q/2 - \tau h/(2(\beta+1))},$$

for the fourth line use Lemma 6.4 and the Borel–Cantelli Lemma. The rest of the proof is analogous to section 5. □

7. Central Limit Theorems for Strongly Mixing Random Fields

In this section we collect some lemmas which will be needed in the proofs of the Theorems 3, 4 and 5. In the proof of Theorem 3 we will use the following lemma which specifies the dependence of the constants in [BM, Lemma 5] of the dimension N of the range of the random vectors. Its proof may be obtained through a detailed analysis of the proofs of [BM, Lemma 5] and [KPh, Proposition 2.1, p. 1011].

Lemma 7.1. Let $(x_j, j \in \mathbb{N}^q)$ be an $\mathbb{R}^N$–valued strongly mixing, weakly stationary random field satisfying $Ex_j = 0$ for all j and the conditions (1.1), (1.7) and (1.8), and that in (1.8)

$$s \geq (1+\epsilon)(1 + 2(17/2)^{q-1}/\min(1,\delta)) \tag{7.1}$$

for some $\epsilon \in (0,1]$. For $n \in \mathbb{N}^q$, $u \in \mathbb{R}^N$ let $f_n(u) := E \exp(i\langle u, \Pi(n)^{-\frac{1}{2}} \sum_{1\leq j\leq n} x_j\rangle)$, let T be defined as in Theorem 1. For $d \in (0, 1/(q-1)]$ define G_d as in [BM], namely

$$G_d := \{n \in \mathbb{N}^q\colon n_k^{1+d} \geq \Pi(n)^d \text{ for } k = 1,\dots,q\}. \tag{7.2}$$

Then there is a $t > 0$ and a $C_3 < \infty$ such that for $\Pi(n)^t \geq N$, $n \in G_d$

$$\sup_{\|u\|\leq\Pi(n)^t} |f_n(u) - \exp(-\langle u,Tu\rangle/2)| \leq C_3\, \Pi(n)^{-t}.$$

Here t depends only on ϵ, δ, d and q, e.g. we can choose $t = 100^{-q-3}\epsilon^2\delta d^2$. C_3 depends only on ϵ, δ, d, q and C_1 and C_2 in (1.7) and (1.8) (but not on N). Of course, $\mathbb{R}^N$ is here equipped with the standard scalar product $\langle\, .\, ,\, .\,\rangle$ and norm $\|.\|$.

We will also make use of the following classical result of F. John which says that the Banach–Mazur distance of an N–dimensional Banach space from the N–dimensional Euclidean space does not exceed $N^{\frac{1}{2}}$:

Lemma 7.2. ([Jo, p. 203], see also [Le, p. 13, Theorem 8]) Let $(E,\|.\|)$ be an N–dimensional Banach space. Then there is a Euclidean norm $\|.\|_1$ on E such that for all $x \in E$

$$\|x\|_1 \leq \|x\| \leq N^{\frac{1}{2}}\|x\|_1. \tag{7.3}$$

The following lemma is a consequence of Theorem 1 of Dehling(1983). Let π denote the Prokhorov distance.

Lemma 7.3. Let $q \geq 2$ and $(x_j, j \in \mathbb{Z}^q)$ be a strongly mixing, weakly stationary family of centered r.v.'s with values in an N–dimensional normed space B such that (1.7) and (1.8) hold with rate (7.1) for some $\epsilon \in (0,1]$. Let $0 < d \leq 1/(q-1)$ and G_d be defined by (7.2). Then the infinite sum

(7.4)
$$T(f,g) = \sum_{j \in \mathbb{Z}^q} E(f(x_0)g(x_j))$$

is absolutely convergent for $f, g \in B'$ (the dual space of B), and if $N(0,T)$ denotes the centered normal distribution with covariance function T then there are constants $C, \lambda \in (0,\infty)$ such that for $n \in G_d$

(7.5)
$$\pi(\mathcal{L}(\Pi(n)^{-\frac{1}{2}} \sum_{j \leq n} x_j) , N(0,T)) \leq C\, \Pi(n)^{-\lambda} N^2.$$

Here λ depends only on ϵ, δ, d and q; C only on ϵ, δ, q, d and C_1 and C_2 in (1.7) and (1.8). If B is a Euclidean space (7.5) can be replaced by the sharper inequality

(7.6)
$$\pi(\mathcal{L}(\Pi(n)^{-\frac{1}{2}} \sum_{j \leq n} x_j) , N(0,T)) \leq C\, \Pi(n)^{-\lambda} N^{3/2}$$

for $n \in G_d$.

<u>Proof.</u> In the first step assume B Euclidean, hence w.l.o.g. $B = \mathbb{R}^N$ with the usual scalar product. The convergence of (7.4) follows from Theorem 1 of [BM]. For easier notation let now T denote the covariance operator defined by

$$\langle u,Tv\rangle = \sum_{j \in \mathbb{Z}^q} E(\langle u,x_0\rangle \langle v,x_j\rangle) ,$$

$u, v \in B$. Let $n \in G_d$. Use Lemmas 4.1 and 4.2 to obtain a uniform bound for the $(2+\alpha)$–th moments of the random vectors

$$Z(j_q) := \left(\prod_{k=1}^{q-1} n_k\right)^{-\frac{1}{2}} \sum_{j_1=1}^{n_1} \cdots \sum_{j_{q-1}=1}^{n_{q-1}} x_{(j_1,\ldots,j_{q-1},j_q)} ,$$

$j_q = 1,\ldots,n_q$, for $\alpha = (2/17)^{q-1}\min(1,\delta) \leq 2/3$. By Theorem 1(a) of [De] and its proof (especially notice the next to last line of his section 6) we now have for constants $C, \lambda' \in (0,\infty)$

(7.7) $$\pi(\mathcal{L}(\Pi(n)^{-\frac{1}{2}} \sum_{j\leq n} x_j) , N(0,T_n)) = \pi(\mathcal{L}(n_q^{-\frac{1}{2}} \sum_{j_q=1}^{n_q} Z(j_q)) , N(0,T_n))$$
$$\leq C N^{3/2} n_q^{-\lambda'} \leq C N^{3/2} \Pi(n)^{-d\lambda'/(1+d)} ,$$

where T_n denotes the covariance operator of $n_q^{-\frac{1}{2}} \sum_{j_q=1}^{n_q} Z(j_q) = \Pi(n)^{-\frac{1}{2}} \sum_{j\leq n} x_j$. For a self–adjoint operator $R: B \to B$ define $\|R\|_s := \sup_{(e_i)} \sum_{i=1}^{N} |\langle e_i, Re_i\rangle|$, where the supremum is taken over all orthonormal bases $(e_i) = (e_1,...,e_N)$ of B. For such (e_i) we have by Lemma 3 of [BM], applied to the random field $(y_j, j \in \mathbb{Z}^q) = (\langle e_i, x_j\rangle, j \in \mathbb{Z}^q)$ (i fixed), that $|\langle e_i,(T-T_n)e_i\rangle| \leq c\Pi(n)^{-\epsilon d/2}$. Summation over i yields

$$\|T - T_n\|_s \leq c\, N\, \Pi(n)^{-\epsilon d/2} ,$$

hence by Theorem 7 of [De]

$$\pi(N(0,T) , N(0,T_n)) \leq c\, N^{1/2}\, \Pi(n)^{-\epsilon d/8}\, N^{1/5}.$$

This together with (7.7) proves the assertion in the Euclidean case.

Now if $(B,\|\cdot\|)$ is a normed space, then apply Lemma 7.2, let $\|\cdot\|_1$ be a Euclidean norm on B with (7.3), let π, π_1 be the associated Prokhorov distances. Now use the result already proved for $(B,\|\cdot\|_1)$. (7.3) ensures that (1.7) is also satisfied for this space. Then note $\pi \leq N^{\frac{1}{2}}\pi_1$. □

8. Some Moment and Maximum Inequalities

We assume that the random variables x_j, $j \in \mathbb{N}$ resp. $j \in \mathbb{N}^q$, occurring in this section are always defined on a perfect probability space (Ω,Σ,P) and take values from some measurable space $(E,\mathcal{F})$ with countably generated σ–algebra $\mathcal{F}$ which contains all singletons from E. Hence by Lemma 2.1 $(x_j, j \in J)$ is a P–perfect E^J–valued r.v. for each $J \subset \mathbb{N}$ resp. $J \subset \mathbb{N}^q$. Let S always be an $\mathbb{R}$–vector space with seminorm $\|.\|$ and let $h, h_j : E \to S$ $(j \in \mathbb{N}$ resp. $j \in \mathbb{N}^q)$ be arbitrary mappings. An analysis of the paper of Dudley and Philipp (1983) shows that their assumption that the x_j are coordinate projections of a product space, may be replaced by the above perfectness hypothesis without affecting their results. (Of course the other hypotheses, in particular independence and identical distribution of the x_j, have to be retained.) The next lemma is a variant of [DuPh, Lemma 3.1] for the case of not identically formed random elements:

Lemma 8.1. Let $(x_j)_{j\in\mathbb{N}}$ be a sequence of i.i.d. r.v.'s, $X_j := h_j(x_j)$ for $j \in \mathbb{N}$, $S_n := \Sigma_{j=1}^{n} X_j$. Assume that $\sup_{j\in\mathbb{N}} E\|X_j\|^{*2} \leq A < \infty$ and for some $\alpha > 0$

(8.1) $$P^*(\|S_j - S_i\| > \alpha n^{\frac{1}{2}}) < 10^{-3}$$

for all i, j with $n_1 > j-i \geq n_0$. Then we have for all n with $n_1 > n \geq \max(n_0, 10^6 n_0^2 A/\alpha^2)$

(8.2) $$E\|S_n\|^{*2} \leq 112\, n\, (4\alpha^2 + A/4).$$

Reason. In the seventh and eighth lines of the proof of [DuPh, Lemma 3.1] replace $nPr(\|X_1\|^* \geq 2t^{\frac{1}{2}})$ by $\Sigma_{i=1}^{n} P(\|X_i\|^* \geq 2t^{\frac{1}{2}})$. Since

$$\int_u^\infty P(\|X_i\|^* \geq 2t^{\frac{1}{2}})\, dt \leq E\|X_i\|^{*2}/4 \leq A/4,$$

following that proof we obtain the inequality

$$10^{-2}\, E\|S_n\|^{*2} \le 4\alpha^2 n + 4\cdot 10^{-3}\cdot\tfrac{1}{4}\, E\|S_n\|^{*2} + nA/4$$

and from this the assertion. □

For the strongly mixing case we want to prove results analogous to this lemma. For this we first of all give a variant of [Ph84, Lemma 2.1]:

Lemma 8.2. Let $(x_j)_{j\in\mathbb{N}}$ be strongly mixing with rate

(8.3) $$\alpha(n) \le C_2\, n^{-(2+\epsilon)(1+2/\delta)}$$

for $\delta, \epsilon \in (0,1]$, $C_2 < \infty$. Let $X_j := h_j(x_j)$ for $j \in \mathbb{N}$ and assume

(8.4) $$\sup_{j\in\mathbb{N}} E\|X_j\|^{*2+\delta} \le M < \infty.$$

Define $S_j := \Sigma_{i=1}^{j} X_i$, $S_0 := 0$, let $n \in \mathbb{N}$ be fixed. Assume that for some $\alpha_1 \in (0,\infty)$

$$c := \max_{j\le n} P(\|S_n - S_j\|^* > \alpha_1/2) < 1.$$

Then there are ρ, $\alpha_0 \in (0,\infty)$ such that for all $\alpha \ge \max(\alpha_0,\alpha_1)$

$$P(\max_{j\le n}\|S_j\|^* > 2\alpha) \le 2(1-c)^{-1}(P(\|S_n\|^* > \alpha) + n\alpha^{-2-\rho}).$$

α_0 depends only on C_2, ϵ, δ, c and M (but not on n and α_1), ρ depends only on δ and ϵ.

The proof can be given by a modification of the proof of Lemma 2.1 of [Ph84]: Instead of distinguishing between the cases $P(t = j) >$ or $\le n^{-1-\rho}$ consider the cases $P(t = j) >$ or $\le \alpha^{-2-\rho}$. Also, the numbers β, ρ, p have to be chosen in another way, namely such that $\frac{2}{(2+\epsilon)(1+2/\delta)} < \beta < \frac{\delta}{2+\delta} = \frac{1}{1+2/\delta}$ and $2+\rho < \beta(2+\epsilon)(1+2/\delta)$

and $-\delta + \rho + (2+\delta)\beta < 0$ and $p := \min([\alpha^\beta], n-j)$. The rest is completely analogous to [Ph84].

Now we give a variant of Lemma 2.8 of [DuPh] for the strongly mixing case:

Lemma 8.3. Let $(x_j)_{j\in\mathbb{N}}$, $(X_j)_{j\in\mathbb{N}}$ and S_j be as in Lemma 8.2, satisfying (8.3) and (8.4). Let $n \in \mathbb{N}$ be fixed. For some $K > 0$ assume $P(\|S_n - S_j\|^* > K/2) \leq 1/2$ for all $j = 0,\dots,n$. Then there are t_0 and $\rho > 0$ such that for all $t > \max(t_0,K)$ and all $s \geq 0$

$$P(\|S_n\|^* \geq 4t+s) \leq 4 \max_{0\leq r\leq n} P(\|S_n - S_r\|^* \geq t)\, P(\|S_n\|^* \geq t) + 5\, n\, t^{-2-\rho} + P(\max_{m\leq n}\|X_m\|^* \geq s).$$

For t_0 resp. ρ we can choose the numbers α_0 (for $c = 1/2$) resp. ρ from Lemma 8.2.

<u>Proof.</u> Define $T = \min\{j\geq 1: \|S_j\|^* \geq 2t\}$ as in the proof of Lemma 2.8 of [DuPh]. As there we have

$$P(\|S_n\|^* \geq 4t+s) \leq P(\max_{m\leq n} \|X_m\|^* \geq s) + \sum_{m=1}^{n} P(T=m, \|S_n - S_m\|^* \geq 2t).$$

With β and ρ as in the sketched proof of Lemma 8.2 and $p = [t^\beta]$ we have

$$\begin{aligned} &P(T = m, \|S_n - S_m\|^* \geq 2t) \\ &\leq P(T = m, \|S_n - S_{m+p}\|^* \geq t) + P(\|S_{m+p} - S_m\|^* \geq t) \\ &\leq P(T = m)\, P(\|S_n - S_{m+p}\|^* \geq t) + \alpha(p) + 8\, M\, t^{-2-\delta}\, p^{2+\delta} \\ &\leq P(T = m) \max_{0\leq r\leq n} P(\|S_n - S_r\|^* \geq t) + o(1)\, t^{-2-\rho} \quad \text{as } t \to \infty \\ &=: A. \end{aligned}$$

In fact, we have with α_0 from Lemma 8.2 for $t \geq t_0 = \alpha_0$

$$A \le P(T=m) \max_{0\le r\le n} P(\|S_n - S_r\|^* \ge t) + t^{-2-\rho}.$$

Hence by Lemma 8.2 for $t \ge \max(t_0, K)$

$$\sum_{m=1}^{n} P(T=m, \|S_n - S_m\|^* \ge 2t)$$

$$\le n\,t^{-2-\rho} + \max_{0\le r\le n} P(\|S_n - S_r\|^* \ge t)\, P(\max_{1\le j\le n} \|S_j\|^* \ge 2t)$$

$$\le n\,t^{-2-\rho} + \max_{0\le r\le n} P(\|S_n - S_r\|^* \ge t)\cdot 4(P(\|S_n\|^* \ge t) + n\,t^{-2-\rho}). \ \square$$

The next lemma is a variant of Lemmas 3.1 and 6.2 of [DuPh] and of the above Lemma 8.1 for the strongly mixing case:

Lemma 8.4. Again let $(x_j)_{j\ge1}$ and $(X_j)_{j\ge1}$ be as in Lemmas 8.2 and 8.3, satisfying (8.3) and (8.4). For some $\alpha > 0$ and $n_0 \in \mathbb{N}$ assume that for all $j, n \in \mathbb{N}_0$ with $n-j \ge n_0$

$$P^*(\|S_n - S_j\| \ge \alpha n^{\frac{1}{2}}) < 10^{-3}.$$

Then there is an n_1 depending only on n_o, α, M, C_2, ϵ, δ such that for all $n \ge n_1$

$$E\|S_n\|^{*2} \le 500\,\alpha^2\, n.$$

More exactly, we can choose $n_1 = C_4\, n_0^2\, \alpha^{-C_3}$ with $C_3, C_4 \in (0,\infty)$ where C_4 depends only on C_2, ϵ, δ, M, and C_3 depends only on ϵ and δ.

Proof. As in [DuPh, p. 527] we have for all j, n with $j \le n$

$$P(\|S_n - S_j\|^* \ge 2\alpha n^{\frac{1}{2}}) < 10^{-3},$$

if

$$n \ge 10^6\, n_0^2\, M^{2/(2+\delta)}\, \alpha^{-2}. \tag{8.5}$$

By Lemma 8.3 with $K = 4\alpha n^{\frac{1}{2}}$ and the Markov inequality we obtain for all t with $2t^{\frac{1}{2}} >$

$\max(t_0,K)$ (with t_0 from Lemma 8.3) that

$$P(\|S_n\|^* \geq 10t^{\frac{1}{2}}) \leq 4\cdot 10^{-3} P(\|S_n\|^* \geq 2t^{\frac{1}{2}}) + 5nt^{-1-\rho/2} + P(\max_{m\leq n} \|X_m\|^* \geq 2t^{\frac{1}{2}})$$

$$\leq 4\cdot 10^{-3} P(\|S_n\|^* \geq 2t^{\frac{1}{2}}) + 5nt^{-1-\rho/2} + nM\cdot 2^{-2-\delta} t^{-1-\delta/2}.$$

If

$$n \geq t_0^2/(16\alpha^2)\,, \tag{8.6}$$

then $\max(t_0,K) = K = 4\alpha n^{\frac{1}{2}}$, so we obtain with $u = 4\alpha^2 n$ (note that for $t > u$ we then have $2t^{\frac{1}{2}} > 4\alpha n^{\frac{1}{2}} = K$)

$$10^{-2} E\|S_n\|^{*2} = \int_0^\infty P(\|S_n\|^* \geq 10t^{\frac{1}{2}})\, dt$$

$$\leq u + 4\cdot 10^{-3} \int_u^\infty P(\|S_n\|^* \geq 2t^{\frac{1}{2}})\, dt + 5n \int_u^\infty t^{-1-\rho/2} dt + \frac{nM}{4} \int_u^\infty t^{-1-\delta/2} dt$$

$$\leq 4\alpha^2 n + 4\cdot 10^{-3}\cdot \frac{1}{4} E\|S_n\|^{*2} + 5n \cdot \frac{2}{\rho} u^{-\rho/2} + \frac{nM}{4} \cdot \frac{2}{\delta} u^{-\delta/2}\,,$$

hence

$$E\|S_n\|^{*2} \leq (10^{-2} - 10^{-3})^{-1}\cdot n\cdot (4\alpha^2 + 10\rho^{-1}(4\alpha^2 n)^{-\rho/2} + \frac{M}{2\delta}(4\alpha^2 n)^{-\delta/2}) \tag{8.7}$$

$$\leq 500\, \alpha^2\, n$$

for sufficiently large n. The assertion on the size of n_1 can be obtained from (8.5), (8.6) and (8.7). □

This lemma can be generalized for random fields with q–dimensional index set:

Lemma 8.5. Assume that $(x_j, j \in \mathbb{N}^q)$ satisfies a strong mixing condition (1.1) with rate (2.4) and $(X_j, j \in \mathbb{N}^q) = (h_j(x_j), j \in \mathbb{N}^q)$ satisfies (2.3) for certain $\epsilon, \delta \in (0,1]$, $C_1, C_2 < \infty$. Assume further that for some $\alpha > 0$ and $n_0 \in \mathbb{N}$ for all rectangles R in $\mathbb{N}^q$ with $\operatorname{card} R \geq n_0$

$$P^*(\|\sum_{j\in R} X_j\| \geq \alpha(\operatorname{card} R)^{\frac{1}{2}}) < 10^{-3}.$$

Then there are $n_1 = n_1(n_0, \alpha, C_1, C_2, \epsilon, \delta, q) \in \mathbb{N}$, $C_5 = C_5(n_0, \alpha, C_1, C_2, \epsilon, \delta, q) \in \mathbb{R}_+$ and $\eta > 0$ such that for all rectangles R with $\text{card}\, R \geq n_1$

$$(8.8) \qquad E\|\sum_{j\in R} X_j\|^{*2} \leq 500\, \alpha^2 \,\text{card}\, R$$

and for all rectangles R (even for small ones)

$$(8.9) \qquad E\|\sum_{j\in R} X_j\|^{*2+\eta} \leq C_5\, (\text{card}\, R)^{1+\eta/2}.$$

η can be chosen to be any number with $0 \leq \eta \leq (17/2)^{-q}\delta$, and we can take $n_1 = C_4 n_0^{2q}\, \alpha^{-C_3}$ where C_4 depends only on $C_1, C_2, \epsilon, \delta, q$, and C_3 depends only on ϵ, δ, q.

Proof. We prove (8.8) and (8.9) simultaneously by induction on q. For $q = 1$ (8.8) holds by Lemma 8.4 and then (8.9) by Lemma 4.2. Now assume that the hypotheses are satisfied for some $Q > 1$ and that the Lemma is already proved for all $q < Q$. So there exists $C_5(Q-1) := C_5(n_0, \alpha, C_1, C_2, \epsilon, \delta, Q-1)$. Define $n_2 := n_1(n_0, \alpha, C_5(Q-1), C_2, \epsilon, (2/17)^{Q-1}\delta, 1)$. For $R = \times_{r=1}^{Q} (a_r, a_r+b_r] \cap \mathbb{N}^Q$ $(a_r, b_r \in \mathbb{N})$ with $\text{card}\, R \geq n_2^Q$ we have $b_r \geq n_2$ for at least one $r \leq Q$, e.g. for $r = Q$, so $b_Q \geq n_2$. Then set $R' := \times_{r=1}^{Q-1} (a_r, a_r+b_r] \cap \mathbb{N}^{Q-1}$. Apply Lemma 8.4 to the sequence

$$((\text{card}\, R')^{-\frac{1}{2}} \sum_{k\in R'} X_{(k,j)})_{j\in\mathbb{N}}\,,$$

since $b_Q \geq n_2$ we obtain

$$(8.10) \qquad E\|\sum_{j=a_Q+1}^{a_Q+b_Q} (\text{card}\, R')^{-\frac{1}{2}} \sum_{k\in R'} X_{(k,j)}\|^{*2} \leq 500\, \alpha^2\, b_Q\,,$$

hence

$$E\|\sum_{k\in R} X_k\|^{*2} \leq 500\, \alpha^2\, b_Q\, \text{card}\, R' = 500\, \alpha^2\, \text{card}\, R,$$

which proves (8.8). The assumption of the induction step says

$$\sup_{j\geq 1} \|(\text{card } R')^{-\frac{1}{2}} \sum_{k\in R'} X_{(k,j)}\|_{*(2+(2/17)^{Q-1}\delta)} \leq c.$$

This and (8.10) imply the assertion (8.9) by Lemma 4.2.

Now the assertion on the exact choice of n_1 can be obtained as follows. In the first step apply what is already proved for some fixed α (and hence for fixed $n_0(\alpha)$), e.g. for $\alpha = 1/2000$. So we get a constant $C_5' = C_5'(C_1, C_2, \epsilon, \delta, q)$ in (8.9) for $\eta = (2/17)^q\delta$. (8.9) holds for all rectangles, in particular for low–dimensional ones. Now follow the above induction argument anew for α and n_0 as assumed and use (8.9) with C_5'. This proves the assertion on the choice of n_1 if we note the corresponding statement on n_1 in Lemma 8.4. □

9. Proof of Theorem 3

In the first part of the proof we will generalize the proof of Theorem 2 of [Ph84] to random fields with q–dimensional index set.

9.0. Notations. Choose β with $\max(1-(1-\vartheta/2)/(3q)\ ,\ 1-\vartheta/4) \leq \beta < 1$. For $k \in \mathbb{N}$ define

(9.1) $\quad N(k) := N_k := [\exp(k^\beta)],\ \ n(k) := n_k := [\exp(k^\beta/4)]$,

(9.2) $\quad t(k) := t_k := \sum_{j=1}^{k} (N_j + n_j) \asymp k^{1-\beta}\exp(k^\beta)$.

For $r \in \mathbb{N}^q$ let

(9.3) $\quad \overline{R}_r := \{v \in \mathbb{R}^q\colon t(r_i) < v_i \leq t(r_i+1) \text{ for } i = 1,\dots,q\}$,

$\quad R_r := \overline{R}_r \cap \mathbb{N}^q$,

(9.4) $\quad H_r := \{j \in \mathbb{N}^q\colon t(r_i) < j_i \leq t(r_i) + N(r_i+1) \text{ for } i = 1,\dots,q\}$

(the "large blocks"),

(9.5) $\quad I_r := R_r \setminus H_r$ (the "gaps"),

(9.6) $\quad h_r := \text{card } H_r = \Pi_{i=1}^{q} N(r_i+1) \asymp \exp(\Sigma_{i=1}^{q} r_i^{\beta})$ as $\Pi(r) \to \infty$,

(9.7) $\quad m(r) := m_r := [\Sigma_{i=1}^{q} r_i^{1-\vartheta/2}]$,

(9.8) $\quad V_j := \Lambda_{m(r)} X_j$ for $j \in R_r$,

(9.9) $\quad \xi_r := \Sigma_{j \in H_r} V_j, \; \zeta_r := \Sigma_{j \in I_r} V_j$.

Let G be defined by (5.18) and

(9.10) $\quad L := \{r \in \mathbb{N}^q: (t(r_1),\dots,t(r_q)) \in G\}$.

By (2.7) and (9.7) we have

(9.11) $\quad N_0(m(r)) \le C_4 \exp((\Sigma_{i=1}^{q} r_i^{1-\vartheta/2})^{1-\vartheta}) = o(h_r)$ for $\Pi(r) \to \infty$.

9.1. Gaussian approximation over the large blocks in finite–dimensional subspaces. For $r \in \mathbb{N}^q$ we have

(9.12) $\quad d(H_r, \cup_{r' \neq r} H_{r'}) \ge \min_{1 \le i \le q} n(r_i) \ge c \min_i \exp(r_i^{\beta}/4)$.

By (2.6) and Lemma 7.2 there exists on $\Lambda_{m(r)} S$ a Euclidean norm $\|.\|_1$ originating from a scalar product $\langle .,. \rangle$ such that

(9.13) $\quad \|x\|_1 \le \|x\| \le c\, m(r)^{\frac{1}{2}} \|x\|_1$ for $x \in \Lambda_{m(r)} S$.

Let $\varphi: \mathbb{N} \to L$ be a bijection, $\varphi(p) = (\varphi_1(p),\dots,\varphi_q(p))$, $\mathcal{F}_p := \sigma(\xi_{\varphi(1)},\dots,\xi_{\varphi(p)})$, $\mathcal{F}_0 := \{\phi, \Omega\}$. By a lemma of Dvoretzky (see e.g. [BM, Lemma 1]) and by (9.12) and (2.4) we have for $p \in \mathbb{N}$, $u \in \Lambda_{m(\varphi(p))} S$

(9.14)
$$E\left| E(\exp(i\langle u, \xi_{\varphi(p)} h_{\varphi(p)}^{-\frac{1}{2}} \rangle \mid \mathcal{F}_{p-1}) - E(\exp(i\langle u, \xi_{\varphi(p)} h_{\varphi(p)}^{-\frac{1}{2}} \rangle))\right|$$
$$\le 2\pi\, \alpha(\min_{a \le q} \exp(\varphi_a(p)^{\beta}/4)) \le c \max_a \exp(-7^{q-1} \varphi_a(p)^{\beta})$$
$$\le c\, \Pi_{a=1}^{q} \exp(-(9q)^{-1}\, 7^{q-1}\, \varphi_a(p)^{\beta}),$$

where the last inequality holds because (9.10), (5.18) and (9.2) imply that

(9.15) $$(\exp(r_1^{\beta}),\dots,\exp(r_q^{\beta})) \in G'(c')$$

for all $r \in L$ with some fixed $c' \in (0,\infty)$ where

(9.16) $$G'(c) := \{j \in \mathbb{N}^q\colon j_a^{9q} \geq c\Pi(j) \text{ for } a = 1,\dots,q\}.$$

By (2.3), (2.5), (9.13) and [DuPh, Lemma 2.2] we have

(9.17) $$\sup_{j\in\mathbb{N}^q} E\|V_j\|_1^{*2+\delta} < \infty.$$

(9.7), (9.6) and (2.6) imply

(9.18) $$\dim \Lambda_{m(\varphi(p))}S \leq c \log h_{\varphi(p)}.$$

Hence by Lemma 7.1 there is a $t > 0$ such that

(9.19) $$\sup_{\|u\|_1\leq h_{\varphi(p)}^t} \left| E\exp(i\langle u, \xi_{\varphi(p)}h_{\varphi(p)}^{-\frac{1}{2}}\rangle) - \exp(-\langle u, \Gamma_{m(\varphi(p))}u\rangle/2)\right|$$

$$<< h_{\varphi(p)}^{-t} << \exp(-t\,\Sigma_{a=1}^{q}\,\varphi_a(p)^{\beta}) \quad \text{as } p \to \infty.$$

Here Γ_m $(m \in \mathbb{N})$ is defined by the following positive semidefinite bilinear form:

(9.20) $$(u,v) \mapsto \langle u,\Gamma_m v\rangle := \Sigma_{j\in\mathbb{Z}^q} E(\langle u,\Lambda_m X_0\rangle\, \langle v,\Lambda_m X_j\rangle)$$

which exists by [BM, Theorem 1]. Using (9.19) and (9.14) we infer from Theorem 1 of Berkes and Philipp (1979) that the $\xi_{\varphi(p)}$ $(p \geq 1)$ can be redefined on some probability space without changing their joint distribution, together with centered Gaussian random vectors $\eta_{\varphi(p)}$ $(p \in \mathbb{N})$ whose distribution depends only on $m(\varphi(p))$, say

(9.21) $$\mathcal{L}(\eta_{\varphi(p)}) = \mu_{m(\varphi(p))}\,,$$

and such that

(9.22) $$P(\|\xi_{\varphi(p)}\, h_{\varphi(p)}^{-\frac{1}{2}} - \eta_{\varphi(p)}\|_1 \geq \alpha_p) \;\leq\; \alpha_p\,,$$

where with $T_p := 10^8(\Sigma_a\, \varphi_a(p))^{4+q}$ and some $\delta_p << \exp(-(\Sigma_a\, \varphi_a(p))^{4+q})$, by (2.6) and (9.7)

$$\alpha_p << m(\varphi(p))\, T_p^{-1} \log T_p + \exp(-c \sum_a \varphi_a(p)^{\beta}) (\sum_a \varphi_a(p))^{cm(\varphi(p))}$$
$$+ \exp(-(\sum_a \varphi_a(p))^{4+q})$$
$$<< (\sum_a \varphi_a(p)^{1-\vartheta/2}) (\sum_a \varphi_a(p))^{-3.9-q} + \exp\Big(-c_1 \sum_a \varphi_a(p)^{\beta}$$
$$+ c_2(\sum_a \varphi_a(p)^{\beta})^{(1-\vartheta/2)/\beta} \log(\sum_a \varphi_a(p))\Big)$$
$$<< (\sum_a \varphi_a(p))^{-2.9-q}\,.$$

Hence $\sum_{p\in\mathbb{N}} \alpha_p < \infty$. By the Borel–Cantelli Lemma this implies for P–almost all $\omega \in \Omega$, with $c(\omega) < \infty$, that $\| \xi_{\varphi(p)}\, h^{-\frac{1}{2}}_{\varphi(p)} - \eta_{\varphi(p)}\|_1 < c(\omega)\, \alpha_p$ for all $p \in \mathbb{N}$, therefore by (9.13) and (9.7)

$$\| \xi_{\varphi(p)}\, h^{-\frac{1}{2}}_{\varphi(p)} - \eta_{\varphi(p)}\| \le c(\omega) \sum_a \varphi_a(p)^{-2-q}\,. \tag{9.22}$$

By applying [BPh, Lemma A.1] several times we infer that there exists (maybe on some new, richer probability space) a family $(W_j, j \in \mathbb{N}^q)$ of independent Gaussian random vectors together with the V_j, $\eta_{\varphi(p)}$ and the ξ_r and ζ_r defined by (9.9), such that

$$h^{-\frac{1}{2}}_{\varphi(p)} \sum_{j\in H_{\varphi(p)}} W_j = \eta_{\varphi(p)} \quad (p \in \mathbb{N}) \tag{9.23}$$

and

(9.24) (a) $$\mathcal{L}(W_j) = \mu_{m(r)}$$

for $j \in R_r$, $r \in \mathbb{N}^q$. Here for $m \in \mathbb{N}$, μ_m denotes the centered Gaussian measure on $\Lambda_m S$ with covariance function

(9.24) (b) $$(f,g) \mapsto \sum_{j\in\mathbb{Z}^q} E(f(\Lambda_m X_0) g(\Lambda_m X_j)) \quad (f, g \in (\Lambda_m S)').$$

Note that this notation is consistent with (9.21). (9.22) and (9.23) imply

$$(9.25)\qquad \|\sum_{j\in H_{\varphi(p)}}(V_j-W_j)\|\le c(\omega)\,h_{\varphi(p)}^{-\frac{1}{2}}\,(\sum_a \varphi_a(p))^{-2-q}\quad \text{a.s. for all } p\in\mathbb{N}.$$

9.2. Partial sums over the gaps. Now we are going to give an upper bound for the ζ_r with $r \in L$. Here we will assume $q \geq 2$; the case $q = 1$ is easier to be dealt with. Set $\gamma := 1/(120q)$. For $r \in \mathbb{N}^q$, I_r is the disjoint union $I_r = \cup_{i=1}^q I_{r,i}$ of rectangles $I_{r,i}$ one edge of which has length at most $n(r_i+1)$ and whose $q-1$ other edges have length no more than $N(r_a+1) + n(r_a+1)$ $(a \neq i)$, hence

$$\text{card } I_{r,i} \leq n(r_i+1) \prod_{a\neq i} (N(r_a+1) + n(r_a+1)).$$

By Chebyshev's inequality, Lemmas 4.1 and 7.2, (9.7), (9.15) and (9.1) we have for $r \in L$

$$\begin{aligned}
P(\|\Sigma_{j\in I_{r,1}} V_j\| \geq h_r^{\frac{1}{2}-\gamma}) &\leq c\,\text{card } I_{r,1}\, h_r^{-1+2\gamma} \dim \Lambda_{m(r)}S \\
&\leq c \prod_{i=1}^q N_{r(i)+1}^{-1+2\gamma} \cdot n_{r(1)+1} \cdot \prod_{i=2}^q (N_{r(i)+1} + n_{r(i)+1}) \dim \Lambda_{m(r)}S \\
&\leq c \prod_{i=1}^q N_{r(i)}^{-1+3\gamma} \cdot N_{r(1)}^{1/4} \cdot \prod_{i=2}^q N_{r(i)} \\
&= c\, N_{r(1)}^{-1+1/4+3\gamma} \Big(\prod_{i=2}^q N_{r(i)} \Big)^{3\gamma} \\
&\leq c\, N_{r(1)}^{-3/4+3\gamma} \cdot N_{r(1)}^{3\gamma(9q-1)} \\
&\leq c\, N_{r(1)}^{-3/4+1/2} \\
&\leq c \Big(\prod_{i=1}^q N_{r(i)} \Big)^{-\frac{1}{4}\cdot\frac{1}{9q}},
\end{aligned}$$

similarly for $I_{r,2}, \ldots, I_{r,q}$. Hence $\Sigma_{r\in L} P(\|\Sigma_{j\in I_r} V_j\| \geq h_r^{\frac{1}{2}-\gamma}) \leq c \sum_{r\in\mathbb{N}^q} \exp(-\frac{1}{36q} \Sigma_{i=1}^q r_i^{\beta}) < \infty$, and so by the Borel–Cantelli Lemma $\|\Sigma_{j\in I_r} V_j\| \leq c(\omega)\, h_r^{\frac{1}{2}-\gamma}$ a.s. The same holds for W_j instead of V_j. This and (9.25) imply

$$\|\Sigma_{j\in R_r}(V_j - W_j)\| \le c(\omega)\, h_r^{\frac{1}{2}} \Big(\sum_{a=1}^{q} r_a\Big)^{-2-q} \quad \text{a.s.,}$$

hence for $M \in \mathbb{N}$, $M \to \infty$

$$\sum_{\substack{r\in L\\ r\le M\mathbb{1}}} \|\sum_{j\in R_r}(V_j - W_j)\| << h_{M\mathbb{1}}^{\frac{1}{2}}(qM)^{-2-q}\, M^q << h_{M\mathbb{1}}^{\frac{1}{2}} M^{-2} \quad \text{a.s.} \tag{9.26}$$

9.3. Partial sums near coordinate planes. Similarly as in section 5 we now need an upper bound for the partial sums of the V_j and W_j over those R_r with $r \notin L$. Let $M, n \in \mathbb{N}$ be linked by

$$t_{M-1} < n \le t_M . \tag{9.27}$$

As in section 5, (5.20), it can be proved that for $r \in \mathbb{N}^q \setminus L$ with $t(r_i) \le n$ $(i = 1,\dots,q)$

$$\text{card } R_r \le n^{q-7/8}. \tag{9.28}$$

Furthermore $\{k \in \mathbb{N}: t_k \le n\} \subset \{k: \exp(k^\beta) \le n\} \subset \{k: k \le (\log n)^{1/\beta}\}$, hence

$$\text{card}\{r \in \mathbb{N}^q: t(r_i) \le n \ \ \forall i\} \le c(\log n)^{q/\beta}. \tag{9.29}$$

Using Lemmas 4.1 and 7.2, (9.7) and (2.6) we infer

$$\begin{aligned}
&P\Big(\sum_{r\in\mathbb{N}^q\setminus L,\ r\le M\mathbb{1}} \|\sum_{j\in R_r} V_j\| \ge n^{q/2-1/10}\Big)\\
&\le \sum_{r\notin L,\ r\le M\mathbb{1}} P\Big(\|\sum_{j\in R_r} V_j\| \ge c n^{q/2-1/8}\Big)\\
&\le c(\log n)^{q/\beta}\, n^{q-7/8}\, n^{-q+2/8} \sup_{r\notin L,\ r\le M\mathbb{1}} m(r)\\
&\le c n^{-3/5}\\
&\le c \exp(-M^\beta/2)\ ,
\end{aligned}$$

therefore by the Borel–Cantelli Lemma

$$\sum_{r\notin L,\ r\le M\mathbb{1}} \|\sum_{j\in R_r} V_j\| << n^{q/2-1/10}$$

a.s. as $n \to \infty$. The same holds for W_j instead of V_j. This together with (9.26), (9.2) and

(9.6) implies

(9.30) $$\sum_{r\in\mathbb{N}^q,\ r\le M\mathbb{1}} \|\sum_{j\in R_r} (V_j - W_j)\| << n^{q/2}\, M^{(\beta-1)/2}\, M^{-2}$$

$$<< n^{q/2}\, (\log n)^{-(2/\beta - 1/2)} \quad \text{a.s.}$$

9.4. Construction of the Gaussian random vectors Y_j. Here we will proceed similarly as in [DuPh], [Ph84] and [MPh]. Let $a, b, d \in \mathbb{N}$. For $n \in \mathbb{N}^q$ we define the finite–dimensional random vector

(9.31) $$U(n;a,b,d) := \Pi(n)^{-\frac{1}{2}} \sum_{j\le n} (\Lambda_a X_j\, ,\, \Lambda_b X_j\, ,\, \Lambda_d X_j)\, .$$

The random field $((\Lambda_a X_j\, ,\, \Lambda_b X_j\, ,\, \Lambda_d X_j),\ j \in \mathbb{N}^q)$ satisfies the assumptions of Theorem 1 of [BM], therefore especially the Central Limit Theorem holds for its partial sums $U(n;a,b,d)$. So there exists a Gaussian measure μ_{abd} such that for the Prokhorov distance π (w.r.t. the norm $\|(u,v,w)\| = \|u\| + \|v\| + \|w\|$ on $\Lambda_a S \times \Lambda_b S \times \Lambda_d S$) we have

(9.32) $$\pi(\mathcal{L}(U(n;a,b,d)),\ \mu_{abd}) < \epsilon_1$$

for all $n = n_1\mathbb{1}$ with $n_1 \ge n_0(\epsilon_1)$. Here $\epsilon_1 > 0$ was given arbitrarily. Now (2.8) implies that for all $m, a, d \in \mathbb{N}$ with $d \ge a \ge m$ and all $n \in \mathbb{N}^q$ with $\Pi(n) \ge \max(N_0(a), N_0(d))$

(9.33) $$P(\Pi(n)^{-\frac{1}{2}} \|\sum_{j\le n} (\Lambda_a X_j - \Lambda_d X_j)\| \ge 2m^{-\frac{1}{8}}) \le 2m^{-q(1+\vartheta)}\, .$$

As in section 3 of [DuPh], (9.32) (with $\epsilon_1 = m^{-2q}$ e.g.) and (9.33) imply for all $d \ge a \ge m$

(9.34) $$\mu_{ad}(\{(u,w) \in T \times T\colon \|u - w\| \ge 3m^{-\frac{1}{8}}\}) \le 3m^{-q(1+\vartheta)}\, ,$$

where μ_{ad} is a marginal of μ_{abd} and these measures are now regarded as living on $T \times T$ resp. $T \times T \times T$. Also as in [DuPh] this implies the existence of Gaussian measures $\mu_{a\infty}$ on $T \times T$ with $\mu_{ad} \to \mu_{a\infty}$ (weakly) as $d \to \infty$ and

(9.35) $$\mu_{a\infty}(\{(u,v) \in T \times T\colon \|u - v\| \ge 3m^{-\frac{1}{8}}\}) \le 3m^{-q(1+\vartheta)}$$

for $a \geq m$ and for some Gaussian measure μ_∞ on T

(9.36) $$\mu_a \to \mu_\infty \text{ for } a \to \infty$$

where μ_a is a marginal of μ_{abd}. (Note that this notation μ_a is consistent with (9.21) and (9.24).)

Now let $((W_j', Z_j), j \in \mathbb{N}^q)$ be a family of independent random vectors with $\mathcal{L}(W_j', Z_j) = \mu_{m(r),\infty}$ for $j \in R_r$. By [BPh, Lemma A.1] we can set $W_j = W_j'$ (maybe after changing the underlying probability space). Then we have

(9.37) $$P((\text{card } R_r)^{-\frac{1}{2}} \| \sum_{j \in R_r} (W_j - Z_j) \| \geq 3m(r)^{-\frac{1}{2}}) \leq 3m(r)^{-q(1+\vartheta)}$$

$$\leq 3[\sum_{i=1}^{q} r_i^{1-\vartheta/2}]^{-q(1+\vartheta)} \leq c(\Sigma\, r_i)^{-(1-\vartheta/2)(1+\vartheta)q}$$

$$\leq c(\Sigma\, r_i)^{-q(1+\vartheta/4)}.$$

Here we used Hölder's inequality. The sum of the last terms over $r \in \mathbb{N}^q$ is finite, so we have by the Borel–Cantelli Lemma

(9.38) $$\| \sum_{j \in R_r} (W_j - Z_j) \| \leq c(\omega)\, (\text{card } R_r)^{\frac{1}{2}}\, m(r)^{-\frac{1}{2}} \quad \text{a.s.}$$

Now if $n \in \mathbb{N}$ and M is defined by (9.27), we obtain

$$\sum_{r \leq M\mathbb{1}} \| \sum_{j \in R_r} (W_j - Z_j) \| \ll \sum_{r \leq M\mathbb{1}} (\text{card } R_r)^{\frac{1}{2}}\, m(r)^{-\frac{1}{2}}$$

$$\ll \sum_{\frac{1}{2}M\mathbb{1} < r \leq M\mathbb{1}} \exp(\tfrac{1}{2} \sum_{i=1}^{q} r_i^{\beta})\, [\sum_{i=1}^{q} r_i^{1-\vartheta/2}]^{-\frac{1}{2}}$$

$$\ll M^{-(1-\vartheta/2)/2} (\sum_{k \in \mathbb{N},\, M/2 < k \leq M} \exp(k^{\beta}/2))^q$$

$$\ll M^{-(1-\vartheta/2)/2}\, M^{q(1-\beta)} (\sum_{M/2 < k \leq M} k^{\beta-1} \exp(k^{\beta}/2))^q$$

$$\ll M^{-(1-\vartheta/2)/2}\, M^{q(1-\beta)} \exp(\tfrac{q}{2} M^{\beta})$$

$$\ll M^{-(1-\vartheta/2)/2}\, M^{q(1-\beta)/2}\, t_M^{q/2}$$

$$<< n^{q/2} (\log n)^{-\alpha_2}$$

for some $\alpha_2 > 0$ by the choice of β. This and (9.30) imply that for some $\alpha > 0$

$$\sum_{r \leq M\mathbb{1}} \| \sum_{j \in R_r} (V_j - Z_j) \| << n^{q/2} (\log n)^{-\alpha} \tag{9.39}$$

a.s. as $n \to \infty$.

Now for $j \in \mathbb{N}^q$ define $\rho(j) = m(r)$ if $j \in R_r$. Then we have by (9.11), (2.8) and [DuPh, Lemma 2.5]

$$P(\| \sum_{j \in R_r} (X_j - \Lambda_{\rho(j)} X_j) \|^* \geq 2m(r)^{-\frac{1}{2}} (\text{card } R_r)^{\frac{1}{2}}) \leq m(r)^{-q(1+\vartheta)}$$

for all except finitely many $r \in \mathbb{N}^q$. As above, this implies by the Borel-Cantelli Lemma

$$\sum_{r \leq M\mathbb{1}} \| \sum_{j \in R_r} (X_j - \Lambda_{\rho(j)} X_j) \|^* << n^{q/2} (\log n)^{-\alpha_2} \quad \text{a.s.} \tag{9.40}$$

By construction we have $\mathcal{L}(V_j, j \in \mathbb{N}^q) = \mathcal{L}(\Lambda_{\rho(j)} X_j, j \in \mathbb{N}^q)$. (Note that (9.8) does now not hold necessarily, because we changed the underlying probability space several times and redefined the V_j.) So as in [DuPh, bottom of p. 525] Lemma 2.11 of [DuPh] implies that on the original probability space (Ω, Σ, P) there exists a family $(Y_j, j \in \mathbb{N}^q)$ such that

$$\begin{aligned} &\mathcal{L}((\Lambda_{\rho(j)} X_j, j \in \mathbb{N}^q), (Y_j, j \in \mathbb{N}^q)) \\ &\quad = \mathcal{L}((V_j, j \in \mathbb{N}^q), (Z_j, j \in \mathbb{N}^q)). \end{aligned} \tag{9.41}$$

This and (9.39) and (9.40) imply for some $\alpha > 0$

$$\sum_{r \leq M\mathbb{1}} \| \sum_{j \in R_r} (X_j - Y_j) \| << n^{q/2} (\log n)^{-\alpha} \tag{9.42}$$

a.s. as $n \to \infty$. The assertion on the expectation and covariance function of $Y_{\mathbb{1}}$ is obtained as in [DuPh, p. 526].

Let us mention that up to this point we only used the strong mixing condition for the family $(x_j, j \in \mathbb{Z}^q)$, but not absolute regularity.

9.5. Approximation of partial sums by sums over rectangles (chaining). Now similarly as in sections 5 and 6 we will approximate the partial sums of the X_j over the nA ($A \in \mathcal{A}$, $n \in \mathbb{N}$) by partial sums over unions of the rectangles $\overline{R}_r$. In order to apply Lemma 2.6 of [DuPh] to blockwise independent random elements (as in the proof of Lemma 6.3) we first of all need bounds for the first and second absolute moments of partial sums. Since here we cannot use Lemma 4.1 (except if S is a separable Hilbert space and $h: E \to S$ measurable), we will apply Lemma 8.5.

The hypotheses of Theorem 3 imply that there are $\tau, r, z \in (0,\infty)$ with (6.2), (6.3) and

(9.43) $$r < 1 - uz/h$$

(with h from (2.11) and (2.12b)), and as in section 6 we note that τ may later be replaced by some arbitrary $\tau' \in (0,\tau)$. Define for $j \in \mathbb{N}^q$

(9.44) $$X_j' := X_j \cdot 1\{\|X_j\|^* \le \Pi(j)^{(1+\tau)/(2+\delta)}\},$$

(9.45) $$X_j'' := X_j - X_j'.$$

We note that [MPh, Lemma 2.3] and (5.6) here also hold if x_j is replaced by X_j and $\|.\|$ by $\|.\|^*$. The following is a generalization of [DuPh, Lemma 6.1]:

Lemma 9.1. Under the hypotheses of each of the Theorems 3, 4, and 5 there is a constant c such that for each $J \subset \mathbb{N}^q$ which is admitted in (2.8) resp. (2.17) and satisfies $\text{card } J \ge N_0(m)$, we have for all $d_j \in [-1, 1]$ ($j \in \mathbb{N}^q$)

(9.46) $$P^*((\text{card } J)^{-\frac{1}{2}} \|\sum_{j\in J} d_j(X_j' - \Lambda_m X_j')\| \ge m^{-\frac{1}{2}})$$
$$\le m^{-1} + cm^{\frac{1}{2}} (\text{card } J)^{\chi} \quad \text{where } \chi = \frac{1}{4} - \frac{(1+\tau)(1+\delta)}{2(2+\delta)} < 0.$$

Proof. By (2.5), (2.8), the linearity of the Λ_m, (5.6) and [DuPh, Lemma 2.5] we have with $C := \sup_m \|\Lambda_m\|$ for $\text{card } J \ge N_0(m)$

$$P^*((\text{card } J)^{-\frac{1}{2}} \|\sum_{j\in J} d_j(X_j' - \Lambda_m X_j')\| \ge m^{-\frac{1}{2}})$$

$$\leq P^*((\text{card } J)^{-\frac{1}{2}} \|\sum_{j\in J} d_j(X_j - \Lambda_m X_j)\| \geq \frac{m^{-\frac{1}{2}}}{2})$$

$$+ P^*((\text{card } J)^{-\frac{1}{2}} \|\sum_{j\in J} d_j(X''_j - \Lambda_m X''_j)\| \geq \frac{m^{-\frac{1}{2}}}{2})$$

$$\leq 1/m + P((\text{card } J)^{-\frac{1}{2}} (1+C) \sum_{j\in J} \|X''_j\|^* > m^{-\frac{1}{2}}/4)$$

$$\leq 1/m + 4(1+C)m^{\frac{1}{2}} (\text{card } J)^{-\frac{1}{2}} \sum_{j\in J} E\|X''_j\|^*$$

$$\leq 1/m + 4(1+C)m^{\frac{1}{2}} (\text{card } J)^{-\frac{1}{2}} \sum_{j\in J} c\Pi(j)^{-(1+\tau)(1+\delta)/(2+\delta)}$$

$$\leq 1/m + cm^{\frac{1}{2}} (\text{card } J)^{-\frac{1}{2}} \Big(\sum_{j\in J} \Pi(j)^{-p(1+\tau)(1+\delta)/(2+\delta)} \Big)^{1/p} (\text{card } J)^{1-1/p}$$

where $\frac{1}{p} := \frac{1}{4} + \frac{(1+\tau)(1+\delta)}{2(2+\delta)} > \frac{1}{2}$. Here we used Hölder's inequality. □

Lemma 9.2. There is a constant C_5 such that for all rectangles $R \subset \mathbb{N}^q$ and all $d_j \in [-1,1]$

(9.47) $$E\|\sum_{j\in R} d_j X'_j\|^{*2} \leq C_5 \text{ card } R ,$$

(9.48) $$E\|\sum_{j\in R} d_j X_j\|^{*2} \leq C_5 \text{ card } R .$$

Proof. From Lemmas 8.5 and 9.1 we obtain for $m = 10^4$ $E\|\Sigma_{j\in R} d_j(X'_j - \Lambda_m X'_j)\|^{*2} << \text{card } R$, from Lemmas 4.1 and 7.2 $E\|\Sigma_{j\in R} d_j \Lambda_m X'_j\|^2 << \text{card } R$. This proves (9.47). For (9.48) use the assumption (2.8) directly instead of Lemma 9.1. □

Now let $n \in \mathbb{N}$ and $p := [n^r]$. Let $D_1,\dots,D_Q$ and H_k^i $(i = 0,\dots,2^q-1,\ k = 1,\dots,Q)$ be as in the proof of Lemma 6.3; $H_k = H_k^0$, $H := \cup_{k=1}^Q H_k$, $\overline{H}_k := H_k \cap \mathbb{N}^q$, $\overline{H} := H \cap \mathbb{N}^q$; as there define E–valued random variables $z_j\ (j \in \overline{H})$ on a probability space $(\overline{\Omega},\overline{\Sigma},\overline{P})$ such that $\mathcal{L}(x_j, j \in \overline{H}_k) = \mathcal{L}(z_j, j \in \overline{H}_k)$ and $(z_j, j \in \overline{H}_{k+1})$ is independent of

$(z_j, j \in \cup_{t=1}^{k} \overline{H}_t)$ for $k = 1,\dots,Q-1$. Here $(\overline{\Omega},\overline{\Sigma},\overline{P})$ can be chosen as the Q–fold product of the space $(E^{(p^q)}, \mathcal{F}^{(p^q)}, \mathcal{L}(x_j, j \in \overline{H}_1))$ with itself and the $(z_j)_{j\in\overline{H}_k}$, $k = 1,\dots,Q$, as the coordinate projections, so $(z_j)_{j\in\overline{H}}$ is the identity on $\overline{\Omega}$. Write $\xi := (x_j, j \in \overline{H})$, $\zeta := (z_j, j \in \overline{H})$. Then by (6.8) we have for some $\gamma > 0$, $P_\xi(B) \le \overline{P}_\zeta(B) + O(n^{-1-\gamma})$ uniformly in $B \in \overline{\Sigma} = \mathcal{F}^{Qp^q}$, hence $(P_\xi)^*(G) \le \overline{P}^*(G) + O(n^{-1-\gamma})$ for all $G \subset \overline{\Omega}$. By [An, Theorem 2.2, p. II.13] this implies $P^*(\xi \in G) \le \overline{P}^*(\zeta \in G) + O(n^{-1-\gamma})$ for all $G \subset \overline{\Omega}$, similarly $\overline{P}^*(\zeta \in G) \le P^*(\xi \in G) + O(n^{-1-\gamma})$, hence

(9.49) $$|P^*(\xi \in G) - \overline{P}^*(\zeta \in G)| << n^{-1-\gamma}.$$

Now let

(9.50) $$Z_j = h \circ z_j \quad \text{for } j \in \overline{H}.$$

(Of course, this notation Z_j has nothing to do with the one introduced above.) Writing $\overline{H}(I) := \cup_{k\in I} \overline{H}_k$, we have by (9.49) and (2.8) for sufficiently large n and for $I \subset \{1,\dots,Q\}$ with $p^q \operatorname{card} I \ge N_0(3000)$

(9.51) $$\overline{P}^*((p^q \operatorname{card} I)^{-\frac{1}{2}} \|\sum_{j\in\overline{H}(I)} d_j(Z_j - \Lambda_{3000} Z_j)\| \ge 1) < 10^{-3}$$

for all $d_j \in [0,1]$, if (2.9) is also satisfied, i.e. if $\beta(p) \operatorname{card} I \le p^{-b}$ with some fixed $b \in (0,1]$. But this is true here, see the last three inequalities in (6.8).

Put $\eta_k := \sum_{j\in\overline{H}_k} d_j Z_j$ for $k = 1,\dots,Q$. By Lemma 9.2 and its proof we have $E\|\eta_k - \Lambda_{3000}\eta_k\|^{*2} \le cp^q$. Now Lemma 8.1, applied to $\Sigma(\eta_k - \Lambda_{3000}\eta_k)$ with $\alpha = p^{q/2}$, and (9.51) imply that for all $I \subset \{1,\dots,Q\}$ with $\operatorname{card} I \ge \max\{p^{-q} N_0(3000), 10^6 p^{-q} c p^{-2q} p^q\}$, i.e. for $\operatorname{card} \overline{H}(I)$ sufficiently large,

(9.52) $$E\|\sum_{j\in\overline{H}(I)} d_j(Z_j - \Lambda_{3000} Z_j)\|^{*2} = E\|\sum_{k\in I} (\eta_k - \Lambda_{3000}\eta_k)\|^{*2} \le \operatorname{card} I \cdot c\, p^q .$$

As in the proof of Lemma 9.2 this implies

(9.53) $$E\|\sum_{j\in\overline{H}(I)} d_j Z_j\|^{*2} \le c \operatorname{card} \overline{H}(I) ,$$

hence

$$E\|\sum_{j\in\overline{H}(I)} d_j Z_j\|^* \le c(\text{card } \overline{H}(I))^{\frac{1}{2}} . \tag{9.54}$$

Writing

$$Z_j' := Z_j \cdot 1\{\|Z_j\|^* \le \Pi(j)^{(1+\tau)/(2+\delta)}\} , \tag{9.55}$$

by (5.6) this implies

$$E\|\sum_{j\in\overline{H}(I)} d_j Z_j'\|^* \le c(\text{card } \overline{H}(I))^{\frac{1}{2}} \tag{9.56}$$

as in the end of the proof of [MPh, Lemma 3.1] (with their x_j replaced by $d_j Z_j$).

Let $t_k = t(k)$, N_k, $\overline{R}_r$ be defined by (9.1) to (9.3) and β be chosen as above. Put for $m \in \mathbb{N}_0$

$$\psi(m) := [\exp(m^{(1-\beta)/2})]. \tag{9.57}$$

For $n \in \mathbb{N}$ given let $m, k \in \mathbb{N}_0$ be defined by

$$\psi(m) < n \le \psi(m+1) , \tag{9.58}$$

$$t(k) < \psi(m) \le t(k+1) . \tag{9.59}$$

Let a, b, d be the smallest integers with

$$2^d \ge m^{h(1-\beta)^2/(8\beta)} , \tag{9.60}$$

$$2^a \ge m^{(1+4\tau)(1-\beta)/2} , \tag{9.61}$$

$$2^b \ge n^z . \tag{9.62}$$

For $A \in \mathcal{A}$ let

$$S_n'(A) := \sum_{j\in\mathbb{N}^q} \mu(nA \cap C_j)\, X_j' , \tag{9.63}$$

$$\overline{S}_n'(A) := \sum_{j\in\overline{H}} \mu(nA \cap C_j)\, Z_j' .$$

Furthermore write for $i = 1,\dots,b$

$$w_n := w_n(i) := S_n'(A(2^{-i})) - S_n'(A(2^{-i+1})) , \tag{9.64}$$

$$\overline{w}_n := \overline{w}_n(i) := \overline{S}_n'(A(2^{-i})) - \overline{S}_n'(A(2^{-i+1})) .$$

Let $c_0 \in (0,\infty)$ be arbitrary (but fixed).

Lemma 9.3. For τ sufficiently small we have for some $\gamma > 0$

$$\overline{P}(\sup\{\|\overline{w}_n(i)\|^* \cdot 2^{i\tau} \colon A(2^{-i}), A(2^{-i+1}), a \le i \le b\} > c_0 n^{q/2}) << n^{-1-\gamma}.$$

Proof. For the time being let i, $A(2^{-i})$, $A(2^{-i+1})$ be fixed. Each cube H_λ $(\lambda = 1,\dots,Q)$ which is needed to cover $B := nA(2^{-i}) \,\Delta\, nA(2^{-i+1})$, i.e. for which $H_\lambda \cap B \neq \phi$, satisfies $H_\lambda \subset B$ or $H_\lambda \cap \partial B \neq \phi$. Since $\partial(C \,\Delta\, D) \subset \partial C \cup \partial D$ for arbitrary sets C, D, every cube H_λ with $H_\lambda \cap B \neq \phi$ satisfies $H_\lambda \subset nA(2^{-i}) \,\Delta\, nA(2^{-i+1})$ or $H_\lambda \cap \partial nA(2^{-i}) \neq \phi$ or $H_\lambda \cap \partial nA(2^{-i+1}) \neq \phi$. Therefore by (1.5), (1.6) and (2.11) the μ–measure and hence by (1.4) also the $\mathbb{N}^q$–counting measure of the union of those H_λ which satisfy $H_\lambda \cap B \neq \phi$, does not exceed

$$F \le 2n^q\, b(\mu,\tfrac{p}{n}) + 2n^q \cdot 2^{-i+1} \le cn^q(n^{-(1-r)h} + 2^{-i}) . \tag{9.65}$$

We intend to apply Lemma 2.6 of [DuPh] to the independent random elements

$$\sum_{j \in H_\lambda \cap \mathbb{N}^q} (\mu(nA(2^{-i}) \cap C_j) - \mu(nA(2^{-i+1}) \cap C_j))\, Z_j', \quad \lambda = 1,\dots,Q.$$

By (6.10) they vanish if $H_\lambda \cap B = \phi$. Let $K := n^{q/2}\, 2^{-i\tau}$. By choosing τ sufficiently small, we can achieve that $2^{-i\tau} \ge 2^{-b\tau} \ge cn^{-z\tau} \ge cn^{-h(1-r)/2 + \gamma}$ for some $\gamma > 0$. Then

$$F^{\frac{1}{2}} \le cn^{q/2}(n^{-h(1-r)/2} + 2^{-i/2}) = o(K) \quad \text{as } n \to \infty. \tag{9.66}$$

Hence for $c_1 \in (0,\infty)$ and $c_2 \in (0,1)$ given, we have $K - c_1 F^{\frac{1}{2}} \ge c_2 K$ for n sufficiently large. Noting (9.65), (9.66), (6.10), (9.56) and (9.47) we now infer from Lemma 2.6 of [DuPh]

$$\overline{P}(\|\overline{w}_n(i)\|^* > c_0 K)$$

$$<< \exp\Big(-c\,\frac{n^q\,2^{-2i\tau}}{n^q(n^{-h(1-r)}+2^{-i})}\Big) + \exp\Big(-c\,\frac{n^{q/2}\;2^{-i\tau}}{p^q\;n^{q(1+\tau)/(2+\delta)}}\Big)$$

$$<< \exp(-c\,2^{-2i\tau}\,n^{h(1-r)}) + \exp(-c\,2^{i(1-2\tau)}) + \exp\Big(-c\,2^{-i\tau}\,n^{q(\frac{1}{2}-\frac{1+\tau}{2+\delta}-r)}\Big).$$

Now we multiply this with $N(\mu,2^{-i})\cdot N(\mu,2^{-i+1})\cdot b << \exp(c2^{iu})$. Since $u < 1$ and therefore $u < 1-2\tau$ for τ sufficiently small, we obtain from the first summand the bound

$$\exp(-c\,2^{i(1-2\tau)}) << \exp(-c\,2^{a(1-2\tau)}) << \exp(-c(\log n)^{1+\tau}).$$

By (9.43) from the first term we obtain the bound

$$\exp(c_1 2^{iu} - c_2 2^{-2i\tau}\,n^{h(1-r)}) << \exp(c_1 2^{bu} - c_2 2^{-2b\tau}\,n^{h(1-r)})$$

$$<< \exp(c_1 n^{zu} - c_2 n^{-2z\tau+h(1-r)}) << \exp(-cn^{\gamma})$$

for some $\gamma > 0$ if τ is sufficiently small. From the third summand we obtain by (6.3) and (9.62) the bound

$$\exp\Big(c_1 2^{bu} - c_2 2^{-b\tau}\,n^{q(\frac{1}{2}-\frac{1+\tau}{2+\delta}-r)}\Big) << \exp(-cn^{\gamma})$$

for some $\gamma > 0$ if τ is sufficiently small. □

Lemma 9.4. For τ sufficiently small we have for $\alpha := \frac{(1+\beta)h}{4(1-\beta)}$ and some $\gamma > 0$

$$\overline{P}\big(\sup_{A(2^{-a})} \|\overline{S}'_n(A(2^{-a})) - \overline{S}'_{\psi(m)}(A(2^{-a}))\|^* > c_0 n^{q/2}(\log n)^{-\alpha}\big) << n^{-1-\gamma}.$$

Proof. By (6.10), for fixed $B := A(2^{-a})$ the random element

(9.67) $$(\mu(nB\cap C_j) - \mu(\psi(m)B\cap C_j))\,Z'_j\,,$$

vanishes if j is outside the set $D := (nB\,\Delta\,\psi(m)B)^{\epsilon'} \subset (nB)^{n\epsilon} \cap (n\breve{B})^{n\epsilon}$, where $\epsilon' = 1$ and $\epsilon = 1 - \psi(m)/n + 1/n$ and $\breve{B} := \mathbb{R}^q_+ \setminus B$ (see [MPh, proof of Lemma 2.7]). Let F denote the $\mathbb{N}^q$–counting measure of the union of those cubes H_λ $(\lambda = 1,\dots,Q)$ which have non–empty intersection with the set D. Then

$$F \le cn^q\, b(\epsilon + p/n) \le cn^q\left(m^{-\frac{1}{2}-\frac{\beta}{2}} + p/n\right)^h$$

$$\le cn^q\, m^{(-\frac{1}{2}-\frac{\beta}{2})h} \le cn^q(\log n)^{-h(1+\beta)/(1-\beta)}.$$

Put $K := n^{q/2}(\log n)^{-\alpha}$, hence $F = o(K)$ as $n \to \infty$. Then [DuPh, Lemma 2.6] (which is applied to the sums of the random elements (9.67) over the cubes $H_\lambda \cap \mathbb{N}^q$ similarly as in the proof of Lemma 9.3) yields

$$\overline{P}(\|\overline{S}'_n(B) - \overline{S}'_{\psi(m)}(B)\|^* > c_0K) << \exp\left(-\frac{cn^q(\log n)^{-2\alpha}}{n^q(\log n)^{-4\alpha}}\right)$$

$$+ \exp\left(-c\frac{n^{q/2}(\log n)^{-\alpha}}{p^q\, n^{q(1+\tau)/(2+\delta)}}\right)$$

$$<< \exp(-c(\log n)^{2\alpha}) + \exp(-cn^{\gamma'}) << \exp(-c(\log n)^{2\alpha})$$

for some $\gamma' > 0$. The choice of β implies $2\alpha > 2h$, and by (2.12b) we have $u < 2h$, so multiplication with $N(2^{-a}) \le \exp(c2^{au}) \le \exp(c(\log n)^{(1+4\tau)u})$ yields the bound $\exp(-c(\log n)^{2\alpha}) << n^{-1-\gamma}$ for some $\gamma > 0$ if τ is sufficiently small. □

Of course, the arguments following Lemma 9.2 up to and including (9.56) remain valid if there n is always replaced by $\psi(m)$. Define $w_{\psi(m)}$ and $\overline{w}_{\psi(m)}$ analogously by (9.64).

Lemma 9.5. For $\tau > 0$ sufficiently small we have for some $\gamma > 0$

$$\overline{P}(\sup\{\|\overline{w}_{\psi(m)}(i)\|^* \cdot 2^{i\tau}: A(2^{-i}), A(2^{-i+1}), d \le i \le a\} > c_0\exp(\tfrac{q}{2}\, m^{(1-\beta)/2}))$$

$$<< \exp(-cm^{\gamma}) \quad \text{as } m \to \infty.$$

<u>Proof.</u> As in the proof of Lemma 9.3 we have for fixed i, $A(2^{-i})$, $A(2^{-i+1})$

$$\overline{P}(\|\overline{w}_{\psi(m)}(i)\|^* > c_0 2^{-i\tau}\exp(\tfrac{q}{2}\, m^{(1-\beta)/2})) << \exp(-c2^{-2i\tau}\, \psi(m)^{h(1-r)})$$

$$+ \exp(-c2^{i(1-2\tau)}) + \exp\left(-c2^{-i\tau}\, \psi(m)^{q(\frac{1}{2} - \frac{1+\tau}{2+\delta} - r)}\right)$$

$$\begin{aligned} &<< \exp(-c(\log \psi(m))^{-2\tau(1+4\tau)} \psi(m)^{h(1-r)}) + \exp(-c2^{i(1-2\tau)}) \\ &\qquad + \exp(-c(\log \psi(m))^{-2\tau(1+4\tau)} \psi(m)^{\gamma'}) \\ &<< \exp(-c2^{i(1-2\tau)}) \end{aligned}$$

for some $\gamma' > 0$. We multiply both sides of the inequality with $N(2^{-i}) \cdot N(2^{-i+1}) \cdot a << \exp(c2^{iu})$ and obtain the bound $\exp(c_1 2^{iu} - c_2 2^{i(1-2\tau)}) << \exp(-c2^{i(1-2\tau)}) << \exp(-c2^{d(1-2\tau)}) << \exp(-cm^{\gamma})$ for some $\gamma > 0$ if τ is sufficiently small. □

For $A \subset \mathbb{R}_+^q$ write $A_* := \cup_{\overline{R}_r \subset A} \overline{R}_r$, where $\overline{R}_r$ is defined in (9.3). For $A \in \mathcal{B}([0,1]^q)$ put

$$V_n'(A) := \sum_{j \in \mathbb{N}^q} \mu(((nA) \setminus (nA)_*) \cap C_j)\, X_j' ,$$

$$\nabla_n'(A) := \sum_{j \in \overline{H}} \mu(((nA) \setminus (nA)_*) \cap C_j)\, Z_j' .$$

Lemma 9.6. For $\alpha := \frac{1-\beta}{4\beta} h$ and some $\gamma > 0$ we have

$$\overline{P}(\sup_{A(2^{-d})} \|\nabla_{\psi(m)}'(A(2^{-d}))\|^* > c_0 \psi(m)^{q/2} (\log \psi(m))^{-\alpha}) << \exp(-cm^{\gamma}) .$$

<u>Proof.</u> For the time being let $B := A(2^{-d})$ fixed. The $\mathbb{N}^q$–counting measure F of the union of those cubes H_λ $(\lambda = 1,\dots,Q)$ which have a non–empty intersection with $(\psi(m)B) \setminus (\psi(m)B)_*$, satisfies

$$\begin{aligned} F &\le c\psi(m)^q\, b(\tfrac{1}{\psi(m)} (2p + t(k+1) - t(k))) \\ &\le c\psi(m)^{q-h} (p + \psi(m) (\log \psi(m))^{-(1-\beta)/\beta})^h \\ &\le c\psi(m)^q (\log \psi(m))^{-4\alpha} . \end{aligned}$$

Hence we have by Lemma 2.6 of [DuPh]

$$\overline{P}(\|\nabla'_{\psi(m)}(B)\|^* > c_0\psi(m)\,(\log \psi(m))^{-\alpha})$$

$$<< \exp\left(-c\,\frac{\psi(m)^q\,(\log\,\psi(m))^{-2\alpha}}{\psi(m)^q\,(\log\,\psi(m))^{-4\alpha}}\right) + \exp\left(-c\,\frac{\psi(m)^{q/2}\,(\log\,\psi(m))^{-\alpha}}{p^q\,\,\psi(m)^{q(1+\tau)/(2+\delta)}}\right)$$

$$<< \exp(-c(\log\,\psi(m))^{2\alpha}).$$

Now we obtain the assertion by multiplying with $N(2^{-d}) << \exp(c(\log\,\psi(m))^{\alpha u})$. □

Now note that by (6.8), the definition of the Z_j $(j \in \overline{H})$, Lemma 2.1 and the remarks following it, the Lemmas 9.3 through 9.6 remain valid if $\overline{P}$, $\overline{w}_n$, $\overline{S}'_n(A)$, $\nabla'_{\psi(m)}(A)$ are replaced by P, w_n, $S'_n(A)$, $V'_{\psi(m)}(A)$. Furthermore, these lemmas hold analogously with Y_j instead of X_j $(j \in \mathbb{N}^q)$. Now in each instance apply the Borel-Cantelli Lemma, note (9.42) and argue analogously as in the end of the proofs of Theorems 1 and 2 in this paper and of Theorems 1 to 3 of [MPh]. Then the assertion of Theorem 3 follows. □

Let us now indicate the changes necessary for the proofs of Theorems 4 and 5. (The details may be found in [Str].) For Theorem 5, the blocks H_r and R_r are chosen with polynomial (instead of subexponential) length. The finite-dimensional ingredient is Lemma 7.3 instead of Lemma 7.1, and the approximation theorem used is [Ph86, Theorem 3.4] (an improvement of Theorem 2 of [BM]) instead of [BM, Theorem 1].

For the proof of Theorem 4, the blocks H_r and R_r are in turn of subexponential size, and again the approximation theorem [Ph86, Theorem 3.4] is used. Here there is an additional complication, namely that the conclusion from (9.37) to (9.38) using the Borel-Cantelli Lemma is not possible in the analogous situation, because the bounds on the right-hand side turn out not to be summable. Instead, Fernique's (1970, p. 1699) inequality can be applied. Also, the easy proof of (9.40) cannot be adapted here for the same reason. Here the proof goes as follows. The x_j are replaced by blockwise independent variables z_j and so the

$X_j^! - \Lambda_m X_j^!$ by some $U_j^! - \Lambda_m U_j^!$ where $U_j^!$ is obtained from $U_j := h \circ z_j$ by truncation. Then the exponential inequality [DuPh, Lemma 2.6] is applied to the independent blocks of random elements. For checking the hypotheses of this Lemma, Lemma 8.5 is needed. In fact, at this point the exact statements on the possible choice of the constants in Lemma 8.5 are essential. The rest of the proof is the same as for Theorem 3.

10. Proof of Theorems 6 and 7

In this section we will give a variant of Theorem 5.1 of [Du78] (resp. Theorem 6.2.1 of [Du84]), of the proof of Theorem 7.1 of [DuPh] and of Proposition 6.1 of [Ph84].

Let $(x_j, j \in \mathbb{N}^q)$ be a strictly stationary family of random variables on a probability space (Ω,Σ,P) and with values in a measurable space $(E,\mathcal{F})$ and distribution $\mathcal{L}(x_j) = P'$ for all $j \in \mathbb{N}^q$. For a family $\gamma = (\gamma_j, j \in \mathbb{N}^q)$ of real numbers with $|\gamma_j| \leq 1$ for all j, for bounded measurable $f: E \to \mathbb{R}$ and a set $J \subset \mathbb{N}^q$ put

$$\nu(J,f,\gamma) := (\text{card } J)^{-\frac{1}{2}} \sum_{j\in J} \gamma_j(f(x_j) - P'(f)), \tag{10.1}$$

where $P'(f) := \int f dP'$.

Theorem 10.1. Let $(x_j, j \in \mathbb{N}^q)$ be as above and assume that it satisfies the condition of absolute regularity (1.2) with rate

$$\beta(t) \leq ct^{-s'} \quad \text{for some } s' > 2q-1. \tag{10.2}$$

Let $\mathcal{G}$ be a class of $\mathcal{F}$–measurable functions $f: E \to [0,1]$ satisfying (3.11) and (3.12). Then there are constants K, L, c_1, $c_2 \in (0,\infty)$ such that for $\epsilon > 0$ given there exist numbers $\delta \geq c_1\epsilon^K$ and $n_0 \leq c_2\epsilon^{-L}$ with the following property: For all families of numbers $\gamma = (\gamma_j, j \in \mathbb{N}^q)$ with $|\gamma_j| \leq 1$ and all rectangles $J \subset \mathbb{N}^q$ with card $J \geq n_0$ we have

(10.3) $$P^*(\sup\{|\nu(J,f,\gamma) - \nu(J,g,\gamma)| : f,g \in \mathcal{G}, P'(|f-g| < \delta\} > \epsilon) < \epsilon.$$

Proof. Write $N_I(y) := N_I(y,\mathcal{G},P')$. By (3.12) there are a, a' with

(10.4) $$\tau < a < a' < 1 - 2q/(s'+1) .$$

As in the proof of Theorem 7.1 of [DuPh] one can show that for some K depending only on τ and a the following holds: For every ϵ there is a $\delta_0 = 2^{-r} >> \epsilon^K$ (with $r \in \mathbb{N}$) such that

(10.5) $$y \log N_I(y) \leq \epsilon^2/2 \text{ for } 0 < y \leq \delta_0^a ,$$

(10.6) $$\exp(-\epsilon^2/\delta_0^a) < \epsilon ,$$

(10.7) $$\sum_{i \geq r} (2^{-ia} \log N_I(2^{-i}))^{\frac{1}{2}} \leq \sum_{i \geq r} 2^{-ia/2} \cdot c \cdot 2^{i\tau/2} \leq c2^{-r(a-\tau)/2} < \epsilon/96 ,$$

(10.8) $$\sum_{i \geq 0} \exp\left(-\frac{2^{a(i+r)}\ \epsilon^2}{1025\ (i+1)^4}\right) < \epsilon.$$

In the proof of Theorem 5.1 of [Du78] independence is only needed where Bernstein's inequality is applied. Here we proceed as there up to (5.10), however with the following changes:

(10.9) $$b_k := (2^{-ka} \log m(k))^{\frac{1}{2}} ,$$

where $\delta_k := \delta(k) := \delta_0/2^k = 2^{-r-k}$ and $m(k) := N_I(\delta_k)$ as in [Du78]. So we have $\log m(i) = 2^{ia}\, b_i^2$. Furthermore put

(10.10) $$d_i := \max((i+1)^{-2}\epsilon/32,\ b_{i+1}\, 2^{-ra/2}) .$$

Therefore we have by (10.7) as in [Du78, (5.10)]

(10.11) $$\sum_{i \geq 0} d_i < \epsilon/8 .$$

Similarly as in sections 6 and 9 we will replace the x_j, $j \in J$, by blockwise independent random variables so that it will be possible to apply Bernstein's inequality. With the a' from (10.4) put $z := \frac{1-a'}{2q}$. For the rest of this proof (and only here!) write $n := \text{card } J$ for short, and $p := p(n) := [n^z]$. With this p we adopt the notations and arguments in section 9 following Lemma 9.2 up to and including the definition of the z_j. Let $H_1,\dots,H_Q$ be those cubes which have non–empty intersection with J. For all except no more than 2^q of the cubes $\bar{H}_1,\dots,\bar{H}_Q$ at least one edge is completely contained in J. W.l.o.g. we assume that this is true for $\bar{H}_1,\dots,\bar{H}_{Q'}$. So $Q-Q' \le 2^q$ and for $\lambda = 1,\dots,Q'$ we have

$$\text{card}(\bar{H}_\lambda \cap J) \ge p\,.$$

Hence

$$Q \le cQ' \le c(\text{card } J)/p = cn/p. \tag{10.12}$$

Writing $\bar{H} := \cup_{\lambda=1}^{Q} \bar{H}_\lambda = \cup_{\lambda=1}^{Q} H_\lambda \cap \mathbb{N}^q$ we have analogously to (6.8)

$$\begin{aligned} \tfrac{1}{2}\|\mathcal{L}(x_j,\ j \in \bar{H}) - \mathcal{L}(z_j,\ j \in \bar{H})\| &\le (Q-1)\,\beta(p) \\ &\le c\, n\, p^{-1}\, p^{-s'} \le c\, n^{1-z(1+s')}\,. \end{aligned} \tag{10.13}$$

Define $n_0 = n_0(\epsilon)$ and $k = k(n)$ as in [Du78, p. 915]. By (5.6) of [Du78] and the lines preceding it we have $1/(16\delta_0 2^{-k}) < n^{\frac{1}{2}}/\epsilon$ and $\delta_0 \le \epsilon$, hence

$$k < c \log(n^{\frac{1}{2}}\delta_0/\epsilon) \le c \log n.$$

Therefore, by (10.4) and the choice of z there are $b_1, b_2 \in (0,\infty)$ (which depend only on c in (10.13) and on s' and z) such that for $n \ge n_0'(\epsilon) := \max(n_0(\epsilon), b_1\epsilon^{-b_2})$ and for the same c as in (10.13)

$$c\, n\, p^{-(s'+1)} < \epsilon/k\,. \tag{10.14}$$

For every $i < k = k(n)$ we have by [Du78, p. 915] (note (10.11)) $d_i < 2n^{\frac{1}{2}}\delta_i$, hence (note that $d_i \le 1$)

(10.15) $$d_i\, n^{qz-\frac{1}{2}} \le d_i^{1-2qz}\, n^{qz-\frac{1}{2}} = (d_i\, n^{-\frac{1}{2}})^{1-2qz} \le 2\delta_i^{1-2qz} .$$

Now let $g\colon E \to [-1,1]$ be measurable. Define τ by $2/(2+\tau) = a'$, note that $P'(|g - P'(g)|^{2+\tau}) \le cP'(|g|)$ since $|g| \le 1$, then by (10.2) we can apply Lemma 4.1 with τ instead of δ there (that δ there has nothing to do with the one here) and obtain an upper bound $O(P'(|g|)^{a'}\mathrm{card}(H_\lambda \cap J))$ for the variance of the real random variables $\Sigma_{j\in H_\lambda\cap J}\, \gamma_j(g(z_j) - P'(g))$, $\lambda = 1,\dots,Q$. Then Bernstein's inequality (see Bennett (1962)) applied to this sequence yields by (10.15) and (10.4) for $g \in \mathcal{G}_i$ (where $\mathcal{G}_i$ is defined as in [Du84, p.45])

(10.16) $$\begin{aligned} &P(|n^{-\frac{1}{2}} \sum_{j\in\cup H_\lambda\cap J} \gamma_j(g(z_j) - P'(g))| > d_i) \\ &= P(|\sum_{\lambda=1}^{Q} n^{-\frac{1}{2}} \sum_{j\in H_\lambda\cap J} \gamma_j(g(z_j) - P'(g))| > d_i) \\ &\le 2\exp(-d_i^2/(2cP'(|g|)^{a'} + d_i p^q n^{-\frac{1}{2}})) \\ &\le 2\exp(-cd_i^2/(\delta_i^{a'} + \delta_i^{1-2qz})) \\ &\le 2\exp(-cd_i^2/\delta_i^{a'}) . \end{aligned}$$

Put

(10.17) $$M_i := 2\, m(i)\, m(i+1) \le 2\, m(i+1)^2 = 2\exp(2^{(i+1)a+1}\, b_{i+1}^2) .$$

Then by (10.13), (10.14) and [Du84, p. 45, (6.2.10)] we have

(10.18) $$\begin{aligned} P_{iJ\gamma} &:= 2^{-q}\, P(|\nu(J,g,\gamma)| > 2^q d_i \text{ for some } g \in \mathcal{G}_i) \\ &\le \epsilon/k + M_i \exp(-cd_i^2/\delta_i^{a'}) \\ &\le \epsilon/k + 2\exp(2^{ia+2}\, b_{i+1}^2 - cd_i^2\, 2^{ia'}/\delta_0^{a'}) \\ &\le \epsilon/k + 2\exp(2^{ia+2}\, d_i^2/\delta_0^{a} - cd_i^2\, 2^{ia'}/\delta_0^{a'}) , \end{aligned}$$

since by (10.10) we have $b_{i+1}^2\, 2^{-ra} \le d_i^2$ and therefore $b_{i+1}^2 \le d_i^2/\delta_0^a$. We have $\delta_0 \le \epsilon$ (see [Du78, p.915]), hence $\delta_0 \to 0$ as $\epsilon \to 0$, and so we have for all ϵ sufficiently small

(10.19) $$P_{iJ\gamma} \leq \epsilon/k + 2\exp(-2^{ia}d_i^2/\delta_0^a)$$
$$\leq \epsilon/k + 2\exp(-2^{(i+r)a}\,\epsilon^2\,(i+1)^4\,32^{-2})\,,$$

hence by (10.8)

(10.20) $$\sum_{0\leq i\leq k} P_{iJ\gamma} \leq \epsilon + 2\epsilon = 3\epsilon\,.$$

Furthermore, with the same arguments as earlier, namely approximation by blockwise independent random variables, and writing $k := k(n)$ and

$$V(J,\gamma) := \max\{|\nu(J,f_{ku}-f_{kt},\gamma)| : f_{kt}\leq f_{ku}\,,\ P'(f_{ku}-f_{kt}) < \delta_k\,,\ t,u = 1,\dots,m(k)\},$$

we have

(10.21) $$Q(J,\gamma) := 2^{-q}\,P(V(J,\gamma) \geq 2^q\epsilon/8)$$
$$\leq \epsilon + m(k)^2\exp(-\epsilon^2\,64^{-1}/(c\delta_k^{a'} + \tfrac{\epsilon}{8}\,n^{qz-\frac{1}{2}}))$$
$$\leq \epsilon + m(k)^2\exp(-c\epsilon^2/(\delta_k^{a'} + \delta_k^{a'}))$$
$$\leq \epsilon + \exp(2^{ka+1}\,b_k^2 - c2^{ka'}\epsilon^2/\delta_0^{a'})\,.$$

Here we used that by [Du78, (5.6)] $n^{\frac{1}{2}}\,\delta_k > \epsilon/16$, hence $\delta_k^{1-2qz} > (\epsilon/16)^{1-2qz}\,n^{qz-\frac{1}{2}} > \frac{\epsilon}{16}\,n^{qz-\frac{1}{2}}$. Now we have by (10.5) for $j := k + r$, hence $k = j - r$,

$$2b_k^2 = 2\cdot 2^{-ka}\log N_I(2^{-j}) = 2\cdot 2^{ra}\cdot 2^{-ja}\log N_I(2^{-j}) \leq 2^{ra}\,\epsilon^2 = \delta_0^{-a}\,\epsilon^2\,.$$

Hence we have for ϵ sufficiently small (since $\delta_0^{-a} \to \infty$ as $\epsilon \to 0$)

(10.22) $$Q(J,\gamma) \leq \epsilon + \exp(2^{ka+ra}\,\epsilon^2 - 2^{ka'+ra'}\epsilon^2) \leq \epsilon + \exp(-\delta_0^{-a}\,\epsilon^2) \leq 2\epsilon$$

by (10.6). Similarly we can show

(10.23) $$P_0(\gamma) :=$$

$$2^{-q}\,P(\sup\{|\nu(J,f_{0i},\gamma) - \nu(J,f_{0j},\gamma)| : P'(|f_{0i}-f_{0j}|) < 3\delta_0\} > 2^q\epsilon/4)$$

$$\leq \epsilon + 2\, m(0)^2 \exp\Big(-\frac{c\epsilon^2}{\delta_0^{a'} + c\epsilon n^{qz-\frac{1}{2}}}\Big)$$

$$\leq \epsilon + 2\, m(0)^2 \exp(-c\epsilon^2 \delta_0^{-a'}) .$$

By (10.5) we have $m(0)^2 \leq \exp(\epsilon^2/\delta_0^a)$, hence by (10.23) and (10.6) for ϵ sufficiently small

(10.24) $$P_0(\gamma) \leq \epsilon + \exp(-\epsilon^2/\delta_0^a) \leq 2\epsilon .$$

Finally we obtain that for all ϵ sufficiently small, all rectangles J with $\operatorname{card} J \geq n_0'(\epsilon)$ and all $\gamma{:}\mathbb{N}^q \to [-1,1]$

$$P^*(\sup\{|\nu(J,f-h,\gamma)|:\ f,h \in \mathcal{G},\ P'(|f-h|) < \delta_0\} > 100^q\epsilon) < 100^q\epsilon. \ \square$$

Remark 10.2. Theorem 10.1 remains valid if for J we allow not only rectangles in $\mathbb{N}^q$ but also disjoint unions of Q_0 cubes in $\mathbb{N}^q$ with length of edges p_0, where Q_0 and p_0 satisfy the condition

(10.25) $$Q_0\beta(p_0) \leq c p_0^{-b}$$

with some fixed $b, c \in (0,\infty)$. (Q_0 and p_0 should not be confused with Q and p in the proof of Theorem 10.1.)

Reason. Put $n := \operatorname{card} J$ and $p := p(n) := [n^z]$ with the same z as in the proof of Theorem 10.1. As there cover J with cubes of length $2p$ and let Q be the numbers of the cubes necessary for this. Only for (10.12) we used the special shape of J. Instead of this we now argue as follows: In the case $p \leq p_0$ we have $\operatorname{card} \cup_{\lambda=1}^{Q} \bar{H}_\lambda \leq cQ_0 p_0^q = c \operatorname{card} J$, hence in this case (10.12) may be replaced by the stronger inequality

(10.26) $$Q \leq cn/p^q .$$

Now let $p > p_0$. Then we have card $\cup_{\lambda=1}^{Q} H_\lambda \leq cQ_0 p^q$, hence $Q \leq cQ_0$. Then in (10.13) we have

(10.27) $$(Q-1)\beta(p) \leq cQ_0\beta(p) \leq cQ_0\beta(p_0) \leq cp_0^{-b} < \frac{\epsilon}{\log n}$$

for p_0 sufficiently large since by (10.25),

(10.28) $$n = \text{card } J = Q_0 p_0^q \leq cp_0^{-b}\, p_0^q\, \beta(p_0)^{-1}.$$

This inequality also tells us that card $J = n \to \infty$ implies $p_0 \to \infty$. Hence we have that (10.27) holds for all sufficiently large n, i.e. for all J (of the described shape) with card $J \geq n_0''(\epsilon)$ for some suitable $n_0''(\epsilon)$ which obviously can be chosen to be a polynomial in $\frac{1}{\epsilon}$. So use (10.27) instead of (10.14). No other changes are necessary in the proof of Theorem 10.1. □

Proof of Theorems 6 and 7. Assume the hypotheses of Theorem 6 (resp. 7) to be satisfied. For $m \in \mathbb{N}$ put $\epsilon := 100^{-q} m^{-1}$. By Theorem 10.1 and Remark 10.2 (for Theorem 7 with $\tau := \frac{1}{2}(1 - \frac{2q}{s'+1}) > 0$ by (3.6)) we have as in the proof of Theorem 7.1 of [DuPh] and with the same mappings Λ_m as there that the hypotheses of Theorem 4 (resp. Theorem 5) are satisfied with $N_0(m) \leq n_0(\epsilon) \leq cm^L$ and $\dim \Lambda_m S \leq N_I(\delta(\epsilon), \mathcal{G}, P') \leq \exp(c\delta^{\tau}) < \exp(cm^{K\tau})$ in the case of Theorem 6 resp. $\dim \Lambda_m S \leq c\delta^{-H} \leq cm^{KH}$ in the case of Theorem 7. It is easily seen that the Λ_m are linear with $\|\Lambda_m\| = 1$. Hence Theorem 6 follows from Theorem 4; Theorem 7 from Theorem 5. (Here (2.3) is satisfied for arbitrarily large δ.) (3.14) can be deduced from (2.14) by noting that in (2.14) the sum over j converges uniformly in m, so that there the limit over m and the sum over j may be interchanged. The assertion on the almost sure continuity of the paths of the Y_j follows from Lemma 1.4 of [DuPh], which is valid here, too (with obvious changes). This is easily seen by an inspection of its proof. □

Acknowledgments. This work is based on the author's Ph.D. thesis. I am grateful to Professors Ernst Eberlein and Walter Philipp for their constant encouragement and advice. I would also like to thank Professors Charles Goldie and Gregory Morrow for some useful discussions.

References

[AlPy] Alexander, K., Pyke, R.: A uniform central limit theorem for set-indexed partial-sum processes with finite variance. Ann. Probab. **14**, 582 – 597 (1986)

[An] Andersen, N.T.: The calculus of non–measurable functions and sets. Aarhus Universitet. Matematisk Institut. Various publications series 36 (1985)

[Ba] Bass, R.F.: Law of the iterated logarithm for set–indexed partial sum processes with finite variance. Z. Wahrscheinlichkeitstheorie verw. Geb. **70**, 591 – 608 (1985)

[BaPy] Bass, R.F., Pyke, R.: Functional law of the iterated logarithm and uniform central limit theorem for partial sum processes indexed by sets. Ann. Probab. **12**, 13 – 34 (1984)

[Bn] Bennett, G.: Probability inequalities for the sum of independent random variables. J. Am. Stat. Assoc. **57**, , 33 – 45 (1962)

[Bk] Berkes, I.: Gaussian approximation of mixing random fields. Acta Math. Hung. **43**, 153 – 185 (1984)

[BM] Berkes, I., Morrow, G.: Strong invariance principles for mixing random fields. Z. Wahrscheinlichkeitstheorie verw. Geb. **57**, 15 – 37 (1981)

[BPh] Berkes, I., Philipp, W.: Approximation theorems for independent and weakly dependent random vectors. Ann. Probab. **7**, 29 – 54 (1979)

[Br] Bradley, R. C.: A caution on mixing conditions for random fields. Preprint 1988

[De] Dehling, H.: Limit theorems for sums of weakly dependent Banach space valued random vectors. Z. Wahrscheinlichkeitstheorie verw. Geb. **63**, 393 – 432 (1983)

[DePh] Dehling, H., Philipp, W.: Almost sure invariance principles for weakly dependent vector valued random variables. Ann. Probab. **10**, 689 – 701 (1982)

[Dh] Dhompongsa, S.: A note on the almost sure approximation of the empirical process of weakly dependent random vectors. Yokohama Math. J. **32**, 113 – 121 (1984)

[Du78] Dudley, R.M.: Central limit theorems for empirical measures. Ann. Probab. **6**, 899 – 929 (1978). Corrections, ibid. **7**, 909 – 911 (1979)

[Du84] Dudley, R.M.: A course on empirical processes. In: Hennequin, P.L. (ed.) Ecole d'Eté de probabilités de Saint–Flour XII–1982. (Lect. Notes Math., vol. 1097, pp. 1 – 142) Berlin: Springer 1984

[DuPh] Dudley, R.M., Philipp, W.: Invariance principles for sums of Banach space valued random elements and empirical processes. Z. Wahrscheinlichkeitstheorie verw. Geb. **62**, 509 – 552 (1983)

[Eb] Eberlein, E.: Weak convergence of partial sums of absolutely regular sequences. Statist. Probab. Lett. **2**, 291 –293 (1984)

[Fe] Fernique, X.: Intégrabilité des vecteurs gaussiens. C. R. Acad. Sci., Paris, Sér. A **270**, 1698 – 1699 (1970)

[Gä] Gaenssler, P.: Limit theorems for empirical processes indexed by classes of sets allowing a finite–dimensional parametrization. Probab. Math. Statist. (Wroclaw) **4**, 1 – 12 (1984)

[GG] Goldie, C.M., Greenwood, P.E.: Variance of set–indexed sums of mixing random variables and weak convergence of set–indexed processes. Ann. Probab. **14**, 817 – 839 (1986)

[GM] Goldie, C.M., Morrow, G.J.: Central limit questions for random fields. In: Eberlein, E., Taqqu, M.S. (eds.) Dependence in probability and statistics. Proceedings, Oberwolfach 1985, pp. 275 – 289. Boston: Birkhäuser 1986

[GR] Guyon, X., Richardson, S.: Vitesse de convergence du théorème de la limite centrale pour des champs faiblements dépendants. Z. Wahrscheinlichkeitstheorie verw. Geb. **66**, 297 – 314 (1984)

[Jo] John, F.: Extremum problems with inequalities as subsidiary conditions. Courant Anniversary Volume, 187 – 204 (1948) (= John, F.: Collected papers, vol. II, pp. 543 – 560. Boston: Birkhäuser 1985)

[KPh] Kuelbs, J., Philipp, W.: Almost sure invariance principles for partial sums of mixing B–valued random variables. Ann. Probab. **8**, 1003 – 1036 (1980)

[Le] Lekkerkerker, G.C.: Geometry of numbers. Groningen: Wolters-Noordhoff, and Amsterdam: North-Holland 1969

[MPh] Morrow, G.J., Philipp, W.: Invariance principles for partial sum processes and empirical processes indexed by sets. Probab. Th. Rel. Fields **73**, 11 – 42 (1986)

[Pl] Pachl, J.K.: Two classes of measures. Colloq. Math. **42**, 331 – 340 (1979)

[Pt] Parthasarathy, K.R.: Probability measures on metric spaces. New York: Academic Press (1967)

[Ph77] Philipp, W.: A functional law of the iterated logarithm for empirical distribution functions of weakly dependent random variables. Ann. Probab. **5**, 319 – 350 (1977)

[Ph84] Philipp, W.: Invariance principles for sums of mixing random elements and the multivariate empirical process. In: Révész, P. (ed.) Limit theorems in probability and statistics. Proceedings, Veszprém 1982. (Colloq. Math. Soc. János Bolyai, vol. 36, pp. 843 – 873) Amsterdam: North–Holland 1984

[Ph86] Philipp, W.: Invariance principles for independent and weakly dependent random variables. In: Eberlein, E., Taqqu, M.S. (eds.) Dependence in probability and statistics. Proceedings, Oberwolfach 1985, pp. 225 – 268. Boston: Birkhäuser 1986

[Py] Pyke, R.: A uniform central limit theorem for partial–sum processes indexed by sets. In: Kingman, J.F.C., Reuter, G.E.H. (eds.) Probability, statistics and analysis. (London Math. Soc. Lecture Note Series, vol. 79, pp. 219 – 240) Cambridge: Cambridge University Press 1983

[SG] Sotres, D.A., Ghosh, M.: Strong convergence of linear rank statistics for mixing processes. Sankhya Ser. B 39, 1 – 11 (1977)

[Se] Serfling, R.J.: Contributions to central limit theory for dependent variables. Ann. Math. Stat. **39**, 1158 – 1175 (1968)

[St] Stout, W.F.: Almost sure convergence. New York: Academic Press (1974)

[Str] Strittmatter, W.: Starke Approximationen für durch Mengen indizierte Partialsummenprozesse und empirische Prozesse von mischenden random fields. Dissertation, Universität Freiburg i. Br. 1986

[VR] Volkonskii, V.A., Rozanov, Yu.A.: Some limit theorems for random functions II. Theory Probab. Appl. 6, 186 – 198 (1961)

The Law of The Iterated Logarithm for Subsequences in Banach Spaces

MICHEL WEBER

1. Introduction—Main Result

In [W$_2$] and [W$_3$], we have obtained a complete description of the law of the iterated logarithm (LIL) for subsequences in Euclidean spaces. This concludes previous works in [G], [To] and [W$_1$]. Our purpose in this paper will be to give a partial extension of this result in arbitrary separable Banach spaces. We will use a procedure recently developed by Ledoux and Talagrand in [LT$_2$], that is an efficient tool for such question. In order to set concretely the problem which interests us, we need to recall some definitions used in [W$_2$] and [W$_3$]. Let $\mathfrak{s}$ be any increasing infinite sequence of positive integers and $M > 1$. Consider the intervals of type $I_n(M) = [M^n, M^{n+1}[$, $n \geq 1$, with $I_0(M) = [0, M[$, which have a nonempty intersection with $\mathfrak{s}$. They define an increasing subsequence that we denote

$$(1.1) \qquad \forall p \geq 1, \qquad I_{k_p}(M) = I_{k_p}(\mathfrak{s}, M).$$

We then define for each element n of $\mathfrak{s}$,

$$\varphi(n) = \varphi(\mathfrak{s}, M, n) = \begin{cases} 1 & \text{if} \quad n \in I_{k_1}, \\ \sqrt{2 \log p} & \text{if} \quad n \in I_{k_p}, \quad p \geq 2. \end{cases}$$

We define in that way a *triple* $(\mathfrak{s}, M, \varphi)$. We recall, for the convenience of the reader, the characterization of the LIL for subsequences in Euclidean spaces.

Theorem 1.1 ([W$_3$], Theorem 2.1). *Let $(\mathfrak{s}, M, \varphi)$ be any triple, and consider a random variable X with values in a m-dimensional Euclidean space $\mathbb{R}^m$. Let $X_1, X_2, \ldots$ be a sequence of independent copies of X and set*

(1.3) $$\forall n \geq 1, \qquad S_n(X) = X_1 + \cdots + X_n.$$

Let B_m be the closed Euclidean unit ball of $\mathbb{R}^m$. Then,

(1.4) $$P\left\{C\left(\left\{\frac{S_n(x)}{\sqrt{n}\varphi(n)}, n \in \mathfrak{s}\right\}\right) = B_m\right\} = 1,$$

if and only if

(1.5) Cov(X) *is the identity matrix of* $\mathbb{R}^m$ *and* $E(X)$ *the zero vector of* $\mathbb{R}^m$,

where in (1.4), $C(\{a_n\})$ *denotes the set of cluster points in* $\mathbb{R}^m$ *of the sequence* $\{a_n\}$.

The justification of the construction in (1.2) lies in the fact that it allows us to get a characterization of the LIL for subsequences. In particular, the LIL for subsequences on the real line may be expressed as follows:

(1.4′) $$P\left\{C\left(\left\{\frac{S_n(X)}{\sqrt{n}\varphi(n)}, n \in \mathfrak{s}\right\}\right) = [-1,1]\right\} = 1,$$

if and only if

(1.5′) $$E(X) = 0 \qquad \text{and} \qquad E(X^2) = 1.$$

Since there are easy examples of sequences $\mathfrak{s}$ and real random variables X such that

(a) $$E(X) = 0 \qquad \text{and} \qquad E(X^2) = \infty,$$

and

(b) $$P\left\{\limsup_{\mathfrak{s}\ni n\to\infty} \frac{|S_n(X)|}{\sqrt{2n\log\log n}} < \infty\right\} = 1,$$

one can conclude that it is an adequate approach for describing the LIL for subsequences.

Now let B be a real separable Banach space with topological dual B^*. Let X be a random variable with values in B; and consider $X_1, X_2, \ldots$ any sequence of independent copies of X. We set again $S_n(X) = X_1 + \cdots + X_n$, $n \geq 1$. Consider also any increasing sequence $\mathfrak{s}$ of positive integers. In the sequel, it will be convenient to fix the value of M in (1.1) and (1.2), say

$M = e$, although it is not needed for proving our main result. We then define, according to (1.2),

$$\forall n \in \mathfrak{s}, \qquad a(n) = \sqrt{n}\varphi(\mathfrak{s}, e, n). \tag{1.6}$$

In particular, $a(n) = \sqrt{2n \log\log n}$, $n \geq 3$, when $\mathfrak{s} = \mathbb{N}$. We say that the random variable X satisfies the bounded, (resp. compact), law of the iterated logarithm (BLIL), (resp. CLIL), for the sequence $\mathfrak{s}$, if for any sequence $X_1, X_2, \dots$ of independent copies of X, $\{S_n(X)/a(n), n \in \mathfrak{s}\}$ is bounded in B almost surely, (resp. relatively compact in B almost surely). To the sequence $\mathfrak{s}$, we associate the following subsequence,

$$\mathfrak{s}^* = \{m_p = \sup\{\mathfrak{s} \cap I_{k_p}(\mathfrak{s}, e), p \geq 1\}\}, \tag{1.7}$$

and we set,

$$\forall p \geq 1, \qquad a_p = a(m_p). \tag{1.7'}$$

When $\mathfrak{s} = \mathbb{N}$, the LIL property has been recently characterized by Ledoux and Talagrand [LT$_1$], who reduce the problem from one of the almost sure behavior to one of the in-probability behavior. More explicitly, one has

Theorem 1.2 ([LT$_1$], Theorem 2.1). (a) (BLIL) *X satisfies the* bounded LIL *if and only if the following three conditions hold*

(1.8) $\quad E(\|X\|^2/\log\log\|X\|) < \infty,$

(1.9) $\quad$ *for each x^* in B^*, $E(\langle x^*, X\rangle^2) < \infty$,*

(1.10) $\quad$ *$\{S_n(X)/a(n), n = 1, 2, \dots,\}$ is bounded in B in probability.*

(b) (CLIL) *X satisfies the* compact LIL *if and only if the following three conditions hold*

(1.8) $\quad E(\|X\|^2/\log\log\|X\|) < \infty,$

(1.11) $\quad$ *$\{\langle x^*, X\rangle^2, x^* \in B^*, \|x\| \leq 1\}$ is uniformly integrable,*

(1.12) $\quad$ *$S_n(X)/a(n) \underset{n\to\infty}{\longrightarrow}$ in probability.*

The main feature of the LIL for subsequences is that condition (1.8) is no longer necessary when $\mathfrak{s}$ is arbitrary. In that case one has instead of (1.8), with the notations (1.7) and (1.7′),

$$\exists \varepsilon < \infty : \sum_{p \geq 1} m_p P\{\|x\| > \varepsilon a_p\} < \infty, \tag{1.13}$$

which reduces to (1.8) when the sequence $\mathfrak{s}$ grows " at most geometrically ", that is to say, when $m_{p+1} \leq Cm_p$, where $C < \infty$ is some constant. Note that (1.13) is easily expressible as an expectation. Indeed, letting

$$(1.14) \qquad P_{\mathfrak{s}}(x) = \begin{cases} 0 & \text{if} \quad 0 \leq x < a_1, \\ \sum_{i=1}^{q} m_i & \text{if} \quad a_q \leq x < a_{q+1}, \quad q \geq 1, \end{cases}$$

then, (1.13) is equivalent to

$$(1.13') \qquad E(P_{\mathfrak{s}}(\varepsilon\|X\|)) < \infty \qquad \text{for some} \qquad \varepsilon > 0.$$

Taking account of the remarks following Theorem 1.1, it becomes clear that one need to use an intrinsic sequence of normalization constants, defined in (1.6), for the partial sums, instead of the usual sequence, namely, $\alpha_n = \sqrt{2n \log\log n}$, $n \geq 3$, in order to obtain the best possible conditions for the LIL for subsequences to hold. Indeed, otherwise the one-dimensional conditions (1.9) and (1.10) are no longer necessary. Before stating our result, we define

$$(1.15) \qquad \Phi_{\mathfrak{s}}(x) = \begin{cases} 1 & \text{if} \quad 0 \leq x < a_1, \\ \log p & \text{if} \quad a_p \leq x < a_{p+1}, \quad p \geq 1. \end{cases}$$

It is plain that

$$\limsup_{x \to \infty} \frac{\Phi_{\mathfrak{s}}(x)}{\log\log x} \leq 1,$$

and

$$(1.16) \qquad \limsup_{x \to \infty} \frac{\Phi_{\mathfrak{s}}(x)}{\log\log x} > 0,$$

if and only if

$$(1.17) \qquad \Lambda(\mathfrak{s}) > 0,$$

where

$$\Lambda^2(\mathfrak{s}) = \limsup_{j \to \infty} \frac{\log \#\{i \leq j : \mathfrak{s} \cap [e^i, e^{i+1}[\neq \emptyset\}}{\log j}.$$

We recall that (1.17) precisely characterizes the LIL behavior ([W3], Theorem 2.3). We also define,

(1.18)

$$\forall\, x \geq 0, \; \Gamma(x) = \frac{x^2 P\{\|x\| > x\}}{\Phi_{\mathfrak{s}}(x)}, \qquad \text{and} \quad \Gamma^*(x) = \sup\{\Gamma(y), y \geq x\}.$$

We note that these functions depend on X and $\mathfrak{s}$.

Theorem 1.3. (a) (BLIL for subsequences). *X satisfies the bounded* LIL *relatively to the sequence $\mathfrak{s}$ whenever the following three conditions are fulfilled:*

(1.19) $\sum_{p\geq 1}\Gamma^*(a_p) < \infty$,

(1.9) $E(\langle x^*, X\rangle^2) < \infty$, *for each x^* in B^*,*

(1.20) $\{S_n(X)/a(n), n \in \mathfrak{s}\}$ *is bounded in B in probability.*

(b) (CLIL for subsequences). *X satisfies the compact* LIL *relatively to the sequence $\mathfrak{s}$ whenever the following three conditions are fulfilled:*

(1.19) $\sum_{p\geq 1}\Gamma^*(a_p) < \infty$,

(1.10) $\{\langle x^*, X\rangle^2, x^* \in B^*, \|x^*\| \leq 1\}$ *is uniformly integrable,*

(1.21) $\{S_n(X)/a(n), n \in \mathfrak{s}\}$ *tends to zero in B in probability.*

Remarks. (1) Apart from condition (1.19), the other sufficient conditions are obviously necessary. We thus will comment on condition (1.19) only. It is strictly stronger than condition (1.13), as one can see by means of the following example: let X be such that

$$\Gamma(a_p) = \begin{cases} k^{-2} & \text{if} \quad p = 2^k, \quad k = 1, 2, \dots, \\ \omega_p & \text{otherwise,} \end{cases}$$

where $\{\omega_p, p \geq 1\} \in l_1$, and $\sup\{\omega_p, 2^k < p < 2^{k+1}\} = o(k^{-2})$, $k \to \infty$. Then (1.13) holds, but (1.19) does not hold since $\Gamma^*(a_p) \geq (k+1)^{-2}$ for each p in $]2^k, 2^{k+1}[$, $k \geq 1$.

(2) Condition (1.19) also implies

$$\forall \varepsilon > 0, \quad \sum_{p\geq 1}\Gamma^*(\varepsilon a_p) < \infty, \text{ and } \Gamma^*(a_p) \leq C/a_p, p \geq 1, \text{ for some } C < \infty.$$

(3) If $\Gamma(\cdot)$ is nonincreasing, then (1.13) and (1.19) are equivalent. As a typical case, one has the following partial characterization.

Theorem 1.4. *Assume that*

(1.22) $x^2 P\{\|X\| > x\}$ *is nonincreasing for large values of x.*

(a) (BLIL for subsequences). *X satisfies the bounded* LIL *relatively to the sequence $\mathfrak{s}$, if and only if the conditions* (1.9), (1.13) *and* (1.20) *are fulfilled.*

(b) (CLIL for subsequences). *X satisfies the compact* LIL *relatively to the sequence $\mathfrak{s}$, if and only if the conditions* (1.10), (1.13) *and* (1.21) *are fulfilled.*

2. Auxiliary Results

We use the notations introduced in the previous section. We collect here for the convenience of the reader some essential lemmas. Let B be a real separable Banach space and $X_1, X_2, \dots$ a sequence of independent, identically distributed random variables with values in B.

Lemma 2.1 ([Ta$_2$]). *Let N be any integer and let $A \subset B^N$ be such that*

$$P\{(X_i)_{i \le N} \in A\} \ge 1/2.$$

Set for any k, q integers

$$H(A,k,q) = \{x \in B^N : \exists\, x^1, \dots, x^q \text{ in } A : \#\{i \le N : x_i \notin \{x_i^1, \dots, x_i^q\}\} \le k\}. \tag{2.1}$$

Then, for $k \ge q$,

$$P^*\{(X_i)_{i \le N} \notin H(A,k,q)\} \le (K/q)^k, \tag{2.2}$$

where K is some universal constant, and P^* denotes the outer probability measure induced by $P(\cdot)$.

Let now $\varepsilon_1, \varepsilon_2, \dots$ be a Rademacher sequence.

Lemma 2.2 ([LT$_2$]). *Let $x_1, \dots, x_N$ be elements of B. Then*

$$E\Big(\sup\Big\{\Big|\sum_{i=1}^N \varepsilon_i \langle x^*, x_i\rangle^2\Big|,\ \|x^*\| \le 1\Big\}\Big) \le \alpha\Big(\sup_{i=1}^N \|x_i\|\Big) E\Big(\Big\|\sum_{i=1}^N \varepsilon_i x_i\Big\|\Big), \tag{2.3}$$

where α is some universal constant.

Lemma 2.3 ([Ta$_1$]). *Let $x_1, \dots, x_N$ be in B and set*

$$\sigma^2 = \sup\Big\{\sum_{i=1}^N \langle x^*, x_i\rangle^2, \|x^*\| \le 1\Big\}.$$

Then, for every $t \geq 0$,

$$(2.4) \qquad P\left\{\left\|\sum_{i=1}^{N} \varepsilon_i x_i\right\| \geq 2E\left(\left\|\sum_{i=1}^{N} \varepsilon_i x_i\right\|\right) + t\right\} \leq 4\exp(-t^2/8\sigma^2).$$

Lemma 2.4 ([A$_1$]). *Let us assume that the* X_i*'s are symmetric and* $\|X_i\| \leq c < \infty$, a.s., $i = 1, \ldots, N$, *and set* $S_N = X_1 + \cdots + X_N$. *Then, for all* $\lambda \geq 0$, $t \geq 0$

$$(2.5) \qquad \begin{aligned} E(\exp\{\lambda\|S_N\|\}) \leq 2\exp\{\lambda(t+c)\}P\{\|S_N\| \geq t\} \\ E(\exp\{\lambda\|S_N\|\}) + \exp\{\lambda t\}. \end{aligned}$$

The last lemma is nothing but a version of the strong symmetrization inequalities and can be proved in the same way (see [St], Lemma 3.2.1).

Lemma 2.5. *Let* $\{T_n, n \geq 1\}$ *be a sequence of* B*-valued random variables and* $\{\mu_n, n \geq 1\} \subset \mathbb{R}^+$. *Then, for any* $\varepsilon > 0$,

$$(2.6) \qquad P\left\{\sup(\|T_n\| - \mu_n, n \geq 1) \geq \varepsilon\right\} \leq \frac{P\{\sup(\|T_n^s\|, n \geq 1) \geq \varepsilon\}}{\inf(P\{\|T_n\| \leq \mu_n\}, n \geq 1)},$$

where $\{T_n^s, n \geq 1\}$ *is a symmetrized version.*

3. Proofs

This section consists of three short subsections. In the two first subsections we examine the necessary conditions and the sufficient conditions for the LIL for subsequences to hold. In the last subsection, we prove Theorem 1.3 and 1.4. Once again X will denote a random variable with values in a separable Banach space B; $X_1, X_2, \ldots$ is a sequence of independent copies of X and $\mathfrak{s}$ is an arbitrary increasing sequence of positive integers.

3.1. *Necessary conditions.*

Lemma 3.1. *Assume that* $P\{\sup(\|S_n(X)\|/a(n), n \in \mathfrak{s}) < \infty\} = 1$. *Then* (1.13) *holds.*

Proof. Let us first assume that X is symmetric. For any $p \geq 1$

$$m_{2(p-1)}/m_{2p} \leq 2^{k_{2(p-1)} - k_{2p} + 1} \leq 1/2.$$

For M large enough,

$$P\Big\{\sup\Big((a_p)^{-1}\|S_{m_{2p}}(X)-S_{m_{2(p-1)}}(X)\|, p\geq 1\Big)\leq M\Big\}>0,$$

and, by the Borel–Cantelli lemma and Levy's maximal inequality,

$$\sum_{p\geq 1}P\Big\{\max\Big(\|X_i\|, i\in\Big[\frac{1}{2}m_{2p}, m_{2p}\Big]\Big)>2Ma_p\Big\}<\infty.$$

Since $X\in B$ a.s., this implies,

$$\sum_{k=0\,\mathrm{mod}(2)} m_k P\{\|X\|>2Ma_k\}<\infty.$$

Similarly,

$$\sum_{k=1\,\mathrm{mod}(2)} m_k P\{\|X\|>2Ma_k\}<\infty.$$

If X is not symmetric, let X^s be a symmetrized version of X and t a median of $\|X\|$. The above part and the weak symmetrization inequalities imply,

$$\sum_{p\geq 1} m_p P\{\|X\|>2Ma_p+t\}<\infty.$$

This gives the conclusion since the sequence $\{a_n\}_{n\geq 1}$ is unbounded. QED

3.2. *Sufficient conditions.* We prove in this part, a proposition that will play a key role for proving the sufficiency part of Theorem 1.3 and 1.4. We set for $n\geq r\geq 1$, $X_n^{(r)}=X_i$, whenever $\|X_i\|$ is the r-th maximum of the sample $\|X_1\|,\ldots,\|X_n\|$.

Proposition 3.2. *Assume that X is symmetric and,*

$$\sum_{p\geq 1}P\Big\{\sum_{i=1}^{c_p}\|X_{m_p}^{(i)}\|>\lambda a_p\Big\}<\infty, \tag{3.1}$$

for some real λ and some sequence of integers $\{c_p, p\geq 1\}$ satisfying $1\leq c_p\leq m_p$, and $\sum_{p\geq 1}2^{-c_p}<\infty$,

$$\sigma^2=\sup\{E(\langle x^*, X\rangle^2), \|x^*\|\leq 1\}<\infty, \tag{3.2}$$

(3.3) $\quad \sup\{P\{\|S_n(X)\| > \theta a(n)\}, n \in \mathfrak{s}\} \leq 1/8e,$ *for some real* θ.

Let $R_0 = \sqrt{e}\{2\lambda + 20K\theta + 16\sqrt{K}\max(\sigma, 2\theta\sqrt{\alpha})\}$, *where* K *and* α *are constants arising from Lemma 2.1 and 2.2 respectively. Then, for all positive integers* p,

$$P\Big\{\sup_{n\in\mathfrak{s}\cap I_{k_p}} \{\|S_n(X)\|/a(n)\} > R_0\Big\} \leq P\Big\{\sum_{i=1}^{c_p} \|X_{m_p}^{(i)}\| > \lambda a_p\Big\} + 5\max(p^{-2}, 2^{-c_p}). \tag{3.4}$$

Moreover, X *satisfies the bounded LIL relatively to the sequence* $\mathfrak{s}$.

Proof. We note $X^{(\delta)} = XI\{\|X\| \leq \delta\}$, $X_{(\delta)} = X - X^{(\delta)}$, $\delta > 0$. Let $\varepsilon_1, \varepsilon_2, \ldots$ be a Rademacher sequence defined on another probability space $(\Omega_\varepsilon, \mathcal{A}_\varepsilon, P_\varepsilon)$ with E_ε as the symbol of integration. A plain application of the triangular inequality, the symmetry assumption and (3.3) also show,

$$\sup\{P\{\|S_n(X^{(\delta)})\| > \theta a(n)\}, n \in \mathfrak{s}\} \leq 1/4e. \tag{3.3'}$$

Applying this with Lemma 2.4 for $c = 1$, $Y_i = \delta^{-1}X_i^{(\delta)}$, $i = 1, \ldots, m_p$, $t = \theta a_p/\delta$, $\lambda = 1/(t+1)$, yield

$$\sup\Big\{E\Big(\exp\Big\{\frac{\|S_n(X^{(\delta)})\|}{(\theta a(n) + t)}\Big\}\Big), n \in \mathfrak{s}\Big\} \leq 2e. \tag{3.5}$$

Fix now an integer p and set for $i = 1, \ldots, m_p$

$$U_i = X_i^{(\theta a_p/c_p)},$$

$$A_1^p = \Big\{(x_i) \in B^{m_p}: \sup\Big\{\sum_{i=1}^{m_p} \langle x^*, x_i I_{\{\|x_i\| \leq \theta a_p/c_p\}}\rangle^2, \|x^*\| \leq 1\Big\} \leq 4m_p\sigma^2 + 32\alpha\theta^2 a_p^2/c_p\Big\}$$

$$A_2^p = \Big\{(x_i) \in B^{m_p}: E_\varepsilon\Big(\Big\|\sum_{i=1}^{m_p} \varepsilon_i x_i\Big\|\Big) \leq 10\theta a_p\Big\},$$

$$A^p = A_1^p \cap A_2^p,$$

$$H^p = H(A^p, c_p, 2K) = \{(x_i) \in B^{m_p}: \exists\, x^1, \ldots, x^{2K} \in A^p: \#\{i \leq m_p: x_i \notin \{x_i^1, \ldots, x_i^{2K}\} \leq c_p\}\},$$

where the constant K comes from Lemma 2.1,

$$B_p = \{(X_i)_{i\le m_p} \in H^p\},$$

$$C_p = \left\{\sum_{i=1}^{c_p} \|X_{m_p}^{(i)}\| \le \lambda a_p\right\}.$$

Clearly, for any $R \ge 0$,

$$\begin{aligned}P\{\sup\{\|S_n(\varepsilon X)\|, n \in \mathfrak{s} \cap I_{k_p}\} > Ra_p\} &\le P\{B_p^c\} + P\{C_p^c\}\\ &+ P\{B_p \cap C_p \cap \{\sup\{\|S_n(\varepsilon X(\|, n \in \mathfrak{s} \cap I_{k_p}\}\\ &> Ra_p\}\}.\end{aligned} \tag{3.6}$$

We first show $P\{(X_i)_{i\le m_p} \in A^p\} \ge 1/2$. Let $\{U_i', i \le m_p\}$ be an independent copy of the sequence $\{U_i, i \le m_p\}$, also independent of the sequence $\{\varepsilon_i, i \le m_p\}$. By recentering,

$$\begin{aligned}E&\left(\sup_{\|x^*\|\le 1}\left\{\sum_{i=1}^{m_p}\langle x^*, U_i\rangle^2\right\}\right)\\
&\le m_p\sigma^2 + E\left(\sup_{\|x^*\|\le 1}\left|\sum_{i=1}^{m_p}\langle x^*, U_i\rangle^2 - E(\langle x^*, U_i'\rangle^2)\right|\right),\\
&\le m_p\sigma^2 + E\left(\sup_{\|x^*\|\le 1}\left|\sum_{i=1}^{m_p}\langle x^*, U_i\rangle^2 - \langle x^*, U_i'\rangle^2\right|\right),\\
&= m_p\sigma^2 + E\left(\sup_{\|x^*\|\le 1}\left|\sum_{i=1}^{m_p}\varepsilon_i(\langle x^*, U_i\rangle^2 - \langle x^*, U_i'\rangle^2)\right|\right),\\
&\le m_p\sigma^2 + 2E\left(\sup_{\|x^*\|\le 1}\left|\sum_{i=1}^{m_p}\varepsilon_i\langle x^*, U_i\rangle^2\right|\right),\end{aligned}$$

(by Lemma 2.2)

$$\le m_p\sigma^2 + 2\alpha\theta a_p(c_p)^{-1}E\left(\left\|\sum_{i=1}^{m_p}\varepsilon_i U_i\right\|\right),$$

(by (3.5) for $\delta = \theta a_p/c_p$)

$$\le m_p\sigma^2 + 8\alpha\theta^2 a_p^2/c_p.$$

Therefore,

$$P\left\{\sup_{\|x^*\|\le 1}\left(\sum_{i=1}^{m_p}\langle x^*, U_i\rangle^2\right) > 4m_p\sigma^2 + 32\alpha\theta^2 a_p^2/c_p\right\} \le 1/4. \tag{3.7}$$

Moreover, by assumption (1.20) and Levy's maximal inequality

$$\sup\{\inf(1, m_p P\{\|X\| > \theta a_p\}), p \geq 1\} \leq 1/5, \tag{3.8}$$

and thus,

$$\begin{aligned} P\{E_\varepsilon(\|S_{m_p}(\varepsilon X)\|) > 10\theta a_p\} &\leq 1/5 + P\{E_\varepsilon(\|S_{m_p}(\varepsilon X^{(\theta a_p)})\|) > 10\theta a_p\}, \\ &\leq 1/5 + 1/20 = 1/4, \end{aligned}$$

by (3.5) with $\delta = \theta a_p$, and Bernstein inequality.

Therefore, we indeed have that $P\{(X_i)_{i \leq m_p} \in A^p\} \geq 1/2$. By Lemma 2.1,

$$P\{B_p^c\} \leq 2^{-c_p}. \tag{3.10}$$

And, on $B_p \cap C_p$, there exist $x^1, \ldots, x^{2K}$ in A^p and integers $\{i_1, \ldots, i_j\}$ with $j \leq c_p$ such that,

$$\{1, \ldots, m_p\} = \{i_1, \ldots, i_j\} \cup I$$

and

$$I = \bigcup_{v=1}^{2K} I_v \qquad \text{where} \qquad I_v = \{i \leq m_p : X_i = x_i^v\},$$

and, at most $c_p - 1$ X_i's are such that $\|X_i\| > \lambda a_p / c_p$. Thus, for every $m \leq m_p$,

$$\left\|\sum_{i=1}^m \varepsilon_i X_i\right\| \leq 2\lambda a_p + \left\|\sum_{i \in I} \varepsilon_i U_i\right\|,$$

and, by using Levy's maximal inequality for Rademacher averages:

$$\forall a > 0, \quad P_\varepsilon\left\{\sup_{m \in \mathfrak{s} \cap I_{k_p}} \left\|\sum_{\substack{i \leq m \\ i \in I}} \varepsilon_i U_i\right\| > a\right\} \leq 2P_\varepsilon\left\{\left\|\sum_{i \in I} \varepsilon_i U_i\right\| > a\right\},$$

$$\begin{aligned} P\left\{\sup_{m \in \mathfrak{s} \cap I_{k_p}} \left\|\sum_{i=1}^m \varepsilon_i X_i\right\| > R a_p\right\} &\leq 2^{-c_p} + P\{C_p^c\} \\ &+ 2\sup\left\{\int_F P_\varepsilon\left\{\left\|\sum_{i \in I} \varepsilon_i U_i\right\| > (R - 2\lambda a_p)\right\} dP\right\}, \end{aligned} \tag{3.11}$$

where the sup is extended over all measurable sets F such that $F \subset B_p \cap C_p$. By monotonicity of the Rademacher averages and by definition of I and A^p,

$$(3.12)\quad E_\varepsilon\left(\left\|\sum_{i\in I}\varepsilon_i U_i\right\|\right) \le \sum_{v=1}^{2K} E_\varepsilon\left(\left\|\sum_{i=1}^{m_p}\varepsilon_i x_i^v I_{\{\|x_i^v\|\le\theta a_p/c_p\}}\right\|\right) \le 20K\theta a_p.$$

Similarly,

$$(3.12')\qquad \sup\left\{\sum_{i\in I}\langle x^*, U_i\rangle^2, \|x^*\|\le 1\right\} \le 8K(m_p\sigma^2 + 8\theta^2\alpha a_p^2/c_p).$$

Thus, by Lemma 2.3,

$$P_\varepsilon\left\{\left\|\sum_{i\in I}\varepsilon_i U_i\right\| > (R-2\lambda)a_p\right\} \le P_\varepsilon\left\{\left\|\sum_{i\in I}\varepsilon_i U_i\right\| > 2E_\varepsilon\left(\left\|\sum_{i\in I}\varepsilon_i U_i\right\|\right)\right.$$

$$\left. + (R-2\lambda-20K\theta)a_p\right\},$$

$$(3.13)\qquad \le 4\exp\left\{-\frac{(R-2\lambda-20K\theta)^2 a_p^2}{64(m_p\sigma^2+8\alpha\theta^2 a_p^2/c_p)}\right\}.$$

A simple calculation finally shows, for every p, letting $R = R_0/\sqrt{e}$,

$$P\{\sup_{n\in\mathfrak{s}\cap I_{k_p}} \|(S_n(\varepsilon X)\|/a(n)) > R_0\} \le P\left\{\sum_{i=1}^{m_p}\|X_{m_p}^{(i)}\| > \lambda a_p\right\}$$

$$(3.14)\qquad + 5sup[p^{-2}, 2^{-c_p}].\qquad \text{QED}$$

3.3. Proofs of Theorems 1.3 and 1.4

3.3.1. *Proof of theorem* 1.3. (a) (BLIL). Let Z be a symmetrized version of X. We are going to apply Proposition 3.2 to Z. It only remains to show that (1.19) implies that (3.1) is satisfied. We choose $c_p = [2\log p]+1$, $p \ge 1$. First, one has the easy bound,

$$P\left\{\sum_{i=1}^{c_p}\|Z_{m_p}^{(i)}\| > 2\lambda a_p\right\} \le P\{\|Z_{m_p}^{(1)}\| > \lambda a_p\} + P\{\|Z_{m_p}^{(2)}\| > \lambda a_p/c_p\},$$

$$\le 2m_p P\{\|X\| > \lambda a_p/2\} + [2m_p P\{\|X\| > \lambda a_p/c_p\}]^2,$$

$$(3.15)\qquad = (\mathrm{I}) + (\mathrm{II}).$$

One the one hand, assumption (1.19) together with Remark (2) following Theorem 1.3, clearly imply that (I) is the summand of a convergent series. On the other hand,

$$(\mathrm{II}) \leq \Gamma^*(\lambda a_p/2)^2 \left[\frac{\Phi_s(\lambda a_p/2)c_p^2}{\lambda^2 \log p}\right].$$

We now simply observe that $\lambda a_p/c_p \geq 2a_{[p/2]}$ provided that $e^{[p/2]} \geq 16e(\log p)^2/\lambda$. Hence, for some $p(\lambda) < \infty$, depending on λ only,

$$(\mathrm{II}) \leq \Gamma^*(a_{[p/2]})^2[4\Phi_s(\lambda a_p)(\log p)/\lambda^2]^2,$$

a bound, which is clearly by (1.19) and Remark (2) following Theorem 1.3, the summand of a convergent series. Thus, we have shown

(3.16) $\qquad \forall \lambda > 0, \quad \exists p(\lambda) < \infty,$ depending on λ only, such that

$$\sum_{p>p(\lambda)} P\left\{\sum_{i=1}^{c_p} \|Z_{m_p}^{(i)}\| > 2\lambda m_p\right\} \leq \lambda.$$

Therefore (3.1) is satisfied. But also putting together Lemma 2.5, Proposition 3.2 and (3.16)

$$\forall \lambda > 0, \sum_{p>p(\lambda)} P\Big\{\sup_{n \in \mathfrak{s} \cap I_{k_p}} [\|S_n(X)\|/a(n)] > T\}$$

$$\leq (\lambda + 5/p(\lambda))/(1 - 1/8e), \tag{3.17}$$

where $p(\lambda)$ arises from (3.16) and

$$T = \sqrt{e}\{2\lambda + 40K\theta + 32\sqrt{K}\max(\sigma, \theta\sqrt{\alpha})\}, \tag{3.18}$$

where K and α are universal constants, σ is defined in (3.2) and is finite under (1.9) by a closed graph argument (see [LT$_1$], p. 1243), θ is such that (3.3) holds, that is always possible under (1.20).

It is now clear that X satisfies the bounded LIL relatively to the sequence $\mathfrak{s}$. But, we have also prepared, with (3.17) and (3.18), a tool for proving the compact LIL.

(b) (CLIL). The separability assumption on B implies by Ulam's theorem that the law induced by X on B is tight. One can thus define an increasing sequence $\{K_N, N \geq 1\}$ of compact sets of B together with a sequence $\{X^N, N \geq 1\}$ of B-valued random variables having the following properties:

$$-\forall\, N \geq 1,\ P\{\#\text{range}(X^N) > \infty, \quad \text{and } X^N = 0 \text{ on } K_N^c\} = 1,$$
$$-\forall\, N \geq 1,\ P\{\|X - X^N\| \leq 2^{-N}, X \in K_N, X^N \in K_N\} \geq 1 - 2^{-N}. \quad (3.19)$$

For proving the almost sure relative compactness of the sequence $\{S_n(X)/a(n),\ n \in \mathfrak{s}\}$, it is enough to show, thanks to Theorem 1.1,

$$\forall\, \varepsilon > 0, \exists\, N(\varepsilon) < \infty : \forall\, N, N' \geq N(\varepsilon),$$
$$P\Big\{ \sup_{n \in \mathfrak{s} \cap [N', \infty[} \Big\| \frac{S_n(X - X^N)}{a(n)} \Big\| > \varepsilon \Big\} \leq \varepsilon. \tag{3.20}$$

Let $\varepsilon > 0$ be fixed. By (1.10) and (3.19), one has $\lim_{n \to \infty} \sigma(X - X^n) = 0$. By (1.21), (3.19) and since the compact LIL is satisfied for each X^N, by virtue of Theorem 1.1; for each N, for each $\theta > 0$, one may find $q(N, \theta) < \infty$, such that

$$\sup\{P\{\|S_n(X - X^N)\| > \theta a(n)\}, \quad n \in \mathfrak{s} \cap [q(N, \theta), \infty[\leq 1/8e.$$

These observations and (3.17) and (3.18) therefore imply (3.20). QED

3.3.2. **Proof of theorem** 1.4. It suffices to observe that under assumption (1.22), $\Gamma = \Gamma^*$, so that (1.13) and (1.19) are equivalent, and therefore, Theorem 1.4 appears as an immediate corollary of Theorem 1.3.

4. Conclusion

In a recent work [S1], M. Slaby has considered the LIL property for subsequences with the classical sequence of normalization constants, namely, $a(n) = \sqrt{2n \log\log n}$, $n \geq 3$, $n \in \mathfrak{s}$. He has obtained a complete description of the LIL property as well as the cluster set of the sequences $\{S_n(X)/\sqrt{2n \log\log n},\ n \geq 3,\ n \in \mathfrak{s}\}$. His characterization is expressed by means of the same set of conditions as in Theorem 1.2. Therefore, it is the best possible result only for sequences $\mathfrak{s}$ such that $\Lambda(\mathfrak{s}) > 0$, because of Theorem 1.1, Theorem 2.3 in [W3] and the necessary condition (1.13). Unlike his work, our approach is, from the beginning, more intrinsic and we obtain a characterization of the LIL property for any kind of subsequence, with the best possible conditions, when the distribution of $\|X\|$ is sufficiently regular, a point that is expressed by condition (1.22). We would like to conclude by making an observation concerning this point as well as the method of proof itself. For simplicity, we will limit our remark to the bounded LIL. It is clear, from Proposition 3.2, that the cornerstone of the method is condition (3.1). Thus, the question is: are the necessary conditions (1.9), (1.13) and (1.20) strong enough to imply (3.1)? A partial negative answer is given by the lemma below, which seems to show at the same time the limits of the employed method, and those of a possible extension of Theorem 1.2. This is perhaps due to the fact that (1.13) contains less and less information on $||X||$, in proportion that $\mathfrak{s}$ is more and more sparse; a point that features here, the infinite dimensional setting.

Lemma 4.1. *Whatever the sequence* $\{c_p, p \geq 1\}$ *satisfying* $\sum_{p\geq 1} 2^{-c_p} < \infty$, *it is possible to find a sequence* $\mathfrak{s}$ *as well as a random variable* X *with values in* B, *such that*

(1.13) $$\sum_{p\geq 1} m_p P\{||X|| > \varepsilon a_p\} < \infty, \quad \text{for some } \varepsilon < \infty,$$

(4.1) $$\forall p \geq 1, \qquad 1 \leq c_p \leq m_p,$$

(4.2) $$P\left\{\limsup_{p\to\infty} \sum_{i=1}^{c_p} ||X_{m_p}^{(i)}||/a_p = \infty\right\} = 1,$$

and thus (3.1) *does not hold.*

Proof. We choose a sequence $\mathfrak{s}$ such that

(4.3) $$\forall p \geq 1, \qquad m_p \geq m_{p-1} \cdot p^2 \cdot c_p,$$

and, a random variable X with values in B, such that

(4.4) $$\forall t \geq 0, \qquad P\{||X|| > t\} = p^{-2} m_p^{-1}, \quad \text{iff} \quad a_p \leq t < a_{p+1}, \quad p \geq 0,$$

with the convention $a_0 = m_0 = 0$. This is always possible, and (1.13) clearly holds. Let now

(4.5) $$\forall p \geq 1, \qquad Y_p = \#\{i \leq m_p : ||X_i|| \geq b_p\},$$

where the sequence $\{b_p, p \geq 1\}$ is defined by

(4.6) $$\forall p \geq 1, \qquad b_p = a_p d_p / c_p \quad \text{and} \quad \{d_p, p \geq 1\} \quad \text{is such that}$$

$$\lim_{p\to\infty} d_p = \infty \qquad \text{and} \qquad d_p = o(c_p), \quad p \to \infty.$$

By the Paley–Zygmund inequality, for any p,

$$\forall\, 0 \leq \lambda \leq 1, \qquad P\{Y_p \geq \lambda E(Y_p)\} \geq (1-\lambda)^2 [E(Y_p)]^2 / E(Y_p^2).$$

Moreover, because of (4.3) and (4.6), for each p, it is possible to find $q(p) < p$, such that,

$$a_{q(p)} \leq \frac{a_p d_p}{c_p} < a_{q(p)+1}.$$

Hence,

$$E(Y_p) = m_p P\{\|X\| \geq b_p\} = m_p / m_{q(p)} \cdot q(p)^2 \geq m_p / m_{p-1} \cdot p^2 \geq c_p,$$
$$\text{and} \qquad \lim_{p\to\infty} E(Y_p) = \infty.$$

Therefore, by Fatou's Lemma, (4.7) and the $0-1$ law,

$$P\Big\{ \limsup_{p\to\infty} Y_p / c_p \geq 1 \Big\} = 1.$$

This clearly implies (4.2). QED

REFERENCES

[A$_1$] Acosta, A. de, *Exponential moments of vector valued random series and triangular arrays,* Ann. Prob., **8**(1980), 262–280.

[G] Gut, A., *Law of the iterated logarithm for subsequences,* Prob. Math. Stat. **7**(1986), 27–58.

[LT$_1$] Hoffman-Jørgensen, *Sums of independent Banach spaces valued random variables,* Studia Math., **52**(1974), 159–186.

[LT$_1$] Ledoux, M. and Talagrand, M., *Characterization of the law of the iterated logarithm in Banach spaces,* Ann. Prob., **16**(1988), 1242–1264.

[LT$_2$] Ledoux, M. and Talagrand, M., *Comparison theorems, random geometry and some limit theorems for empirical processes,* (1986), Preprint.

[LT$_3$] Ledoux, M. and Talagrand, M., *Some applications of isoperimetric methods to strong limit theorems for sums of independent random variables,* (1988), Preprint.

[St] Stout, W., *Almost sure convergence,* Acad. Press, (1974).

[S1] Slaby, M., *The law of the iterated logarithm for subsequences and characterization of the cluster set of* $\{S_{n_k}/\sqrt{2n_k \log\log n_k},\ k \geq 1\}$, (1988), Preprint.

[Ta$_1$] Talagrand, M., *Isoperimetry and integrability of sums of independent Banach spaces valued random variables,* (1986), Preprint.

[Ta_2] Talagrand, M., *An isoperimetric inequality on the cube and the Khintchin–Kahane inequalities,* (1987), Preprint.

[To] Torrang, I., *The law of the iterated logarithm-cluster points of deterministic and random subsequences,* Uppsala Univ., Report No. 14, (1984).

[W_1] Weber, M., *La loi du logarithme itéré sur les sous-suites,* C.R. Acad. Paris, **303**(1986), 77–80.

[W_2] Weber, M., *La loi du logarithme itéré sur toute sous-suite, caractérisations,* C.R. Acad. Paris, **305**(1987), 835–840.

[W_3] Weber, M., *The law of the iterated logarithm for subsequences – characterizations,* (1987), Preprint.

Acknowledgement: This work was carried out during my stay at Kyoto University, during the Spring 1988; and the author expresses the warmest appreciations to Pr. Norio Kôno and Pr. S. Watanabe for the hospitality he received.

Michel Weber
Institut de Recherche Mathématique
Avancée, Laboratoire associé au CNRS,
Université Louis Pasteur,
Strasbourg

CENTER, SCALE AND ASYMPTOTIC NORMALITY FOR SUMS OF INDEPENDENT RANDOM VARIABLES

Daniel Charles Weiner*
Department of Mathematics
Boston University
Boston, MA 02215
USA

1. Introduction

This paper investigates the related properties of center and scale for distributions and features application to asymptotic normality criteria for sums of independent random variables. Necessary and sufficient conditions are given for convergence in distribution to the standard normal for suitably centered and scaled sums of independent variables in the context of uniform asymptotic negligibility of the summands (Theorem 5.1). New here is the expression of the precise conditions and prior construction of centering and scaling constants directly in terms of the distributions of the original variables, without recourse to symmetrization, and the display of the centered, scaled sums as sums of uniformly asymptotically negligible summands without the necessity of further recentering (cf. Gnedenko and Kolmogorov (1968), Chapter 6, Theorem 2, and Hahn and Klass (1981), Theorem 2 and especially Remark 2 following Theorem 1). The resulting theorem allows a direct and informative expression of Lévy's (1937) profound realization of asymptotic normality for sums as equivalent to negligibility of the summand of maximum modulus (cf. Remark 5 and equation (5.9) here following the statement of Theorem 5.1).

Pursuit of a more unified and complete theory of center and scale for the distribution of sums of independent random variables in the context of asymptotic normality depends upon their simultaneous consideration. While considerable attention has been focused on the problem of scale (e.g., Pruitt (1981)), until recently (Hahn and Klass (1981)) relatively little effort has been devoted to the centering aspect. Traditionally, centerings for sums have been computed only after, and in terms of, scalings. Thus the need for prior construction of the scalings has prompted the technique of symmetrizing the summands in order to eliminate the effect of centrality upon scale. But in fact such symmetrization permanently obscures the mutual effect of centrality and scale upon each other when each is undergoing change, as in, for example, a partial sum of an increasing number of independent summands. Two important points recommend a deeper investigation into the joint behavior of center and scale:

* Research supported in part by NSF Grant DMS-8603188.

(i) In the case of nonidentically distributed summands, the motion of the center may be as important as, or even more important than, the change of scale, in contrast to the i.i.d. case where one has the well-known asymptotic negligibility of the squared truncated mean to the truncated variance (assuming the untruncated mean, if it exists, is zero). (See, e.g., Example 6.3). In particular, in order to determine useful truncation levels for the summands, it is necessary simultaneously to determine useful centers about which to truncate. These centers and scale levels would then be useful in studying sums with respect to general weak convergence, and not just in the context of possible asymptotic normality. (ii) Aesthetics concerning good results on the asymptotic behavior of sums require that conditions on the summands be expressed directly in terms of their original distributions. Without recourse to symmetrization, such expression requires an understanding of the joint effects of centrality and scale.

These points are successfully addressed in our unified approach to centrality and scale as part of a relatively complete treatment of asymptotic normality for sums of independent random variables. The key feature of the development is the reliance on a smoother variation of the truncation operation which we term censoring. The resulting unified approach introduced here should prove useful in the general study of weak convergence of sums; a related approach is being considered and developed elsewhere (Weiner (1989)).

The author wishes to thank Professors J. Kuelbs and M. Hahn for many useful conversations, and for their steady encouragement.

2. Censoring and Associated Moment Functions

In this section we will introduce the censoring and truncating operations and investigate the properties of the associated moment functions.

Let X be a nondegenerate, real random variable with distribution function F.

We define the *censor* $c : [0,\infty) \times (-\infty,\infty) \to (-\infty,\infty)$ by

$$\begin{aligned} c(t,x) &= xI(|x| \le t) + t \text{ sgn } (x)\ I(|x| > t) \\ &= (|x| \wedge t) \text{ sgn } (x), \end{aligned} \tag{2.1}$$

and the *truncator* $\tilde{c} : [0,\infty) \times (-\infty,\infty)$ by

$$\tilde{c}(t,x) = xI(|x| \le t). \tag{2.2}$$

Define the censored and truncated expectation functions $e : [0,\infty) \times (-\infty,\infty) \to (-\infty,\infty)$ and $\tilde{e} : [0,\infty) \times (-\infty,\infty) \to (-\infty,\infty)$ by

$$\begin{aligned} e(t,y) &= Ec(t,X-y) \\ &= E(X-y)I(|X-y| \le t) + t\{P(X-y>t) - P(X-y<-t)\} \end{aligned} \tag{2.3}$$

and

$$\tilde{e}(t,y) = E\tilde{c}(t,X-y) = E(X-y)I(|X-y| \le t). \tag{2.4}$$

Clearly $|e(t,y)| \le t$, and furthermore

$$\lim_{y\to-\infty} e(t,y) = t, \quad \lim_{y\to\infty} e(t,y) = -t. \tag{2.5}$$

Similar facts hold for $\tilde{e}$.

If $E|X| < \infty$, then $\lim_{t\to\infty} tP(|X| > t) = 0$, and we have $\lim_{t\to\infty} e(t,y) = E(X-y)$ $= EX - y$, as well as $\lim_{t\to\infty} \tilde{e}(t,y) = EX - y$.

The next Proposition gives two important explicit representations for e. The proof is an elementary application of Fubini's Theorem.

Proposition 2.1.

For $t \ge 0$ and $-\infty < y < \infty$,

$$e(t,y) = t - \int_{-\infty}^{y} P(|X-s| \le t)\,ds \tag{2.6}$$

$$= \int_0^t \{P(X-y>s) - P(X-y<-s)\}\,ds. \tag{2.7}$$

The following Corollary is immediate:

Corollary 2.2.

For $t \ge 0$ and $-\infty < y < \infty$, the functions $e(t,\cdot)$ and $e(\cdot,y)$ are each absolutely continuous. Each obeys a Lipschitz condition with Lipschitz constant one; thus e is jointly uniformly continuous. Moreover, at every point (t_0,y_0) such that $t_0 > 0$ and $y_0 \pm t_0$ are points of continuity of F, we have

$$e_1(t_0,y_0) = \left.\frac{\partial e(t,y)}{\partial t}\right|_{(t_0,y_0)} = P(X-y_0>t_0) - P(X-y_0<-t_0) \tag{2.8}$$

and

$$e_2(t_0,y_0) = \left.\frac{\partial e(t,y)}{\partial y}\right|_{(t_0,y_0)} = -P(|X-y_0| \le t_0). \tag{2.9}$$

Next, define the censored (truncated) second moment function $m(\tilde{m}) : [0,\infty)\times(-\infty,\infty) \to [0,\infty)$ by

$$\tilde{m}(t,y) = E\tilde{c}^2(t,X-y) = E(X-y)^2 I(|X-y| \le t)$$

(2.10)

$$m(t,y) = Ec^2(t,X-y) = E\tilde{c}^2(t,X-y) + t^2P(|X-y|>t) = E[(X-y)^2 \wedge t^2].$$

Clearly $m(t,y) \le t^2$, $\tilde{m}(t,y) \le t^2$, and

$$\lim_{|y|\to\infty} m(t,y) = t^2, \tag{2.11}$$

while $\tilde{m}(t,y) \to 0$ as $|y| \to \infty$.

If $EX^2 < \infty$, then $\lim_{t\to\infty} t^2 P(|X| > t) = 0$, and we have $\lim_{t\to\infty} m(t,y) = E(X-y)^2 = \lim_{t\to\infty} \tilde{m}(t,y)$.

The next Proposition gives two important explicit representations for m, and relates m to $\tilde{e}$ (recall (2.4)). Its proof is another application of Fubini's Theorem.

Proposition 2.3.

For $t \geq 0$ and $-\infty < y < \infty$,

$$m(t,y) = t^2 - 2\int_{-\infty}^{y} \tilde{e}(t,s)ds \tag{2.12}$$

$$= 2\int_0^t sP(|X-y| > s)ds. \tag{2.13}$$

The following Corollary is immediate:

Corollary 2.4.

For $t \geq 0$ and $-\infty < y < \infty$, the functions $m(t,\cdot)$ and $m(\cdot,t)$ are each absolutely continuous. Furthermore, m is jointly uniformly continuous on sets of form $[0,R] \times (-\infty,\infty)$, for $R < \infty$. If (t_0, y_0) is a point such that $t_0 > 0$ and $y_0 \pm t_0$ are points of continuity of F, then

$$m_1(t_0,y_0) = 2t_0 P(|X-y_0| > t_0) \tag{2.14}$$

and

$$m_2(t_0,y_0) = -2\tilde{e}(t_0,y_0). \tag{2.15}$$

Consider the problem of locating a suitable center for the distribution of X censored at the level t. To motivate the two distinct methods we will propose (one here, one elsewhere–cf. Weiner (1989)) suppose $EX^2 < \infty$ and take $t = \infty$. To define a center y for the distribution of X, one might require

$$\tilde{e}(\infty,y) = e(\infty,y) = E(X-y) = 0, \tag{2.16}$$

in order that the centered variable $X-y$ have zero expectation, or one might require

$$m(\infty,y) = E(X-y)^2 = \inf_s E(X-s)^2 = \inf_s m(\infty,s), \tag{2.17}$$

in order that the centered variable $X-y$ have minimized second moment among all shifted versions $X - s$ of X.

Of course, each of (2.16) and (2.17) leads to the same unique choice of $y = EX$, despite their a priori unrelated motivational origins. However, for finite t, the requirements analogous to (2.16) and (2.17), respectively, lead to distinct results.

Given a scale level $0 < t < \infty$, the attempt to locate the center of the censored distribution of X by solving

$$e(t,y) = Ec(t,X-y) = 0 \tag{2.18}$$

will be called the *Censored Centered Method* (CCM). This smoother but technically involved method is considered elsewhere (Weiner (1989)). The attempt to locate this center by minimizing, i.e., solving

$$m(t, y) = Ec^2(t, X - y) = \inf_s Ec^2(t, X - s) = \inf_s m(t, s) \tag{2.19}$$

will be called *Minimum Moment Method* (MMM).

We will consider the MMM approach to the centering problem in this article. This method is by far the easier of the two, and moreover has the advantage of producing the smallest precentered normalizing constants in a given situation. The benefit when this method is considered in the context of possible statistical inference is clear, since the theoretical margin of error will be proportional to the normalizing constants used.

The MMM method is also the basis for the success of the CCM method being considered elsewhere. As we will see, MMM centerings overcome some of the disadvantages of the classical approach using medians (see Example 6.3), while retaining most of their concreteness, in studying uniform asymptotic negligibility (u.a.n.) conditions. The lack of uniqueness (as with medians) of MMM centerings and other remaining disadvantages of the MMM approach, can be overcome via the CCM approach. This latter, being analytically quite involved and resting upon the MMM approach itself, is thus better considered separately.

3. The Minimum Moment Method

In this section we will construct the functions needed for determining the precentered normalizing constants of the Minimum Moment Method, and demonstrate that they possess the necessary regularity properties.

We will show that given $t > 0$, the function $m(t, \cdot)$ attains a strictly positive absolute minimum, and that if y_0 is the location of one such minimum, then $E\tilde{c}(t, X - y_0) = \tilde{e}(t, y_0) = 0$. It will follow that var $\tilde{c}(t, X - y_0) = E\tilde{c}^2(t, y_0) = \tilde{m}(t, y)$.

Fix $t > 0$, and define $h : (-\infty, \infty) \to [0, \infty)$ by $h(y) = m(t, y)$. Recalling (2.11), we note

$$t^2 > h(y) \to t^2, \tag{3.1}$$

as $|y| \to \infty$. But, as h is a nonnegative, continuous function (cf. (2.12)), (3.1) guarantees that h achieves an absolute minimum. Let y_0 be the location of an absolute minimum for h. We claim $0 < h(y_0) < t^2$. For $0 = h(y_0) = E(X - y_0)^2 \wedge t^2$ would imply $P(X = y_0) = 1$, contrary to X nondegenerate, and $t^2 = h(y_0) = \inf_y E(X - y)^2 \wedge t^2$ would imply, for every $y, t^2 \leq E(X - y)^2 \wedge t^2 \leq t^2$, which would imply, for every y, $P(|X - y| \geq t) = 1$, contrary to $P(-\infty < X < \infty) = 1$.

Next we claim $\tilde{e}(t, y_0) = 0$. Inspection of (2.15) shows this fact is immediate if $y_0 \pm t$ are points of continuity of F, since $h'(y_0) = 0$ if it exists (because y_0 locates the minimum of h). But we will show directly that $\tilde{e}(t, y_0) = 0$, and deduce that $y_0 \pm t$ are points of continuity of F as a byproduct.

Define $g : (-\infty, \infty) \to (-\infty, \infty)$ by $g(s) = \tilde{e}(t, s)$.

Then (2.13) gives

$$h(y) - h(y_0) = -2 \int_{y_0}^{y} g(s)ds, \tag{3.2}$$

for every y.

Let $y > y_0$. Since y_0 locates the absolute minimum for h, we have $h(y) \geq h(y_0)$. From (3.2) we get

$$0 \leq \frac{h(y) - h(y_0)}{y - y_0} = \frac{-2}{y - y_0} \int_{y_0}^{y} g(s)ds, \tag{3.3}$$

or

$$0 \geq \frac{1}{y - y_0} \int_{y_0}^{y} g(s)ds \geq \inf_{y_0 < s < y} g(s). \tag{3.4}$$

Letting $y \downarrow y_0$ in (3.4) gives

$$\liminf_{y \downarrow y_0} g(y) \leq 0. \tag{3.5}$$

Similarly, if $y < y_0$, then $h(y) \geq h(y_0)$, and (3.2) will lead to

$$\limsup_{y \uparrow y_0} g(y) \geq 0. \tag{3.6}$$

But in fact the left- and right-hand limits of g exist: Fix $a < y_0 - t$. Then

$$g(y) = \int_{[a,y+t]} (x - y)dF(x) - \int_{[a,y-t)} (x - y)dF(x) \tag{3.7}$$

for every $y > a+t$, in particular for y near $y_0 > a+t$. Now (3.7) exposes g as the difference of a right-continuous function and a left-continuous function, each of which possesses both one-sided limits at y_0. We find from (3.5)-(3.7)

$$\begin{aligned} 0 &\geq \liminf_{y \downarrow y_0} g(y) = g(y_0+) \qquad (3.8) \\ &= \int_{[a,y_0+t]} (x - y_0)dF(x) - \int_{[a,y_0-t]} (x - y_0)dF(x) \\ &= E(X - y_0)I(|X - y_0| \leq t) + tP(X = y_0 - t) \geq g(y_0) \\ &\geq E(X - y_0)I(|X - y_0| \leq t) - tP(X = y_0 + t) \\ &= \int_{[a,y_0+t)} (x - y_0)dF(x) - \int_{[a,y_0-t)} (x - y_0)dF(x) = g(y_0-) \\ &= \limsup_{y \uparrow y_0} g(y) \geq 0, \end{aligned}$$

which forces $\tilde{e}(t, y_0) = g(y_0) = 0$ along with $P(X = y_0 \pm t) = 0$, as desired.

Thus for each $t > 0$, we can find a y_0 (depending on t) such that $m(t, y_0) = \inf_y m(t, y)$. Unfortunately, given t, y_0 is not necessarily unique, nor is it possible in general to choose y_0 as a continuous function of t (see Example 6.1). Nevertheless one can define a function $\tilde{\gamma} : [0, \infty) \to (-\infty, \infty)$ by selecting, for every $t \geq 0$, a point $\tilde{\gamma}(t)$ minimizing $m(t, \cdot)$. (Note $m(0, \cdot) = 0$, so $\tilde{\gamma}(0)$ is completely arbitrary.) Then the function $\tilde{v} : [0, \infty) \to [0, \infty)$ given by

$$\tilde{v}(t) = m(t, \tilde{\gamma}(t)) = \inf_y m(t, y) \tag{3.9}$$

is well-defined, and strictly positive on $(0,\infty)$. Note that $\tilde{v}(t) = t^2 P(|X - \tilde{\gamma}(t)| > t) + \tilde{m}(t, \tilde{\gamma}(t))$.

Now by (2.14), $m(\cdot, y)$ is nondecreasing, for every y, and thus $\tilde{v}$ is also nondecreasing. If X is essentially unbounded, (2.14) shows $m(\cdot, y)$ is strictly increasing, for every y, and then $\tilde{v}$ will also be strictly increasing.

Previous arguments imply $0 < \tilde{v}(t) < t^2$, for every $t > 0$.

We define $\tilde{w} : [0,\infty) \to [0,1]$ by

$$\tilde{w}(t) = \tilde{v}(t)/t^2, \; t > 0. \tag{3.10}$$

(We will define $\tilde{w}(0)$ later.)
We claim $\tilde{w}$ is strictly decreasing and $\lim_{t\to\infty} \tilde{w}(t) = 0$. Because $\tilde{w}(t) \geq P(|X - \tilde{\gamma}(t)| \geq t)$, we will thus obtain

$$\lim_{t\to\infty} P(|X - \tilde{\gamma}(t)| \geq t) = 0, \tag{3.11}$$

regardless of choice of $\tilde{\gamma}$.

Now for each y, the function $f_y(t) = m(t,y)/t^2$ is nondecreasing, by (2.13) together with the quotient rule. Let $0 < t_1 < t_2$. Because $\tilde{\gamma}(t_1) \pm t_1$ are points of continuity of F and $P(|X - \tilde{\gamma}(t_1)| > t_1) \leq \tilde{w}(t_1) < 1$, (2.14) shows that $f_{\tilde{\gamma}(t_1)}(\cdot)$ is strictly decreasing near t_1. Thus

$$\begin{aligned}\tilde{w}(t_2) &= m(t_2, \tilde{\gamma}(t_2))/t_2^2 \leq m(t_2, \tilde{\gamma}(t_1))/t_2^2 \\ &< m(t_1, \tilde{\gamma}(t_1))/t_1^2 = \tilde{w}(t_1),\end{aligned} \tag{3.12}$$

where the first inequality holds due to the minimizing property of $\tilde{\gamma}$. Finally

$$\tilde{w}(t) = m(t, \tilde{\gamma}(t))/t^2 \leq m(t,0)/t^2 = E(X^2 \wedge t^2)/t^2 \to 0, \tag{3.13}$$

as $t \to \infty$, by the Dominated Convergence Theorem.

Since $\tilde{w}$ is strictly decreasing we can define $\tilde{w}(0) = \lim_{t\downarrow 0} \tilde{w}(t) \leq 1$, and thus $\tilde{w}(0) > 0$.

We claim that $\tilde{v}$ (hence $\tilde{w}$) is continuous. Let $t > 0$, and suppose $0 < t_n \to t$. If $\{\tilde{\gamma}(t_n)\}$ is bounded, we can suppose (by taking subsequences, if necessary) that $\tilde{\gamma}(t_n) \to L$. By joint continuity of m (Corollary 2.4) we have $\tilde{v}(t_n) = m(t_n, \tilde{\gamma}(t_n)) \to m(t, L)$. But, as $\tilde{\gamma}$ minimizes, we have $m(t_n, \tilde{\gamma}(t_n)) \leq m(t_n, \tilde{\gamma}(t)) \to m(t, \tilde{\gamma}(t))$, i.e., $m(t, L) \leq m(t, \tilde{\gamma}(t)) = \inf_y m(t,y)$, forcing $m(t,L) = m(t, \tilde{\gamma}(t)) = \tilde{v}(t)$, i.e., $\tilde{v}(t_n) \to \tilde{v}(t)$.

If $\{\tilde{\gamma}(t_n)\}$ is unbounded, we can suppose $\tilde{\gamma}(t_n) \to \infty$ (the case of $-\infty$ being similar). We claim $\tilde{w}(t_n) \to 1$, which is impossible in light of (3.12) and $\tilde{w}(\cdot) < 1$. We have, via Corollary 2.4 and (2.11), for all sufficiently large n,

$$\begin{aligned}|\tilde{w}(t_n) - 1| &\leq 2t^{-2}|m(t_n, \tilde{\gamma}(t_n)) - t_n^2| \\ &\leq 2t^{-2}\{|m(t_n, \tilde{\gamma}(t_n)) - m(t, \tilde{\gamma}(t_n))| \\ &\quad + |m(t, \gamma(t_n)) - t^2| + |t_n^2 - t^2|\} \\ &\to 0,\end{aligned} \tag{3.14}$$

as $n \to \infty$. Thus $\{\tilde{\gamma}(t_n)\}$ must be bounded, and $\tilde{v}$ and $\tilde{w}$ are indeed continuous.

As we will see, $\tilde{v}$ and $\tilde{w}$ will be useful in determining the precentered normalizing constants for the Minimum Moment Method.

Next we claim

$$\lim_{t\to\infty} \frac{\tilde{\gamma}(t)}{t} = 0, \tag{3.15}$$

regardless of choice of $\tilde{\gamma}$. (See also Example 6.2.)

First, we note that $\overline{\lim_{t\to\infty}}(|\tilde{\gamma}(t_n) - t) < \infty$, for if not, take $t_n \to \infty$ with $\tilde{\gamma}(t_n) - t_n \to \infty$ (the case of $-\tilde{\gamma}(t_n) - t_n \to \infty$ being similar). Then $P(|X - \tilde{\gamma}(t_n)| \leq t_n) \leq P(X \geq \tilde{\gamma}(t_n) - t_n) \to 0$, as $n \to \infty$, which implies $P(|X - \tilde{\gamma}(t_n)| \geq t_n) \to 1$, a contradiction of (3.11).

So fix $C > 0$ with $|\tilde{\gamma}(t)| \leq t + C$ for all sufficiently large t. Now $P(|X - \tilde{\gamma}(t)| \leq t) > 0$ (lest $\tilde{w}(t) = 1$) so that from $\tilde{e}(t, \tilde{\gamma}(t)) = 0$ we obtain

$$\tilde{\gamma}(t) = \frac{EXI(|X - \tilde{\gamma}(t)| \leq t)}{P(|X - \tilde{\gamma}(t)| \leq t)}. \tag{3.16}$$

Thus, for all sufficiently large t, we obtain (since $P(|X - \tilde{\gamma}(t)| \leq t) \to 1$)

$$\begin{aligned} |\tilde{\gamma}(t)|/t &\leq 2E|X|I(\tilde{\gamma}(t) - t \leq X \leq \tilde{\gamma}(t) + t)/t \\ &\leq 2E|X|I(|X| \leq 2t + C)/t \\ &\to 0, \end{aligned} \tag{3.17}$$

as $t \to \infty$, by Dominated Convergence. Thus (3.15) holds, and in particular $\tilde{\gamma}(t) + t \to \infty$ and $\tilde{\gamma}(t) - t \to -\infty$, as $t \to \infty$.

Now we can examine the case $E|X| < \infty$. Letting $t \to \infty$ in (3.16) and applying the previous facts together with the Dominated Convergence Theorem, we obtain

$$\lim_{t\to\infty} \tilde{\gamma}(t) = EX, \tag{3.18}$$

regardless of choice of $\tilde{\gamma}$.

The converse of (3.18), i.e., $\tilde{\gamma}(t) \to L \in (0, \infty)$ implies EX exists, is false (see Example 6.2).

Finally, if $EX^2 < \infty$, we note that $t^2 P(|X - \tilde{\gamma}(t)| \geq t) \leq t^2 P(|X| > t/2) \to 0$, so $\tilde{v}(t) \sim E(X - \tilde{\gamma}(t))^2 I(|X - \tilde{\gamma}(t)| \leq t) \to \operatorname{var} X$, as $t \to \infty$, using (3.18).

We summarize the main results of the MMM in

Proposition 3.1.

There exists a function $\tilde{\gamma} : [0, \infty) \to (-\infty, \infty)$ *such that for every* $t \geq 0$,

$$E(X - \tilde{\gamma}(t))^2 \wedge t^2 = \inf_y E(X - y)^2 \wedge t^2 \tag{3.19}$$

Every such function $\tilde{\gamma}$ *satisfies, for every* $t \geq 0$,

$$E(X - \tilde{\gamma}(t))I(|X - \tilde{\gamma}(t)| \leq t) = 0, \tag{3.20}$$

and

$$\lim_{t\to\infty} \frac{\tilde{\gamma}(t)}{t} = 0. \tag{3.21}$$

Moreover, for every $t > 0$,

$$\tilde{\gamma}(t) \pm t \in C(F), \tag{3.22}$$

where $C(F)$ denotes the set of continuity points of F.

The function $\tilde{v} : [0,\infty) \to [0,\infty)$ given by $\tilde{v}(t) = Ec^2(t, X - \tilde{\gamma}(t)) = t^2 P(|X - \tilde{\gamma}(t)| > t)$ $+E(X - \tilde{\gamma}(t))^2 I(|X - \tilde{\gamma}(t)| \le t)$ is well-defined, continuous and nondecreasing.

The function $\tilde{w} : [0,\infty) \to (0,1]$ given by $\tilde{w}(t) = \tilde{v}(t)/t^2$ for $t > 0$ and defined by continuity at $t = 0$, is continuous, strictly decreasing and satisfies

$$\lim_{t\to\infty} \tilde{w}(t) = 0. \tag{3.23}$$

If $E|X| < \infty$, then each function $\tilde{\gamma}$ satisfies

$$\lim_{t\to\infty} \tilde{\gamma}(t) = EX. \tag{3.24}$$

If $EX^2 < \infty$, then

$$\lim_{t\to\infty} \tilde{v}(t) = \operatorname{var} X. \tag{3.25}$$

Remarks:

1. It is not true that symmetry of X about 0 will allow the choice $\tilde{\gamma} \equiv 0$ (see Example 6.1).

2. The minimum Moment Method for choosing centers may be viewed as a kind of concentration - approach (cf. Lévy (1937)). Instead of maximizing $P(|X - y| \le t)$ with respect to y (the classical concentration function idea), we obtain a smoother solution by minimizing $E(X - y)^2 \wedge t^2 = \int_0^t 2s\ P(|X - y| > s)ds$ (recalling (2.13)), or what is the same, by maximizing $\int_0^t s\ P(|X - y| \le s)ds$ - that is, we consider maximum *averaged* concentration. The linearly weighted averaging adopted here owing to its relevance, via (2.13), to second moment computations, could certainly be replaced for other purposes by any smooth-weighted averaging. For example, the use of constant weights will give rise to a centering concept bearing a relationship to ordinary medians similar to that of the MMM centering idea to ordinary means. In any event, the concentration-like properties of the MMM centerings will be exploited later regarding asymptotic normality in connection with uniform asymptotic negligibility conditions.

3. We use tilde ($\sim$) here so as to distinguish the MMM functions being constructed here from their analogues to be constructed via the CCM approach (in Weiner (1989)).

4. That (3.21) holds uniformly in choice of $\tilde{\gamma}$ is needed and established in Example 6.2.

4. The Precentered Scale Sequence

Let $\{X_j : j \geq 1\}$ be an independent sequence of nondegenerate real random variables. Subscript by j the various functions, introduced in the previous sections, corresponding to X_j.

In this section, assuming the sequence $\{X_j\}$ satisfies a necessary "nondegeneracy" condition (see (4.1), below), we construct the centering and scaling constants to be used in considering affine-normalized asymptotic normality for the partial sums $S_n = X_1 + \ldots + X_n$.

Our only assumption on the sequence $\{X_j\}$, aside from nondegeneracy of each variable, is that

$$\sum_{j=1}^{\infty} \inf_y E(X_j - y)^2 \wedge t^2 = \infty, \tag{4.1}$$

for some (every) $t > 0$. In view of the MMM results of Section 3, we may rewrite (4.1) in the form

$$\sum_{j=1}^{\infty} \tilde{v}_j(t) = \infty, \tag{4.2}$$

for some (every) $t > 0$.

Now if $\{X_j\}$ is i.i.d., then (4.1) always holds due to nondegeneracy and our previous observation that $\tilde{v}_j(t) > 0$ for $t > 0$. To demonstrate that (4.1) is extremely general in the nonidentically distributed case, we will show that the failure of (4.1) for *some* $t > 0$ is equivalent to the almost sure convergence of the random series

$$\sum_{j=1}^{\infty} (X_j - \tilde{\gamma}_j(t)) \tag{4.3}$$

to a nondegenerate random variable for *every* $t > 0$. In this case (and only in this case) there can be little interest in normalizing the partial sums to obtain weak convergence.

We will show that the left member of (4.2) either converges for every $t > 0$ or diverges for every $t > 0$.

So fix $t > 0$ such that (4.2) fails. We have

$$\begin{aligned} &\sum_{j=1}^{\infty} var(X_j - \tilde{\gamma}_j(t)) I(|X_j - \tilde{\gamma}_j(t)| \leq t) \qquad (4.4)\\ &= \sum_{j=1}^{\infty} E(X_j - \tilde{\gamma}_j(t))^2 I(|X_j - \tilde{\gamma}_j(t)| \leq t) \\ &\leq \sum_{j=1}^{\infty} \tilde{v}(t) \\ &< \infty, \end{aligned}$$

$$\sum_{j=1}^{\infty} P(|X_j - \tilde{\gamma}_j(t)| > t) \leq t^{-2} \sum_{j=1}^{\infty} \tilde{v}_j(t) < \infty, \tag{4.5}$$

and (recalling the properties of $\tilde{\gamma}_j$, in particular (3.20)),

$$\sum_{j=1}^{\infty} E(X_j - \tilde{\gamma}_j(t))I(|X_j - \tilde{\gamma}_j(t)| \leq t) = \sum_{j=1}^{\infty} 0 = 0. \tag{4.6}$$

Thus the "three series theorem" of Kolmogorov (see, e.g., Chung (1974), p. 118) applies to give almost sure convergence of the random series (4.3). But now the converse to Kolmogorov's theorem shows that this almost sure convergence of the random series (4.3) for some $t > 0$ implies

$$\sum_{j=1}^{\infty} E(X_j - \tilde{\gamma}_j(t))^2 \wedge s^2 < \infty, \tag{4.7}$$

for every $s > 0$. The minimizing property of $\tilde{\gamma}_j$ gives

$$\sum_{j=1}^{\infty} \tilde{v}_j(s) \leq \sum_{j=1}^{\infty} E(X_j - \tilde{\gamma}_j(t))^2 \wedge s^2 < \infty, \tag{4.8}$$

for every $s > 0$, which leads by the direct part of Kolmogorov's theorem (as above) to the almost sure convergence of (4.3) for ***every*** $t > 0$.

To see that the random variable defined by (4.3) is nondegenerate, note that by independence,

$$\infty \geq var \sum_{j=1}^{\infty}(X_j - \tilde{\gamma}_j(t)) = \sum_{j=1}^{\infty} var\ X_j > 0, \tag{4.9}$$

by nondegeneracy of each X_j.

We remark that the same argument shows that the failure of (4.2) for some $t > 0$ is actually equivalent to the almost sure convergence of some (any) random sequence of the form

$$\sum_{j=1}^{n} X_j - e_n \tag{4.10}$$

for constants $\{e_n\}$.

Thus in the sequel we will assume (4.1) and hence also (4.2) hold for every $t > 0$, and in particular we can assume that no series of form (4.10) converges, even in distribution (cf. symmetrization and Araujo and Giné (1980), p.105), so that scaling is actually required.

Define $\tilde{C}_n : [0, \infty) \to [0, \infty)$ by

$$\tilde{C}_n(t) = \sum_{j=1}^{n} \tilde{w}_j(t) = \sum_{j=1}^{n} t^{-2} E(X_j - \tilde{\gamma}_j(t))^2 \wedge t^2. \tag{4.11}$$

Note that $\tilde{C}_n$ is continuous and strictly decreasing due to Proposition 3.1.

We will show that for all sufficiently large n, we can uniquely solve the equation

$$\tilde{C}_n(t) = 1. \tag{4.12}$$

Then we will examine the solution sequence.

For each $t > 0$, (4.1) guarantees

$$\lim_{n\to\infty} \tilde{C}_n(t) = \sum_{j=1}^{\infty} \tilde{w}_j(t) = \infty > 1, \tag{4.13}$$

so, fixing $t_0 > 0$, there exists $n_0 = n_0(t_0)$ such that for all $n \geq n_0$,

$$\tilde{C}_n(t_0) \geq \tilde{C}_{n_0}(t_0) > 1. \tag{4.14}$$

But, for any fixed n, (3.23) guarantees

$$\lim_{t\to\infty} \tilde{C}_n(t) = 0, \tag{4.15}$$

so given $n \geq n_0$, there exists t such that

$$\tilde{C}_n(t) < 1. \tag{4.16}$$

Using continuity and strict decrease, it follows that for $n \geq n_0$ there exists unique $t \geq t_0$ such that (4.12) holds. For $n \geq n_0$, then, we can uniquely define scaling constants $\tilde{a}_n > 0$ by

$$\tilde{a}_n^{-2} \sum_{j=1}^{n} E(X_j - \tilde{\gamma}_j(\tilde{a}_n))^2 \wedge \tilde{a}_n^2 = 1. \tag{4.17}$$

Because $\tilde{C}_{n+1}(\tilde{a}_n) > \tilde{C}_n(\tilde{a}_n) = 1$, it follows $\tilde{a}_{n+1} > \tilde{a}_n$, so $\{\tilde{a}_n\}$ is strictly increasing. But, given $t \geq t_0$, since $\lim_{n\to\infty} \tilde{C}_n(t) = \infty$ (via (4.13)), it follows there exists n_1 such that $a_{n_1} > t$, using $\tilde{C}_n(\tilde{a}_n) = 1$ and strict decrease of $\tilde{C}_n(\cdot)$; thus, $\tilde{a}_n \longrightarrow \infty$ as $n \to \infty$.

We summarize these results in

Proposition 4.1

Given nondegenerate $X_1, X_2, X_3, \ldots$, such that (4.1) holds for some $t > 0$, there exists n_0 such that for every $n \geq n_0$, there exists unique $\tilde{a}_n > 0$ satisfying

$$\sum_{j=1}^{n} E(X_j - \tilde{\gamma}_j(\tilde{a}_n))^2 \wedge \tilde{a}_n^2 = \tilde{a}_n^2. \tag{4.18}$$

The sequence $\{\tilde{a}_n : n \geq n_0\}$ is strictly increasing, and satisfies

$$\lim_{n\to\infty} \tilde{a}_n = \infty. \tag{4.19}$$

5. Asymptotic Normality

Let $\{X_j\}$ be an independent sequence of nondegenerate, real random variables, with $S_n = X_1 + \ldots + X_n$. In this section we consider the problem of determining when (and how) $\{S_n\}$ can be centered and normalized so as to converge in distribution to the standard normal, yet be composed of uniformly asymptotically negligible (u.a.n.) summands.

(Equivalent to u.a.n. is that the scalings d_n satisfy $d_{n+1}/d_n \to 1$.) The classical solution to this problem (Gnedenko and Kolmogorov, 1968, Chapter 6, Theorem 2) and the recent version (Hahn and Klass, 1981, Theorem 2) share the drawback that the scale sequence which is constructed in order to test for asymptotic normality is computed not directly from the distributions of the summands, but rather from the symmetrizations of these distributions. (See especially Remark 2 following Theorem 1 in Hahn and Klass.) Only after the scale sequence has been constructed can the centering sequence be computed. Both approaches involve the use of medians of the summand distributions as centers of concentration, necessitating an awkward recentering after the expression of the scaled, centered sum as the sum of u.a.n. scaled, precentered summands. Now it is easily seen that if any centerings render the scaled, precentered summands u.a.n., then the medians will do the same; howevber, the use of these medians as unchanging centers of concentration obscures the scale-dependence of the centrality with respect to truncated means, which centrality is of greater computational relevance in applying the criteria for normal convergence. Moreover, use of medians to directly compute scalings can result in an error (toward *standard* normal convergence) of up to a factor of $\sqrt{2}$ (see Example 6.3).

Our approach, utilizing the Minimum Moment Method, simultaneously computes the appropriate centering and scaling sequences, directly and exclusively in terms of the distributions of the original, unsymmetrized summands $\{X_j\}$. (See Section 4.) The u.a.n. condition and the corresponding tail-sum condition are each expressed, not in terms of the classical medians, but rather in terms of the more useful and (especially with respect to truncated means) computationally relevant MMM centerings. These centerings allow the final display of the scaled, centered sum as the sum of u.a.n. scaled, precentered summands *without the necessity of further recentering*; the MMM centerings perform the dual task of centering the sums for the normal convergence and also serving as centers of concentration of the summands for their uniform asymptotic negligibility.

For the Theorem, we recall the notation of Section 4, in particular (4.18).

Theorem 5.1.

The following conditions are equivalent:

(A) There exist constants $\{d_n\}, \{e_{nj}\}$ *and* $\{y_n\}$ *such that*

$$(S_n - y_n)/d_n \xrightarrow{D} N(0,1) \tag{5.1}$$

and for every $\epsilon > 0$,

$$\lim_{n\to\infty} \max_{j\le n} P(|X_j - e_{nj}| > \epsilon d_n) = 0. \tag{5.2}$$

(B) There exist constants $\{d_n\}, \{y_n\}$ *such that (5.1) holds and*

$$\lim_{n\to\infty} d_n = \infty, \qquad \lim_{n\to\infty} d_{n+1}/d_n = 1 \tag{5.3}$$

(C) For some (hence every) $t > 0$,

$$\sum_{j=1}^{\infty} \tilde{v}_j(t) = \sum_{j=1}^{\infty} \inf_y E(X_j - y)^2 \wedge t^2 = \infty. \tag{5.4}$$

and for every $\epsilon > 0$,

$$\lim_{n\to\infty} \sum_{j=1}^{n} P(|X_j - \tilde{\gamma}_j(\tilde{a}_n)| > \epsilon \tilde{a}_n) = 0. \tag{5.5}$$

When (A),(B) and (C) hold,

$$\tilde{a}_n^{-1} \sum_{j=1}^{n} (X_j - \tilde{\gamma}_j(\tilde{a}_n)) \xrightarrow{D} N(0,1) \tag{5.6}$$

$$\{(X_j - \tilde{\gamma}_j(\tilde{a}_n))/\tilde{a}_n : j \leq n\} \text{ is u. a.n.} \tag{5.7}$$

and

$$\lim_{n\to\infty} \tilde{a}_{n+1}/\tilde{a}_n = 1. \tag{5.8}$$

Remarks:

1. If (5.2) is satisfied by some constants $\{e_{nj}\}$, then it is also satisfied by $e_{nj} = med\ X_j$. Part of the proof of Theorem 5.1 involves showing that if (5.2) is satisfied by some $\{e_{nj}\}$, then it is also satisfied by $e_{nj} = \tilde{\gamma}_j(d_n)$. Thus under (5.2), we must actually have $\lim_{n\to\infty} \max_{j\leq n} |\tilde{\gamma}_j(d_n) - medX_j|/d_n = 0$.

2. If (A) holds, the Theorem and a convergence-of-types argument guarantee that

$$d_n/\tilde{a}_n \longrightarrow 1, \quad (y_n - \sum_{j=1}^{n} \tilde{\gamma}_j(\tilde{a}_n))/\tilde{a}_n \longrightarrow 0.$$

3. The Hahn-Klass (1981) method of centering the sums employing t-trimming and medians may be comparéd to the present method by observing that under (A) we will have a posteriori (recalling (2.3)),

$$\lim_{n\to\infty} \tilde{a}_n^{-1} \sum_{k=1}^{n} (e_j(\tilde{a}_n, med\ X_j) - \tilde{\gamma}_j(\tilde{a}_n)) = 0.$$

4. Condition (5.4), which assures the existence of the constants $\{\tilde{a}_n\}$, is truly minimal in this sense: We have seen, in Section 4, that failure of (5.4) gives almost sure convergence of the random series (4.3). Cramer's famous result (e.g., Feller (1971), p.525) shows that the limit random variable cannot be normal unless every summand X_j is already normal. The converse being obvious, we regard the asymptotic normality problem as settled when (5.4) fails.

5. Together, (5.6) and (5.7) give the promised display of the asymptotically normal scaled, centered sums directly as sums of u.a.n. summands, without recentering. We obtain ultimate realization of Lévy's (1937) original idea of asymptotic normality via negligibility when we rewrite (5.5) and (5.6) in the form

$$\sum_{j=1}^{n} \left(\frac{X_j - \tilde{\gamma}_j(\tilde{a}_n)}{\tilde{a}_n} \right) \xrightarrow{D} N(0,1), \qquad \max_{j\leq n} \left| \frac{X_j - \tilde{\gamma}_j(\tilde{a}_n)}{\tilde{a}_n} \right| \xrightarrow{P} 0. \tag{5.9}$$

Proof of the Theorem

(A)⇒ (B): Symmetrize: Let $\{X_j'\}$ be an independent copy of $\{X_j\}$, and write $\tilde{X}_j = X_j - X_j'$, $\tilde{S}_n = \sum_{j\le n}\tilde{X}_j$. Then $\tilde{S}_n/d_n \xrightarrow{D} N(0,2)$, and $\{\tilde{X}_j/d_n : j \le n\}$ is u.a. n. Then

$$N(0,2) \xleftarrow{D} \frac{\tilde{S}_{n+1}}{d_{n+1}} = \frac{\tilde{S}_n}{d_n}\frac{d_n}{d_{n+1}} + \frac{\tilde{X}_{n+1}}{d_{n+1}}. \tag{5.10}$$

Now, by the u.a.n., $\tilde{X}_{n+1}/d_{n+1} \xrightarrow{P} 0$; by convergence-of-types and $\tilde{S}_n/d_n \xrightarrow{D} N(0,2)$, (5.3) follows

(B)⇒(A) Choosing e_{nj} = median of X_j, it is enough to show, having symmetrized, that $\{\tilde{X}_j/d_n : j \le n\}$ is u.a.n. For this, we need only show that given any $1 \le j_n \le n$, $\tilde{X}_{j_n}/d_n \xrightarrow{P} 0$. Now $d_n \to \infty$ so without lost generality $j_n \to \infty$. Writing (5.10) with n replaced by $j_n - 1$, we obtain via (5.3) that $\tilde{X}_{j_n}/d_{j_n} \xrightarrow{P} 0$. But writing

$$N(0,2) \xleftarrow{D} \frac{\tilde{S}_n}{d_n} = \frac{\tilde{S}_{j_n}}{d_{j_n}}\frac{d_{j_n}}{d_n} + \frac{\tilde{S}_n - \tilde{S}_{j_n}}{d_n} \tag{5.11}$$

and using independence and convergence of types shows that $\overline{\lim}_{n\to\infty} d_{j_n}/d_n \le 1$. Thus $\tilde{X}_{j_n}/d_n \xrightarrow{P} 0$, as required.

(C)⇒(A). We note that (5.4) is just (4.1); thus the constants $\{\tilde{a}_n\}$ of (4.18) actually (eventually) exist. Obviously (5.5) implies (5.2) and (5.7). Let us recall (4.18) here: For sufficiently large n,

$$\tilde{a}_n^{-2}\sum_{j=1}^{n} E(X_j - \tilde{\gamma}_j(\tilde{a}_n))^2 \wedge \tilde{a}_n^2 = 1. \tag{5.12}$$

Taking $\epsilon = 1$ in (5.5), we see (5.12) gives

$$\tilde{a}_n^{-2}\sum_{j=1}^{n} E(X_j - \tilde{\gamma}_j(\tilde{a}_n))^2 I(|X_j - \tilde{\gamma}_j(\tilde{a}_n)| \le \tilde{a}_n) \to 1. \tag{5.13}$$

We recall (3.20):

$$\tilde{a}_n^{-1}\sum_{j=1}^{n} E(X_j - \tilde{\gamma}_j(\tilde{a}_n)) I(|X_j - \tilde{\gamma}_j(\tilde{a}_n)| \le \tilde{a}_n) \to 0 \tag{5.14}$$

In conjunction with (5.5), (5.13) and (5.14) provide the usual sufficient conditions (cf., Loève (1977), p.328) for the normal convergence in (5.6). Now (5.1) is obvious, so (A) holds. This part of the proof was easy because of the construction of $\{\tilde{a}_n\}$ and due to property (3.20) of the $\tilde{\gamma}$-function.

(A)⇒(C). We are assuming the asymptotic normality (5.1); in order to derive information from this fact relevant to establishing (5.5), we will need to identify useful $\{e_{nj}\}$, such as in (5.7).

First, we claim (5.4) holds (so we can work with $\{\tilde{a}_n\}$ later). If (5.4) fails, we have seen that for suitable constants $\{e_n\}$, $S_n - e_n \to S$ a.s., where S is a nondegenerate r.v. Then

(5.1) and convergence of types force S to be normal and $d_n^2 \to \text{var } S$. Thus without lost generality we may assume $d_n \equiv 1$. Fix k, and let $\epsilon > 0$. For $n \geq k$, (5.2) implies (using $e_{nj} = med\ X_j$ for now),

$$P(|X_k - med\ X_k| > \epsilon) \leq \max_{j \leq n} P(|X_j - med\ X_j| > \epsilon) \to 0, \tag{5.15}$$

as $n \to \infty$. Let $\epsilon \to 0$, we have X_k degenerate at its median, contrary to the nondegeneracy assumption on the sequence $\{X_j\}$. Thus (5.4) holds, and $\{\tilde{a}_n\}$ is defined for all large n.

We claim, for each $\delta > 0$, $\epsilon > 0$,

$$\lim_{n \to \infty} \max_{j \leq n} P(|X_j - \tilde{\gamma}_j(\delta d_n)| > \epsilon d_n) = 0 \tag{5.16}$$

i.e., the triangular array $\{(X_j - \tilde{\gamma}_j(\delta d_n))/d_n : j \leq n\}$ is u.a.n., for each $\delta > 0$. To establish (5.16), we will need the following easy lemma:

Lemma 5.2

Let $\{u_{nj} : j \leq n\}$ be a u.a.n. triangular array. Then for every $\epsilon > 0$,

$$\lim_{n \to \infty} \max_{j \leq n} E u_{nj}^2 \wedge \epsilon^2 = 0. \tag{5.17}$$

Proof of Lemma:

Fix $\epsilon > 0$, and choose $0 < \beta < \epsilon$. Then

$$\max_{j \leq n} E u_{nj}^2 I(|u_{nj}| \leq \epsilon) \leq \beta^2 + \max_{j \leq n} E u_{nj}^2 I(\beta < |u_{nj}| \leq \epsilon) \tag{5.18}$$

$$\leq \beta^2 + \epsilon^2 \max_{j \leq n} P(|u_{nj}| > \beta).$$

Letting $n \to \infty$, we have

$$\limsup_{n \to \infty} \max_{j \leq n} E u_{nj}^2 \wedge \epsilon^2 \leq \beta^2. \tag{5.19}$$

Letting $\beta \to 0$ gives (5.17), and the lemma is proved.

Recalling the minimizing property (3.19) of $\tilde{\gamma}$, we have from (5.2) and lemma 5.2, that for each $\epsilon > 0$,

$$\lim_{n \to \infty} \max_{j \leq n} d_n^{-2} E(X_j - \tilde{\gamma}_j(\epsilon d_n))^2 \wedge (\epsilon^2 d_n^2) \tag{5.20}$$

$$\leq \lim_{n \to \infty} \max_{j \leq n} d_n^{-2} E(X_j - e_{nj})^2 \wedge (\epsilon^2 d_n^2)$$

$$= 0,$$

since $\{u_{nj} = (X_j - e_{nj})/d_n : j \leq n\}$ is a u.a.n. array.

Let $\delta > 0$. Now (5.20) gives us

$$\lim_{n \to \infty} \max_{j \leq n} P(|X_j - \tilde{\gamma}_j(\delta d_n)| > \delta d_n) = 0. \tag{5.21}$$

Let $0 < \epsilon < \delta$. We see

$$\begin{aligned} 0 &= \lim_{n\to\infty} \max_{j\le n} d_n^{-2} E(X_j - \tilde{\gamma}_j(\delta d_n))^2 \wedge (\delta d_n)^2 \\ &\ge \lim_{n\to\infty} \max_{j\le n} d_n^{-2} E(X_j - \tilde{\gamma}_j(\delta d_n))^2 I(|X_j - \tilde{\gamma}_j(\delta d_n)| \le \delta d_n) \\ &\ge \lim_{n\to\infty} \max_{j\le n} d_n^{-2} E(X_j - \tilde{\gamma}_j(\delta d_n))^2 I(\epsilon d_n < |X_j - \tilde{\gamma}_j(\delta d_n)| \le \delta d_n) \\ &\ge \lim_{n\to\infty} \epsilon^2 \max_{j\le n} P(\epsilon d_n < |X_j - \tilde{\gamma}_j(\delta d_n)| \le \delta d_n). \end{aligned} \tag{5.22}$$

If $0 < \epsilon < \delta$, adding (5.21) and (5.22) gives (5.16), while if $0 < \delta \le \epsilon$, (5.21) renders (5.16) obvious. We have, therefore, established the "flexible" u.a.n. condition we will need, namely (5.16).

Now we appeal to the classical necessity conditions for normal convergence of sums from u.a.n. arrays: For each $\delta > 0, \epsilon > 0$, we have, via (5.1) and (5.16),

$$\lim_{n\to\infty} \sum_{j=1}^{n} P(|X_j - \tilde{\gamma}_j(\delta d_n)| > \epsilon d_n) = 0. \tag{5.23}$$

Since from (3.20) we have, for every $j \le n$, $E(X_j - \tilde{\gamma}_j(\delta d_n))I(|X_j - \gamma_j(\delta d_n)| \le \delta d_n) = 0$, we also have from the normal convergence necessary conditions that for every $\delta > 0$,

$$\begin{aligned} &\lim_{n\to\infty} d_n^{-2} \sum_{j=1}^{n} E(X_j - \tilde{\gamma}_j(\delta d_n))^2 I(|X_j - \tilde{\gamma}_j(\delta d_n)| \le \delta d_n) \\ &= \lim_{n\to\infty} d_n^{-2} \sum_{j=1}^{n} var\ (X_j - \tilde{\gamma}_j(\delta d_n)) I(|X_j - \tilde{\gamma}_j(\delta d_n)| \le \delta d_n) \\ &= 1, \end{aligned} \tag{5.24}$$

which, together with (5.23), gives us, for every $\delta > 0$,

$$\lim_{n\to\infty} d_n^{-2} \sum_{j=1}^{n} E(X_j - \tilde{\gamma}_j(\delta d_n))^2 \wedge (\delta d_n)^2 = 1, \tag{5.25}$$

We now claim

$$\lim_{n\to\infty} d_n/\tilde{a}_n = 1. \tag{5.26}$$

If (5.26) fails, there exist $0 < \beta < 1$ and infinitely many n such that either

$$(1 + \beta)d_n < \tilde{a}_n \tag{5.27}$$

or

$$\tilde{a}_n < (1 - \beta)d_n. \tag{5.28}$$

Recall the functions $\tilde{w}_j(\cdot)$ used in Sections 3 and 4; they are strictly decreasing. Then (5.25) with $\delta = (1 \pm \beta)$, this monotonicity, and (4.18) give, as $n \to \infty$ under (5.27),

$$(1+\beta)^{-2} \sim ((1+\beta)d_n)^{-2} \sum_{j=1}^{n} E(X_j - \tilde{\gamma}_j((1+\beta)d_n))^2 \wedge ((1+\beta)d_n)^2 \tag{5.29}$$

$$= \sum_{j=1}^{n} \tilde{w}_j((1+\beta)d_n) > \sum_{j=1}^{n} \tilde{w}_j(\tilde{a}_n) = 1,$$

a contradiction, with the similar contradiction as $n \to \infty$ under (5.28), of

$$(1-\beta)^{-2} \sim ((1-\beta)d_n)^{-2} \sum_{j=1}^{n} E(X_j - \tilde{\gamma}_j((1-\beta)d_n))^2 \wedge ((1-\beta)d_n)^2 \tag{5.30}$$

$$= \sum_{j=1}^{n} \tilde{w}_j((1-\beta)d_n) < \sum_{j=1}^{n} \tilde{w}_j(\tilde{a}_n) = 1.$$

Thus (5.26) holds.

Now we claim that the array $\{(X_j - \tilde{\gamma}_j(\tilde{a}_n))/\tilde{a}_n\}$ is u.a.n. Choose $0 < \delta < 1$ so that $\delta d_n < \tilde{a}_n$ for all large n, by (5.26). Then, since each $\tilde{w}_j(\cdot)$ decreases, we have, from (5.20) and the definition of $\tilde{w}_j$,

$$\lim_{n\to\infty} \max_{j\le n} \tilde{a}_n^{-2} E(X_j - \tilde{\gamma}_j(\tilde{a}_n))^2 \wedge \tilde{a}_n^2 \tag{5.31}$$

$$= \lim_{n\to\infty} \max_{j\le n} \tilde{w}_j(\tilde{a}_n) \le \lim_{n\to\infty} \max_{j\le n} \tilde{w}_j(\delta d_n)$$

$$= 0.$$

Arguing as in (5.22), we see that (5.31) implies, for each $\epsilon > 0$,

$$\lim_{n\to\infty} \max_{j\le n} P(|X_j - \tilde{\gamma}_j(\tilde{a}_n)| > \epsilon \tilde{a}_n) = 0, \tag{5.32}$$

the required u.a.n. condition. Thus, (5.1), (5.32), and the normal convergence criteria give us (5.5), and hence (C) holds.

Finally the proof of the direct part (C) $\Rightarrow$(A) gives us (5.6), and the proof of Theorem 5.1 is complete.

6. Examples

In this section we will present some examples illustrating the centering/scaling method we have proposed, and clarifying some of the fine points in its execution. One of these examples showcases the main asymptotic normality theorem, with a direct comparison with previous related results and methods.

Example 6.1.

Here we will consider some of the pathologies possible using the Minimum Moment Method.

First, $\tilde{\gamma}(t)$ need not be unique. Consider the simplest example, a r.v. X with $P(X = 1) = P(X = -1) = 1/2$. Consideration of the possibilities for solving (3.20) for $\tilde{\gamma}(t)$, and direct comparison of the results shows

$$\tilde{\gamma}(t) = \begin{cases} -1 \text{ or } 1, & 0 < t < \sqrt{2} \\ -1 \text{ or } 0 \text{ or } 1, & t = \sqrt{2} \\ 0, & t > \sqrt{2} \end{cases} \tag{6.1}$$

(Incidentally, for $0 < t < 1$, the choice $y = 0$ actually *maximizes* $m(t,y)$.)

Second, there need not exist a version of $\tilde{\gamma}$ which is even eventually continuous. Suppose X is a r.v. whose atoms form an unbounded set. Now, if for some number M and version $\tilde{\gamma}_0$ of $\tilde{\gamma}$, we had $\tilde{\gamma}_0$ continuous on $[M,\infty)$, then the functions $a(t) = \tilde{\gamma}_0(t)+t$ and $b(t) = \tilde{\gamma}_0(t)-t$ would each be continuous. By (3.21), we have $a(t) \to \infty$ and $b(t) \to -\infty$ as $t \to \infty$. But then (3.22) implies that every point in the range $[a(M),\infty) \cup (-\infty, b(M)]$ is a continuity point for the distribution of X, contrary to the unboundedness of the set of atoms of X.

Finally, as we have remarked, even symmetry about the origin for X need not imply admissibility of the version $\tilde{\gamma} \equiv 0$. Indeed, modifying the example above to have X symmetric and its set of atoms dense on the line, we see that the set of points t such that the choice $\tilde{\gamma}(t) = 0$ is inadmissible is actually dense, utilizing (3.22).

Example 6.2.

Here we observe that finite limiting behavior of $\tilde{\gamma}$ at ∞ need not imply the existence of EX. Let X be a symmetric r.v. with Lebesgue density

$$f(t) = \begin{cases} 1/2t^2, & |t| > 1 \\ 0, & |t| \leq 1. \end{cases} \tag{6.2}$$

Contrary to the pathology in Example 6.1, here we can show, at least for all sufficiently large t, that $\tilde{\gamma}(t) = 0$. Yet $P(|X| > t) = t^{-1}$ for $t > 1$, so that $E|X| = \infty$.

We need a uniform version of (3.21): Since, for each fixed t, the set of y-values minimizing $m(t,\cdot)$ is compact, we can select, for each t, the value of y with the largest magnitude. Since the resulting "exterior" version of $\tilde{\gamma}$ satisfies (3.21), we see (3.21) is actually uniform in all versions of $\tilde{\gamma}$.

Now, for t fixed, candidates for $\tilde{\gamma}(t)$ must satisfy (3.20). Certainly the choice $\tilde{\gamma}(t) = 0$ works, by symmetry; we will show that for large t, no other solution of (3.20) exists in the interval $[-o(t), o(t)]$ permitted by the uniform version of (3.21). Tedious but direct calculation of $m(\cdot,\cdot)$ shows that for $y - t < -1 < 1 < y + t$, we have

$$\begin{aligned} (y^2 - t^2) m_{22}(t,y) &= t^4 - 2t^2y^2 + y^4 \\ &= t^4(1 + o(1)), \end{aligned} \tag{6.3}$$

as $t \to \infty$ and $|y|/t \to 0$ (here is where the uniformity of (3.21) is used). In particular, $m_{22}(t,y) > 0$ for $y \in [-o(t), o(t)]$, so that $y = 0$ is the unique solution of $m_2(t,y) = 0$ (i.e., of (3.20)) in this interval. Consequently, since $\tilde{\gamma}(t)$ does exist, we must have $\tilde{\gamma}(t) = 0$ (uniquely), for all sufficiently large t.

Example 6.3.

Here we borrow a fine example from Hahn and Klass (1981, Example 2) to demonstrate the need for and applicability of the MMM centerings in Theorem 5.1. Let $\{X_j\}$ be independent with $P(X_j = 0) = p_j$, $P(X_j = c_j + 1) = P(X_j = c_j - 1) = (1 - p_j)/2$, where for $j \geq 1$ we have $p_j = 1/(j+1)^2$, $c_{2j-1} = -c_{2j} = (2j)!$. Hahn and Klass showed indirectly that the partial sums S_n were asymptotically normal, yet given any intervals $\{I_{nj}\}$ (whose lengths depend only on n) such that $\sum_{j=1}^n P(X_j \in I_{nj}) \to 0$, the traditional centerings $\sum_{j=1}^n EX_jI(X_j \in I_{nj})$ and scalings $(\sum_{j=1}^n \ var\ X_jI(X_j \in I_{nj}))^{1/2}$ are of the wrong order to secure asymptotic normality. Moreover, even centering the X_j at their expectation EX_j before normalization will not give a u.a.n. array.

We will show that our Theorem 5.1 nevertheless applies; in particular, one *must* center using the MMM, so that symmetrization followed by "t-trimming" (the Hahn-Klass approach) can be replaced by more direct calculations involving $\tilde{\gamma}$.

It is easy to see in general that if X lives on some interval I of radius r, then $\tilde{\gamma}(t) \in I$, for every $t > 0$; thus, if $t \geq 2r$, we see $P(|X - \tilde{\gamma}(t)| \leq t) = 1$. Hence (3.20) forces the unique choice $\tilde{\gamma}(t) = EX$, whence $\tilde{v}(t) = varX$, when $t \geq 2r$.

In our example, we have, for $t \geq |c_j| + 1$, $\tilde{\gamma}_j(t) = EX_j = c_j(1 - p_j)$ and $\tilde{v}_j(t) = varX_j = (1 - p_j)(1 + p_j\ c_j^2)$.

For $2 \leq t < |c_j| + 1$, we restrict attention to the possibilities for $\tilde{\gamma}(t)$ allowed by (3.20). Now as the desired central interval $[\tilde{\gamma}_j(t) - t, \tilde{\gamma}_j(t) + t]$ must, by a process of elimination, contain either exactly the atoms of X_j at $c_j - 1$ and $c_j + 1$, or else all three atoms. Then (3.20) forces comparison of the centers c_j and $c_j(1 - p_j)$. Direct comparison via m shows that for $t > (1 - p_j)^{1/2}|c_j|$, we have $\tilde{\gamma}_j(t) = EX_j$ uniquely, and for $2 \leq t < (1 - p_j)^{1/2}|c_j|$ we have $\tilde{\gamma}(t) = c_j$ uniquely. For $t = (1 - p_j)^{1/2}|c_j|$, either value of $\tilde{\gamma}_j(t)$ is allowed.

In summary, we find that

$$\tilde{v}_j(t) = \begin{cases} (1 - p_j)(1 + p_j\ c_j^2), & t \geq (1 - p_j)^{1/2}|c_j| \\ (t^2 - 1)p_j + 1 & 2 \leq t \leq (1 - p_j)^{1/2}|c_j|. \end{cases} \tag{6.4}$$

Now $\tilde{v}_j(t) \geq 1$ for $t \geq 2$. So (5.4) is trivially satisfied. Thus $\tilde{a}_n \uparrow \infty$ exist satisfying (4.18). To find them, let $\Delta_n = \sup\{j \geq 1 : (1 - p_j)^{1/2}|c_j| \leq \tilde{a}_n\} \wedge n$. Now Δ_n is even, by the definiton of $\{c_j\}$, and also $\Delta_n \to \infty$, $\Delta_n/n \to 0$.

We have, as $n \to \infty$,

$$\begin{aligned} \tilde{a}_n^2 &= \sum_{j=1}^n \tilde{v}_j(\tilde{a}_n) = \sum_{j \leq \Delta_n} (1 - p_j)(1 + p_jc_j^2) + \sum_{\Delta_n < j \leq n} \{1 + p_j(\tilde{a}_n^2 - 1)\} \\ &= n - \sum_{j=1}^n p_j + \sum_{j \leq \Delta_n} (1 - p_j)p_jc_j^2 + \tilde{a}_n^2 \sum_{\Delta_n \leq j \leq n} p_j \\ &= n + O(1) + 2(1 + o(1))p_{\Delta_n}c_{\Delta_n}^2 + o(\tilde{a}_n^2) = n + o(\tilde{a}_n^2), \end{aligned} \tag{6.5}$$

due to $\sum_{j=1}^\infty p_j < \infty$ and the definition of Δ_n. But now (6.5) forces $\tilde{a}_n \sim \sqrt{n}$.

To verify (5.5), fix $\epsilon > 0$ and consider $\pi_{nj} = P(|X_j - \tilde{\gamma}_j(\tilde{a}_n| > \epsilon\tilde{a}_n))$. Let $\delta_n = \sup\{j \geq 1 : (1 - p_j)|c_j| \leq \epsilon\tilde{a}_n\} \wedge n$. Then as $n \to \infty$, $\Delta_n \geq \delta_n \to \infty$. For $j \leq \delta_n, \tilde{\gamma}_j(\tilde{a}_n) = c_j(1 - p_j)$, so that $\pi_{nj} = 0$, while for $\delta_n < j \leq \Delta_n$, we see $\pi_{nj} \leq P(X_j = 0) = p_j$. For $\Delta_n < j \leq n$, $\tilde{\gamma}_j(\tilde{a}_n) = c_j$, so that $\pi_{nj} \leq p_j$ again. Hence $\sum_{j \leq n} \pi_{nj} \leq \sum_{\delta_n < j \leq n} p_j \to 0$, validating (5.5).

It remains to identify the centerings:

$$\begin{aligned}\sum_{j=1}^{n}\tilde{\gamma}_j(a_n) &= \sum_{j=1}^{\Delta_n} c_j(1-p_j) + \sum_{j=\Delta_n+1}^{n} c_j \\ &= \sum_{j=1}^{n} c_j - \sum_{j=1}^{\Delta_n} c_j p_j = c_n I(n\ odd) + O(c_{\Delta_n} p_{\Delta_n}) \\ &= c_n I(n\ odd) + o(\tilde{a}_n).\end{aligned} \tag{6.6}$$

Thus Theorem 5.1 yields

$$(S_n - nI(n\ odd))/\sqrt{n} \xrightarrow{D} N(0,1), \tag{6.7}$$

where in addition the left side can be written as an unrecentered sum from a u.a.n. array, as in the Theorem, using the values of $\tilde{\gamma}_j(\tilde{a}_n)$ computed above (6.4).

By contrast, the more classical direct use of medians to compute scalings d_n via

$$\sum_{j=1}^{n} E(X_j - medX_j)^2 \wedge d_n^2 = d_n^2 \tag{6.8}$$

would yield (as is checked similarly to the above) $d_n \sim \sqrt{2n}$, erroneous (with respect to *standard* normal convergence) by a factor of $\sqrt{2}$. Thus further recentering/rescaling would be required.

BIBLIOGRAPHY

Araujo, A. and Giné, E. (1980). *The Central Limit Theorem for Real and Banach Valued Random Variables.* Wiley, New York.

Chung, K. (1974). *A Course in Probability Theory.* Academic New York.

Feller, W. (1971). *An Introduction to Probability Theory*, vol. II. Wiley, New York.

Gnedenko, B. and Kolmogorov, A. (1968). *Limit Distributions for Sums of Independent Random Variables.* Addison-Wesley, Reading, MA.

Hahn, M. and Klass, M. (1981). The multidimensional central limit theorem for arrays normed by affine transformations. *Ann. Probab.* **9**, 611-623.

Lévy, P. (1937). *Theorie de l'addition des Variables Aleatoires.* Gauthier-Villars, Paris.

Loève, M. (1977). *Probability Theory.* Springer-Verlag, New York.

Pruitt, W. (1981). General one-sided laws of the iterated logarithm. *Ann. Probab.* **9**, 1-48.

Weiner, D. (1989). Center, scale and CLT behavior for sums of independent random variables, in preparation.

Progress in Probability

Edited by:

Professor Thomas M. Liggett
Department of Mathematics
University of California
Los Angeles, CA 90024-1555

Professor Charles Newman
Department of Mathematics
University of Arizona
Tucson, AZ 85721

Professor Loren Pitt
Department of Mathematics
University of Virginia
Charlottesville, VA 22903-3199

Progress in Probability includes all aspects of probability theory and stochastic processes, as well as their connections with and applications to other areas such as mathematical statistics and statistical physics. Each volume presents an in-depth look at a specific subject, concentrating on recent research developments. Some volumes are research monographs, while others will consist of collections of papers on a particular topic.

Proposals should be sent directly to the series editors or to Birkhäuser Boston, 675 Massachusetts Avenue, Suite 601, Cambridge, MA 02139.

1 ÇINLAR/CHUNG/GETOOR. Seminar on Stochastic Processes, 1981
2 KESTEN. Percolation Theory for Mathematicians
3 ASMUSSEN/HERING. Branching Processes
4 CHUNG/WILLIAMS. Introduction to Stochastic Integration
5 ÇINLAR/CHUNG/GETOOR. Seminar on Stochastic Processes, 1982
6 BLOOMFIELD/STEIGER. Least Absolute Deviation
7 ÇINLAR/CHUNG/GETOOR. Seminar on Stochastic Processes, 1983
8 BOUGEROL/LACROIX. Products of Random Matrices with Application to Schrödinger Operator
9 ÇINLAR/CHUNG/GETOOR. Seminar on Stochastic Processes, 1984
10 KIFER. Ergodic Theory of Random Transformations
11 EBERLEIN/TAQQU. Dependence in Probability and Statistics
12 ÇINLAR/CHUNG/GETOOR. Seminar on Stochastic Processes, 1985
13 ÇINLAR/CHUNG/GETOOR/GLOVER. Seminar on Stochastic Processes, 1986
14 DEVROYE. A Course in Density Estimation
15 ÇINLAR/CHUNG/GETOOR/GLOVER. Seminar on Stochastic Processes, 1987
16 KIFER. Random Pertubations of Dynamical Systems
17 ÇINLAR/CHUNG/GETOOR/GLOVER. Seminar on Stochastic Processes, 1988
18 ÇINLAR/CHUNG/GETOOR/FITZSIMMONS/WILLIAMS. Seminar on Stochastic Processes, 1989
19 ALEXANDER/WATKINS. Spatial Stochastic Processes: A Festschrift in Honor of Ted Harris on His 70th Birthday
20 HAAGERUP/HOFFMANN-JØRGENSEN/NIELSEN. Probability in Banach Spaces 6: Proceedings of the Sixth International Conference, Sandbjerg, Denmark 1986
21 EBERLEIN/KUELBS/MARCUS. Probability in Banach Spaces 7: Proceedings of the Seventh International Conference